AF181460

# Mechanical Alloying

L. Lü and M. O. Lai

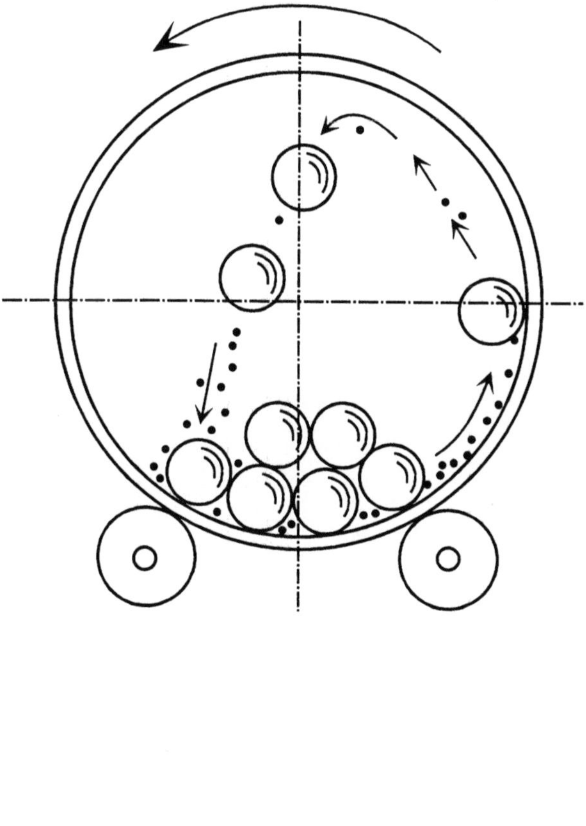

# MECHANICAL ALLOYING

by

**L. Lü and M. O. Lai**

SPRINGER SCIENCE+BUSINESS MEDIA, LLC

**Library of Congress Cataloging-in-Publication Data**

A C.I.P. Catalogue record for this book is available
from the Library of Congress.

ISBN 978-1-4613-7517-3 ISBN 978-1-4615-5509-4 (eBook)
DOI 10.1007/978-1-4615-5509-4

**Copyright** © 1998 by Springer Science+Business Media New York
Originally published by Kluwer Academic Publishers in 1998
Softcover reprint of the hardcover 1st edition 1998
All rights reserved. No part of this publication may be reproduced, stored in a retrieval system or transmitted in any form or by any means, mechanical, photocopying, recording, or otherwise, without the prior written permission of the publisher, Springer Science+Business Media, LLC.

*Printed on acid-free paper.*

# TABLE OF CONTENTS

# PREFACE

Mechanical alloying (or mechanical milling) was invented in the 1970's as a method to develop dispersion-strengthened high temperature alloys with unique properties. With the discovery of formation of amorphous alloys using this technique, it has received new research interest in developing different material systems. Potential applications of this technique have been demonstrated in different areas of materials research.

This book is intended as an introduction to mechanical alloying technique used in difference areas. This book contains basic information on the preparation of materials using the mechanical alloying technique. It is useful not only to undergraduate and post-graduate students, but also to scientists and engineers who wish to gain some understanding on the basic process and mechanisms of the process.

The book begins with a brief introduction to provide a historical background understanding to the development of the mechanical alloying process. The experimental set-up in the alloying process is important. Currently there are different types of ball mills available. Some of them are specially designed for mechanical alloying process. Since the resultant materials are milling intensity and milling temperature dependent, ball mills should be carefully selected in order to obtain the desired materials and structures. This is discussed in chapter 2. The actual mechanical alloying process is being considered in Chapter 3. As it is essential to understand the use of processing control agents, the physical properties of some commonly used processing control agents are listed. Chapter 4 deals with the formation of new materials in which the mechanical alloying of Al, Ti and Mg alloys and their intermetallics is presented. Formation of composite materials using this technique is discussed in detail. Several examples on the mechanical alloying of amorphous materials, an important group of engineering materials where mechanical alloying is commonly employed are given. Chapter 5 looks into the characterization of the alloyed powders. Detailed use of x-ray diffraction to identify crystalline size, the use of thermal analysis and measurement of particle size are provided. Densification methods are discussed in Chapter 6 while strengthening mechanisms of the composite materials by mechanical alloying are considered in Chapter 7. To understand the mechanisms of the alloying technique so that better control of the process could be exercised, the effects of temperature, activation energy, grain size on diffusion during the alloying process are examined in Chapter 8. The last chapter of the book, Chapter 9, deals with the dynamics and modeling of the alloying process together with some experimental outputs.

This book could not be completed without help from friends and students or without referring to the works of other researchers. Sincere thanks are due to all of them in one way or another who have contributed their expertise to this book.

Finally, the authors would like to express their deep appreciation their families for their support during the preparation of this book.

L. Lü
M. O. Lai

# ACKNOWLEDGMENT

We would like to sincerely acknowlelge the use of figures and tables from the following thesis, books, proceedings and journals in the preparation of this book.

***Ph.D. thesis***

Niu, X.P., *Ph.D. Thesis at KULeuven*, Belgium, (1991).

***Books***

Aldrich, *Handbook of Fine Chemicals*, Aldrich Chemical Company, Inc., WI 53233, USA (1992).

Allen, T., *Particle Size Measurement*, Publ. Chapman & Hall, (1990), 125.

Atkinson, H.V. and Rickinson, B.A., *Hot Isostatic Processing*, The Adam Hiler Series on New Manufacturing Processes and Materials, ed. J. Wood, Adam Hilger, Bristol, Philadelphia and New York, (1991), 34.

***Proceedings and journals***

Aoki, A., Itoh, Y. and Masumoto, T., *Scripta Metall. Mater.*, Vol.31 (1994), 1271.

Arnhold, V. and Hummert, K., *New Meter. by Mechanical Alloying*, DGM Confer., Calw-Hirsau, Germany, Oct. 1988, Ed. E. Arzt and L. Schultz, Publ. Deutsche Gesellschaft für Metallkunde eV., (1989), 263.

Benjamin, J.S. and Bomford, M.J., *Metall. Trans.*, Vol.8A (1977), 1301.

Bhattachary, A.K. and Artz, E., *Scripta Metall. Mater.*, Vol.28 (1993), 395.

Burgio, N., Guo, W., Magini, M. and Padella,F., *Structureal Applications of Mechanical Alloying*, Proc. an ASM Intern. Confer., Myrtle Beach, South Carolina, 27-29 March 1990, Ed: F.H.Froes and J.J. deBarbadillo, ASM Intern. Mater. Park, Ohio (1990), 175.

Calka, A., Jing, J., Jayasurlya, K.D. and Campbell, S.J., *Proc. of the 2nd Intl' confer. on Structural Appl. of Mechanical Alloying*, Vancouver, British Columbia, Canada, 20-22 Sep. (1993), Ed. J.J. deBarbadillo, F.H. Froes and R. Schwarz, ASM Intern. Mat. Park, OH, USA, 27.

Chen, Y. and Williams, J.S., *J. Alloys & Compounds*, Vol.217 (1995), 181.

Chen, Y., Le Hazif, R. and Martin, G., *Mater. Sci. Forum*, Vols.88-90 (1992), 35.

Cocco, G., Soletta, I, Battezzati, L., Baricco, M. and Enzo, S., *Phil. Mag.* B, Vol.61 (1990), 473.

Creasy, T.,Weertman, J.R. and Fine, M.E., *Dispersion Strengthened Aluminium Alloys*, Proc. of six-session Symposium, TMS Annual Mtg, Phoenix, Arizona, Jan.25-29, 1988, Ed. Y.W. Kim and W.M. Griffith, Publ. TMS, (1988), 539.

Davis, E.W., *Trans. AIMME*, Vol61 (1919), 250.

Delaey, L., *et al*, "Shape Memory Effect, Super-elasticity and Damping in Cu-Zn-Al Alloys" K.U.Leuven, Metaalkunde, Report 78 R, 1978.

Fecht, H.J., *Nanophase Mater.*, Ed. G.C. Hadjipanayis and R.W. Siegel, Publ. Kluwer Academic Publishers, 1994, 125.

Fischmeister, H.F. and Arzt, E., *Powder Metall.*, Vol.26 (1983), 82.

Fritsch GmbH, Industriestr. 8, D-55743 Idar-Oberstein.

Gilman, P.S. and Nix, W.D., *Metall. Trans.*, Vol.12A (1981), 813.

Goodwin, P.S. and Ward-Close, C.M., *Mater. Sci. Forum*, Vols.179-181 (1995), 411.

Goodwin, P.S. and Ward-Close, C.M., *Proc. of the 2nd Intern. confer. on Structural Appl. of Mechanical Alloying*, Vancouver, British Columbia, Canada, 20-22 Sep. (1993), Ed. J.J. deBarbadillo, F.H. Froes and R. Schwarz, ASM Intern. Mat. Park, OH, USA, 139

Harris, A.M., Schaffer, G.B. and Page, N.W., *Proc. of the 2nd Intern. confer. on Structural Appl. of Mechanical Alloying*, Vancouver, British Columbia, Canada, 20-22 Sep. (1993), Ed. J.J. deBarbadillo, F.H. Froes and R. Schwarz, ASM Intern. Mat. Park, OH, USA, 15.

Janng, G., *New Materials by Mechanical alloying Techniques*, DGM Confer., Calw-Hirsau (FRG), Oct. 1988, Ed. E. Arzt and L. Schultz, Publ. Informationsgesellschaft Verlag, (1989), 39.

Johnson, W.L., *Mater. Sci. Eng.*, Vol.97 (1988), 1.

Kim, H.S. Kim, G. and Kum, D.W. *Design Fund. of High Tem. Comp. Interm. and Metal-Crem. Systems*, Ed: R.Y. Liu, Y.A. Change, R.G. Reddy and C.T. Liu, the Mineral, Metals & Materials Society, (1996), 223.

Kumpfert, J., Staniek, G and Kleinekathofer, W., *Structural Applications of Mechanical Alloying*, Proc. of An ASM Inter. Confer., Myrtle Beach, South Carolina, 27-29 March 1990, Ed. F.H. Froes and J.J. deBarbadillo, Publ. ASM Inter., Mater. Park, OH, 163.

Le Brun, B., Froeyen, L. and Delaey, L., *Mater. Sci. Eng.*, A161 (1993), 75.

Le Brun, P., Froyen, L. and Delaey, L., *Structural Applications of Mechanical Alloying*, Proc. of an ASM Intern.. Confer., Myrtle Beach, South Carolina,

27-29 March 1990, Ed. F.H. Froes and J.J. deBarbadillo, Publ. ASM Intern.., Mater. Park, Ohio, 155.

Lu, L., Lai, M.O. and Zhang, S., *Mater. Design*, Vol.15 (1994), 79.

Lu, L, Lai, M.O. and Zhang, S., *J. Mater. Proc. Tech.*, Vol.67 (1997), 100.

Magini, M. and Iasonna, A., *Mater. Trans. JIM*, Vol.36 (1995), 123.

Malchere, A. and Gaffet, E., *Proc. of the 2nd Intern. Confer. on Structural Applications of Mechamical Alloying*, Vancouver, Canada, Ed.J.J. deBarladillo, F.H. Froes and R. Schwarz, Pul. ASM Intern., Materials Park, 20-22 Sep. (1993), 297.

Martin, G. and Gaffet, E., *Coll. Phys.*, C4 (1990), 71.

Matsuura, K. and Matsuuda, N., *Sintering'87*, Proc. of The Intern. Inst. for The Sci. of Sintering (IISS), Tokyo, Japan, Nov. 4-6, 1987, 587.

Miki, M., Yamasaki, T. and Ogino, Y., *Mater. Sci. Forum*, Vols.179-181, 307.

Miki, M., Yamasaki, T. and Ogino, Y., *Mater. Trans. JIM,* Vol.34 (1993), 952..

Morris, M.A. and Morris, D.G., *Mater. Sci. Forum*, Vols.88-90 (1992), 529.

Mukhopadhyay, D.K., Suryanarayana, C. and Froes, F.H., *Proc. of the 2nd Intern. Confer. on Structural Applications of Mechamical Alloying*, Vancouver, Canada, Ed.J.J. deBarladillo, F.H. Froes and R. Schwarz, Pul. ASM Intern., Materials Park, 20-22 Sep. (1993), 131.

Niu, X.P., Froyen, L., Delaey, L. and Peytour, C., *Scripta Metall. Mater.*, Vol.30 (1994), 13.

Park, Y.H., Hashimoto, H., Watanabe, R., Ahn, J.H. and Chung, H.S., *Mater. Sci. Forum*, Vols.88-90 (1992), 155.

Schaffer, G.B. and McCormick, P.G., *Metall. Trans. A*, Vol.21A (1990), 2789.

Schaffer, G.B. and McCormick, P.G., *Metall. Trans. A*, Vol.22A (1991), 3019.

Suryanarayana, C., Sundaresan, R. and Froes, F.H., *Mater. Sci. Eng.*, Vol.A150 (1992) 117.

Weber, J.H. and Schelleng, R.D., *Dispersion Strengthened Aluminium Alloys*, Proc. of six-session Symposium, TMS Annual Mt, Phoenix, Arizona, Jan.25-29, 1988, Ed. Y.W. Kim and W.M. Griffith, Publ. TMS, (1988), 467.

Zbiral, J., Jangg, G. and Korb, G., *Mechanical Alloying for Structural Appls.*, Proc. of the 2nd Intern. Confer. on Structural Appls. of Mechanical alloying, 20-22 Sep. 1993, Vancouver, British Columbia, Canada, Ed. J.J. deBarbadillo, F.H. Froes and R. Schwarz, Publ. ASM Intern. Mater. Park, Ohio, (1993), 59.

Zoz, H., *Mater. Sci. Forum*, Vols.179-181 (1995), 419.

# 1

# INTRODUCTION TO MECHANICAL ALLOYING

## 1.1 Mechanical alloying

Mechanical alloying is a ball milling process where a powder mixture placed in the ball mill is subjected to high energy collision from the balls. The process is usually carried out in an inert atmosphere. It is an alternative technique for producing metallic and ceramic powder particles in the solid state. The two most important events involved in mechanical alloying are the repeated welding and fracturing of the powder mixture. The alloying process can only be continued if the rate of welding balances that of fracturing and the average particle size of the powders remains relatively coarse [(1)]. Alloys with different combination of elements have been successfully synthesized due to the uniqueness of the process being able to produce new materials from the bottom of the phase diagrams [(2)]. Since mechanical alloying is a solid state process, it provides a means to overcome the drawback of formation of new alloys using a starting mixture of low and high melting temperature elements. Although in general, the raw materials used in mechanical alloying should include at least one fairly ductile metal to act as a host or binder to hold together the other ingredients [(3)], a lot of studies have confirmed that brittle metals can also be mechanically alloyed to form solid solution [(4, 5)], intermetallics [(6)] and amorphous alloys as well [(7)].

## 1.2 History and development

The mechanical alloying process was invented at INCO's Paul D. Merica Research Laboratory around 1966. The initial attempt was to develop a material by combining oxide dispersion strengthening with gamma prime precipitation hardening in a nickel-based superalloy for gas turbine applications and other heat resistant alloys [(3-12)]. As the oxides could not be dispersed in the liquid phase, a solid state processing technique was therefore needed. In the early 1960s, INCO's Pual D. Merica Research Laboratory had successfully coated nickel onto graphite surface using this technique. The coated graphite particulates were injected into molten aluminium alloys using an argon sparging gas [13]. At this stage, this process was referred to as "milling/mixing". The term "mechanical alloying" was later introduced by E. C. MacQueen in the late 1960's [14].

The applications at the early stage of research after the discovery of the technique had mainly been confined to oxide dispersion strengthened alloys. Typical examples of use are the mechanical alloying of superalloys [15-18] and aluminium alloys [19-22]. The repeated cold welding and fracturing of the powder particles eventually leads to the formation of oxide layers on the surfaces of the powder particles as well as layered structures. With extended mechanical alloying, even the thin oxide layers are continuously broken up to form homogeneously distributed fine particles. By using the appropriate processing parameters, materials containing strengthening particles in the size of nanometer can be produced [21]. Large amount of research funding has been spent to develop this new technique. Most of the work carried out in research laboratories between 1965 and 1973 was expended on small scale processes and alloy development to meet the need of high temperature engine components [3].

Developed by Jangg [23, 24], reaction milling has been employed to produce aluminium strengthened by $Al_4C_3$ and $Al_2O_3$. The process is performed using a chemical reaction between milling additives and the powder particles to be milled. In the presence of carbon or graphite and in a controlled atmosphere of oxygen, both the graphite and the oxygen may serve as processing control agent during the milling. The introduction of carbon and oxygen thus results in $Al_4C_3$ and $Al_2O_3$ [26]. Fine grained microstructures and homogeneous composites can be produced via chemical reactions between intimately mixed elemental powders. Chemical reactions during mechanical alloying occur by solute diffusion. The latter, although is accelerated due to the presence of deformation induced point and lattice defects and also due to localized increase in temperature, normally requires a time duration of the order of hours [25]. This technique has been utilized to produce Al-C-O alloys under the trade name of DISPAL, where the alloys have been dispersion strengthened. One of the major advantages of dispersion strengthened alloys lies in the retention of useful strength up to a relatively high fraction of their melting points (about 90%). However, if strengthening is caused by precipitation or solid solution, strength will rapidly lose effectiveness [21] when the materials are used at a relatively high temperature.

The mechanical alloying process is not limited to the production of complex oxide dispersion strengthened alloys only. It is also a means of producing composite metal powders with controlled, extremely fine microstructures. Composite structures result when ceramics and metals are milled. The method has also been recognized as one of the novel techniques in the synthesis of new alloys. Since the discovery of formation of amorphous materials by Koch *et al.* [27], it is the first time that such alloys were produced when a mixture of elemental metal powders was mechanically alloyed in an inert gas atmosphere. Koch *et al.* prepared amorphous $Ni_{60}Nb_{40}$ by milling elemental powders in a SPEX 8000 mill. It was found that the effective size of the crystallites as calculated based x-ray diffraction pattern decreases rapidly. X-ray measurement of the mechanically alloyed Ni-Nb powder mixture showed similar patterns as those of a liquid quenched sample [(27)]. Amorphisation occurs either by diffusion controlled reaction where the more stable crystalline site is constrained kinetically from forming [(28, 29)], or by defect-induced decomposition of the crystalline state which is analogous to amorphisation by irradiation. Amorphisation by solid state reaction has been observed in many other binary multilayer systems [(30-33)]. Amorphous phases can be formed from bimetallic layers when chemical affinity between the two compounds is strongly negative enthalpy of mixing $\Delta H_{mix}$ and when there is a significant asymmetry in the atomic mobility. During mechanical alloying, the increase in energy is achieved by the generation of chemical disorder, point defects (vacancies and interstitials) and lattice defects (dislocations and disclinations) [(34)]. The driving force for this reaction is the composition-induced destabilization of the crystalline phases in the systems. Ermakov [(35)] observed that an intermediate phase which is an equilibrium phase could also be amorphized. Yeh [(36)] found that amorphisation could take place by a reaction between hydrogen and an intermetallic compound when hydrogen gas reacted with ZrRh below a temperature of 498 K. The product is an amorphous metallic hydride phase with composition $Zr_3RhH_{5.5}$ [(37)]. A wide range of amorphisation has been carried out using Ni-Ti [(38-41)], Ni-Zr [(42-44)], Ni-Nb [(27, 45-47)], Ti-Cu [(48-51)], Co-Zr [(52)], Au-La [(30)], Au-Y [(53)], Fe-Si-Fe [(54-56)] and other materials. A more detailed review has been conducted by Weeber and Bakker [(57)]. The methods by which amorphous alloys are formed have been divided into two main categories [(57)]: (a) mechanical alloying of elemental powders in which the enthalpy of the alloyed materials during amorphisation decreases through diffusion [(27, 30, 36, 58)], and (b) mechanical milling of alloy powders in which the materials milled is thermodynamically unstable due to the increase in its enthalpy [(35, 59)]. The negative mixing enthalpy and thermodynamic instability drive the formation of amorphous alloys.

Depending upon the material system, not all alloys can be synthesized using the mechanical alloying technique. For instance, Nd and Fe continue to remain a mixture of two elements after mechanical alloying [(60)]. Milling of Al and Fe has been found to result in no significant formation of new intermetallic compound if Fe content is low. After the alloying process, Fe is seen to be homogeneously embedded in the Al matrix just as it originally was. To process Al-Fe intermetallic, a heat treatment or thermal process of the mechanically alloyed powders must be followed [(61-64)].

Because of the possibility of utilizing intermetallics as high temperature materials, new research interests on mechanical alloying of intermetallic systems have been evidenced. The formation of Ni-Al intermetallic compounds using mechanical alloying technique, for example, has been widely investigated. Koch *et al.* [(65)]

observed the formation of $Ni_3Al$ compound after milling Ni and Al powders together for 300 min. Ivanov [66] studied the synthesis of $Ni_2Al_3$ using mechanical alloying of mixture of Ni and Al powders at a composition of $Ni_{40}Al_{60}$. It was found that the alloying process led to the formation of metastable β' NiAl phase which reverted to the rhombohedral $Ni_2Al_3$ phase after annealing. Large amount of research works on TiAl [67-70], $Ti_3Al$ [71-73] and $Al_3Ti$ [74-76] have been carried out as a consequence of the great potential in developing mechanically alloyed Ti alloys with enhanced behaviour [74]. A more novel process by Suryanarayana and Froes [77] made use of mechanical alloying and a thermochemical process to synthesize TiAl from two brittle compounds of $Al_3Ti$ and $TiH_2$.

Since large plastic deformation is induced onto the powder particles during mechanical alloying, the crystal is heavily strained and the deformation occurs in a rather inhomogeneous manner. With longer duration of mechanical alloying, shear bands which are about 1 μm thick at the initial milling grow over larger areas and eventually the entire powder particle disintegrates into subgrains with a final grain size of few nanometer [78]. For nanocrystalline powders prepared using conventional processes, when these powder particles are consolidated by warm processing, the compacts are found to consist of nanosized crystalline grains that are nearly dislocation-free. The grains are separated by clean grain boundaries which are only one or two atomic dimensions thick. On the other hand, the nanosized crystalline domains in the powders synthesized using mechanical alloying are separated by thick walls containing a large density of dislocations [81]. Hence, nanocrystalline powders produced by mechanical alloying will possess different properties from those prepared by conventional techniques. Since there are about 50% of the atoms located in the grain boundaries, this new solid state manifests considerable changes in physical and mechanical properties. Fecht [82, 83] and Eckert [84] have successfully synthesized nanocrystalline structures from various elemental bcc materials of Fe, Cr, Nb and W, hcp materials of Co, Zr, Hf and Ru, fcc materials of Al, Cu, Ni, Pd, Rh and Ir, and from CsCl metal of NiTi, CuEr, SiRu and AlRu. Gaffet [85] reported the synthesis of $MoSi_2$ nanophase by a mechanically activated annealing process of combining short durations of mechanical alloying and low temperature isothermal annealing. Other binary systems including Al-Fe [79, 80], Ag-Fe [86], Ni-Al [87], Ti-Mg [88], Al-Ti [89-91], W-Fe [92] and other systems [93-97] have been successfully processed to produce nanocrystalline structure. As mechanical alloying technique is an easy and cheap way to produce large quantity of nanocrystalline materials or amorphous powders, it has induced a great deal of excitement and research interest.

Formation of hard materials of nitrides, carbides, borites and oxides is another interesting area of research. Titanium borites of TiB [98-102] and $TiB_2$ [103-105] have been synthesized directly from mechanical alloying of elemental powders of Ti and B. Besides the formation of titanium borites, formation of titanium carbides have also been reported [106]. Malchere *et al.* [113] found the formation of βSiC by milling mixture of Si and C powders. However, no SiC could be formed if ternary Al-Si-C system was milled since Si and C were diluted by the presence of aluminium. As metal and metalloid nitrides possess considerable hardness, high temperature stability, high thermal conductivity and high corrosion resistance, they are considered to be very important materials in mechanical alloying. Calka *et al.* have produced a group of nitrides including TiN [105, 107-109], ZrN [110, 111], VN [110], BN [110, 112], $Mo_2N$ [111, 114], $Si_3N_4$ [111, 114], $Cu_3N$ [110], $Mg_3N_2$ [110] and WN [110]. These nitrides can be formed by mechanical alloying the appropriate metals in an atmosphere of molecular nitrogen or ammonia [110, 115, 116]. Because high exothermic reaction during mechochemical milling is involved, the temperature of the process may rise to a very high value within a very short time. Extreme care should therefore be exercised to prevent a sudden increase in the internal pressure of the milling container [117].

## 1.3 Applications

Mechanical alloying has the potential to be used in high temperature applications. Commercial products prepared by mechanical alloying include the iron-based MA956, the aluminium-based MA952, and the nickel-based Inconel MA754 and MA6000 [81]. Inconel is one of the largest producer that fabricates mechanical components for aerospace applications. An example of a contrasting high volume small component application is the INCOLOY alloy MA956 which is being used at high temperature in the precombustion chamber of new generation diesel engines [118, 119]. This iron-base alloy has a melting temperature of 1755 K which is well above those of conventional nickel-base alloys. It possesses superior oxidation resistance compared to conventional alloys. Because of its high corrosion resistance, the alloy is being evaluated for the applications of firing-kiln roller, muffle tubes and furnace racks [120].

Another exciting area of application is the extension of mechanical alloying to mechanochemistry from which not only nanocrystalline materials but also nano sized powder particles can be synthesized. Schaffer and McCormick have provided the evidence of successful reduction of oxide from oxides via simple mechanochemistry using ball milling technique [121]. One of the early investigations of reduction of CuO via Ca during milling showed that mechanical alloying technique could be used as a vehicle for solid state reactions [122, 123].

## 1.4 Conclusions

Mechanical alloying, the innovation of Benjamin [8] thirty years ago, has changed the traditional method in which production of materials is carried out by high temperature synthesis. It has attracted much attention and inspired numerous research interests because of its promising results, wide possible applications and potential scientific values. Mechanical alloying has now been recognized as a unique technique. It has been utilized in different areas of material processing and applied to many different material systems. Oxide dispersion strengthened materials, composites, amorphous, nanocrystalline alloys, intermetallic compounds, non-equilibrium materials and ceramics have all been successfully synthesized, an achievement not possible by means of traditional techniques.

## 1.5 References


1. P.S. Gilman and J.S. Benjamin, *Ann. Rev. Mater. Sci.*, Vol.13 (1983), 279.
2. A. Johnson, *New Materials by Mechanical alloying Techniques*, DGM Confer., Calw-Hirsau (FRG), Oct. 1988, Ed. E. Arzt and L. Schultz, Informationsgesellschaft Verlag, 354.
3. J.S. Benjamin, *Novel Powder Processing Adv. In Powder Metall.*, Vol. 7 (1992), Proc. of the 1992 Powder Metallurgy, World Congr., San Francisco, CA, USA, 21-26 June (1992), Publ. Metal Powder Inductries, 155.
4. R.M. Davis and C.C Koch, *Scripta Metall.* Vol.21 (1987), 305.
5. S. Zhang, K.A. Khor and L. Lu, *J. Mater. Proc. Tech.*, Vol.48 (1995), 779.
6. R.M. Davis, B.T. McDermott and C.C. Koch, *Metall. Trans.*, Vol.19A (1988), 2867.
7. D. Lee, J. Cheng, M. Yuan, C.N.J. Wagner and A.J. Ardell, *J. Appl. Phys.*, Vol.64, 4772.
8. J.S. Benjamin, *Met. Constr. Mech.*, Vol.104 (1972), 12.
9. J.S. Benjamin, *Dispersion-strengthened electrical heating alloys by powder metallurgy*, US Patent #US3 660 049, May 2, 1972.
10. J.S. Benjamin, *Powder metallurgical products, British Patent* #1 298 944, December 6, 1972.
11. J.S. Benjamin, R.L. Cairns and J.H. Weber, *Hot working, heat resistant alloys, S. African Patent* #7 104 328, February 21, (1972).
12. J.S. Benjamin, *Dispersion-hardened nickel-chromium-cobalt wrought alloy, German patent #2* 223 715, December 21, (1972).
13. J.S. Benjamin, *Mater. Sci. Forum*, Vol.88-90 (1992), 1.
14. J.S. Benjamin, *New Materials by Mechanical alloying Techniques*, DGM Confer., Calw-Hirsau (FRG), Oct. 1988, Ed. E. Arzt and L. Schultz, Informationsgesellschaft Verlag, 3.
15. J.S. Benjamin, T.E. Volin and J.H. Weber, *High Temp. High Press.*, Vol.6 (1974), 443.
16. R.L. Cairns, *Metall. Trans.*, Vol.5 (1974), 1677.
17. N. Kenyon and R.J. Hrubec, *Welding J.*, Vol.53 (1974), 145.
18. M.S. Grewal, A.S. Sastri and N.J. Grant, *Metall. Trans.*, Vol.6A (1975), 1393.
19. M.J. Bomford and J.S. Benjamin, *U.S. Patent* #US 3 816 080 (1974).
20. G. Jangg, F. Kutner and G. Korb, *Aliminium*, Vol.51 (1975), 641.
21. E. Arzt, *New Materials by Mechanical alloying Techniques*, DGM Confer., Calw-Hirsau (FRG), Oct. 1988, Ed. E. Arzt and L. Schultz, Informationsgesellschaft Verlag, 185-200.
22. J.S. Benjamin and M.J. Bomford, *Metall. Trans.*, Vol.8A (1977), 1301.
23. G. Jangg, F. Kutner and G. Korb, *Powder Metall. Intern.*, Vol.9 (1977), 24.
24. G. Jangg, *New Materials by Mechanical alloying Techniques*, DGM Confer., Calw-Hirsau (FRG), Oct. 1988, Ed. E. Arzt and L. Schultz, Informationsgesellschaft Verlag, 39.
25. N.T. Naresh, *Progress in Mater. Sci.*, Vol.37 (1993), 117.
26. X.P. Niu, *Processing and Characterization of Mechanically Alloyed Aluminium for High Temperature and Wear Resistance Applications*, Ph.D. thesis, K.U.Leuven, Belgium, April 1994.
27. C.C. Koch, O.B. Calvin, C.G. Mckamey and J.O. Scarbrough, *J. Appl. Phys. Lett.*, Vol.43 (1983), 1017.
28. R.B. Schwarz, R.R. Petrich and C.K. Saw, *J. Non-Cryst. Sol*, Vol.76 (1987), 281.
29. E. Hellstern and L. Schultz, *J. Appl. Phys.*, Vol.63 (1988), 1408.
30. R.B. Schwarz and W.L. Johnson, *Phys. Rev. Lett.*, Vol.41 (1983), 415.
31. E. Hellstern and L. Schultz, *Mater. Sci. Eng.*, Vol.93 (1987), 213.
32. J.A. Hunt, I. Soletta, S. Enzo, L. Meiya, R.L. Havill, L. Battezzati, G. Cocco and N. Cowlam, *Mater. Sci. Forum*, Vols.179-181 (1995), 255.
33. T. Nasu, K. Nagaoka, M. Sakurai and K. Suzuki, *Mater. Sci. Forum*, Vols.179-181 (1995), 97.
34. D.L. Beke, H. Bakker and P.I. Loeff, *Coll. Phys*, C4 (1990), 64.
35. A.E. Ermakov, E.E. Yurchikov and V.A. Barinov, *Phys. Met. Metall.*, Vol.52 (1981), 50.
36. X.L. Yeh, K. Samwer and W.L. Johnson, *Appl. Phys. Lett.*, Vol.42 (1983), 242.

37. W.L. Johnson, *Mater. Sci. Eng.*, Vol.97 (1988), 1.
38. R.B. Schwarz, R.R. Petrich and C.K. Saw, *J. Non-Cryst. Solids*, Vol.76 (1985), 281.
39. B.P. Dolgin, *Sci. & Tech. of Rapidly Quenched Alloys*, Ed. M. Tenhover, W.L. Johnson and L.E. Tanner, Publ. Mater. Res. Soc., Pittsburgh, PA, (1987), 447.
40. M.S. Boldrick, D. Lee and C.N.J. Wagner, *J. Non-Cryst. Solids*, Vol.106 (1988), 60.
41. D.G. Morris and M.A. Morris, *J. Less-Commom Metals*, Vol.145 (1988), 277.
42. M. Atzmon, J.D. Verhoeven, E.D. Gibson and W.L. Johnson, *Appl. Phys. Lett.*, Vol.45 (1984), 1052.
43. L. Schultz, *Rapidly Quenched Metals*, Ed. S. Steeb and H. Warlimont, Publ. Elsevier Sci. Publ., B.V., Amsterdam, The Netherlands, Vol.II (1985), 1585.
44. J. Echert, L. Schultz, E. Hellstern and K. Urban, *J. Appl. Phys.*, Vol.64 (1988), 3224.
45. L. Schultz, *J. Less-Common Metals*, Vol.145 (1988), 233.
46. P.Y. Lee and C.C. Koch, *J. Non-Cryst. Solids*, Vol. 94 (1987), 88.
47. A.W. Weeber, P.I. Loeff and H. Bakker, *J. Less-Common Metals*, Vol.145 (1988), 293.
48. C. Plitis and W.L. Johnson, *J. Appl. Phys.*, Vol.60 (1986), 1147.
49. K. Uenishi and P.H. Shingu, *Sintering'87*, Proc. Confer., Ed. S. Somiya, M. Shimada, M. Yoshimura and R. Watanabe, Publ. Elsevier Appl. Sci. Pub., Barking, UK, Vol.1 (1988), 206.
50. G. Cocco, I. Soletta, S. Enzo, M. Magini and N. Cowlam, *J. de Phys.*, Vol51 (1990), C4-181
51. Y.H. Park, H. Hashimoto and R. Watanabe, *Mater. Sci. Forum*, Vols.88-90 (1992), 59.
52. A. Thoma, G.S. Ischenko, L. Schultz and E. Hellstern, *Japaness J. Appl. Phys.*, Vol.26 (1987), 977.
53. R.B. Schwarz, K. Wong and W.L. Johnson, *J. Non-Cryst. Solids*, Vols.61-62 (1984), 129.
54. M.L. Trudeau, R. Schulz, D. Dussault and A. Van Neste, *Phys. Rev. Lett.*, Vol.64 (1990), 99.
55. Z.J. Chui, L. Wang, K.Y. Wang, L. Sun, G.W. Qiao and J.T. Wang, *J. Non-Cryst. Solids*, Vol.150 (1992), 487.
56. K.Y. Wang, A.Q. He, T.D. Shen, M.X. Quan and J.T. Wang, *J. Mater. Res.*, Vol.9 (1994), 866.
57. A.W. Weeber and H. Bakker, *Phys. B*, Vol.53 (1988), 93.
58. L. Schultz, *Mater. Sci. Eng.*, Vol.97 (1988), 15.
59. M. Von Allmen and A. Blatter, *Appl. Phys. Lett.*, Vol.50 (1987), 1873.
60. L. Schultz, K. Schnitzke and J. Wecker, *J. Mag. & Mag. Mater.*, Vol.80 (1989), 115.
61. X.P. Niu, L. Froyen, L. Delaey, C. Peytour, *J. Mater. Sci.*, Vol.29 (1994), 3724.
62. X.P. Niu, P. Le Brun, L. Froyen, C. Peytour and L. Delaey, *Powder Metall. Inter.*, No.3 (1993), 120.
63. P. Le Brun, L. Froyen and L. Delaey, *Mater. Sci. Eng.*, A157 (1992), 79.
64. L. Froyen, L.Delaey, X.P. Niu, P.Le Brun and C. Peytour, *JOM* (1995), 16.
65. C.C. Koch, J.S.C. Jang and P.Y. Lee, *New Materials by Mechanical Alloying*, DGM Confer. Calw-Hirsau, FRG, October (1988), Ed: E. Arzt and L. Schultz, DGM Informationsgesellschaft Verlag, 101.
66. E. Ivanov, T. Grigorieva, G. Golubkova, V. Boldyrev, A.B. Fasman, *Mater. Lett.*, Vol. 1 (1988), 51.
67. C. Suryanarayana, R. Sundaresan and F.H. Froes, *Solid State Powder Processing*, Ed. A.H. Clauer and J.J. deBarbadillo, TMS, Warrendale, PA, USA (1990), 55.
68. M.S. Ei-Eskandarany, K. Aoki and K. Suzuki, *Mater. Sci. Forum*, Vol.88-90 (1992), 81.
69. S. Kobayashi and H. Kimura, *Mater. Sci. Forum*, Vol.88-90 (1992), 97.
70. W. Guo, S. Martelli, F. Padella, M. Magini, N. Burgio, E. Paradiso and U. Franzoni, *Mater. Sci. Forum*, Vol.88-90 (1992), 139.
71. G. Cocco, I. Soletta, L. Battezzati, M. Baricco and S. Enzo, *Phil. Mag.*, Vol.B61 (1990), 473.
72. A. Miyazaki, M. Tokizane and T. Inaba, *J. Japan Inst. Metals*, Vol.54 (1990), 1279.
73. N. Burgio, W. Guo, M. Magini, E. Padella, S. Martelli and I. Soletta, *Structural Appl. of Mechanical Alloying*, Ed. F.H. Froes and J.J. deBarbadillo, ASM Intnl, Mater. Park, OH, USA (1990), 175.
74. C. Suryanarayana, R. Sundaresan and F.H. Froes, *ibid*, 193.
75. R.C. Benn, P.K. Mirchandani and A.S. Watwe, *Solid State Powder Processing*, Ed. A.H. Clauer and J.J. deBarbadillo, TMS, Warrendale, PA, USA (1990), 157.
76. R. Lerf and D.G. Morris, *Mater. Sci. Eng.*, Vol.A128 (1990), 119.

77. C.R. Suryanarayana, F.H. Froes, *Mater. Sci. Eng.*, Vol.150A (1992), 117.
78. H.J. Fecht, *Nanophase Materials, Synthesis-Properties-Applications*, Ed. G.C. Hadjipanayis and R.W. Siegel, Kluwer Academic Publishers, Netherlands, (1994), 145.
79. P.H. Shigu, B. Huang, S.R. Nishitani, S. Nasu, *Suppl. Trans. Japan Inst. Metals*, Vol.29 (1988), 3.
80. M.A. Morris and D.G. Morris, *Mater. Sci. Eng.*, Vol.A136 (1991), 4687.
81. R.B. Schwarz, *Scripta Metall. Mater.*, Vol.34 (1966), 1.
82. H.J. Fecht, Hellstern, Z. Fu and W.L. Johnson, *Adv. Powder Metall.*, Vol.1 (1989), 111.
83. H.J. Fecht, E. Hellstem, Z. Fu and W.L. Johnson, *Metall. Trans.*, Vol.A21 (1990), 1744.
84. J. Eckert, J.C. Holzer, C.E. Krill III and W.L. Johnson, *J. Mater. Res.*, Vol.7 (1992).
85. E. Gaffet, N.M. Gaffet, *J. Alloys. Comp.*, Vol.205 (1994), 27.
86. P.H. Shingu, B. Huang, J. Kuyama, K.N. Ishihara and S. Nasu, *New Materials by Mechanical Alloying Techniques*, Proc. of The Confer., Calw-Hirsau, West Germany, 3-5 October, 1988, Ed. E. Arzt and L. Schultz, Deutsche Gesellschaft fur Metall., Oberursel, Germany (1989), 319.
87. J.S.C. Jang and C.C. Kock, *J. Mater. Res.*, Vol.5 (1990), 498.
88. C. Suryanarayana and F.H. Froes, *J. Mater. Res.*, Vol.5 (1990), 1880.
89. S. Srinivasan, P.B. Desh and R.B. Schwarz, *Sripta Metall. Mater.*, Vol.25 (1991), 2513.
90. C. Suryarayana and F.H. Froes, *Mater. Sci. Eng.*, Vol.A179/190 (1994), 108.
91. T. Christman and M. Jainm, *Scripta Metall. Mater.*, Vol.25 (1991), 767.
92. M.S. Boldrick, E. Yang and C.N.J. Wagner, *J. Non-Cryst. Solids*, Vol.150 (1992), 478.
93. C. Suryanarayana, W. Li and F.H. Froes, *Scripta Metall. Mater.* Vol.31 (1994), 1465.
94. C. Suryanarayana, E. Zhou, E. Peng and F.H. Froes, *Scripta Metall. Mater.*, Vol.30 (1994), 781.
95. A. Teresiak, N. Mattern, H. Kubsch and B.F. Kieback, *Nanostructured Mater.*, Vol.4 (1994), 775.
96. J.C. Rawers, R.D. Govier and G. Korth, *Mater. Sci. Forum*, Vols.179-181 (1995), 363.
97. M. Oehring, F. Appel, Th. Pfullmann and R. Bormann, *Mater. Sci. Forum*, Vols.179-181 (1995),435.
98. Y.H. Park, H. Hashimoto, M. Nakamura, T. Abe and R. Watanabe, *J. Japan Inst. Metals*, Vol.57 (1993), 952.
99. A.A. Popovich, V.P. Reva, V.N. Vasilenko and O.A. Belous, *Mater. Sci. Forum*, Vol.88-90 (1992), 737.
100 T. Takahashi, *Mechanical Alloying for Structural Applications*, Proc. of the 2nd Intl, Confer., Vancouver, British Columbia, Canada, 20-22 September (1993), Ed. J.J. deBarbadillo, F.H. Froes and R. Scharz, ASM Intl., Mater. Park, OH, 307.
101. Y.H. Park, H. Hashimoto, M. Nakamura, T. Abe and R. Watanabe, *Proc. of 1993 Powder Metall. World Congr.*, Kyoto, Japan, 12-15 July (1993), Ed. Y. Bando and K. Kosuge, Japan Society of Powder and Powder Metall., 189.
102. Y.H. Park, H. Hashimoto, T. Abe and R. Watanabe, *Mater. Sci. Eng.*, Vol.181/182 (1994), 1291.
103. D.D. Radev and D. Klissurski, *J. Alloys and Compounds*, Vol.206 (1994), 39.
104. A. Calka and A.P. Radlinski, *J. Less-Common Metals*, Vol.161 (1990), L23.
105. J.L. Hoyer, *Mater. Manufact. Proc.*, Vol.9 (1994), 623.
106. S. Wanikawa and T. Takeda, *J. Japan Soc. Powder Powder Metall.*, Vol.36 (1989), 672.
107. A. Calka, A.P. Radlinski, R.A. Shanks, A.P. Pogany, *J. Mater. Sci. Lett.*, Vol.10 (1991), 734.
108. A. Calka, *Key Eng. Mater.*, Vol.81-83 (1993), 17.
109. W.Y. Lim, M. Hida, A. Sakakibara, Y. Takemoto and S. Yokomizo, *J. Mater. Sci.*, Vol.28 (1993), 3463.
110. A. Calka and J.S. Williama, *Mater. Sci. Forum*, Vol.88-90 (1992), 787.
111. A. Calka, J.I. Nikolov and B.W. Ninham, *Mechanica Alloying for Structural Applications*, Proc. 2nd Intl' Confer. on Structural App. of Mechanical Alloying, Vancouver, British Columbia, Canada, 20-22 Sep. 1993, Ed. J.J. deBarbadillo, F.H. Froes and R. Schwarz, (1993), 189.
112. T. Hagio and H. Yoshida, *J. Mater. Sci. Lett.*, Vo.13 (1994) 653.

113. A. Malchere and E. Gaffet, *Mechanical Alloying for Structural Applications*, Proc. of the 2nd Intl, Confer., Vancouver, British Columbia, Canada, 20-22 September (1993), Ed. J.J. deBarbadillo, F.H. Froes and R. Scharz, ASM Intl., Mater. Park, OH, 297.
114. A. Calka, *Proc. of the 1st Intern. Confer. on Mechanochemistry*, Kosice, Slovak Republic, 23-26 March, 1993, Ed. P. Balaz, B. Plesingerova, V. Sepelak and N. Stevulova, Cambridge Interscience Publishing, Cambridge, England, Vol.2, (1994), 36.
115. M.S. El-Eskandarany, K. Sumiyama, K. Aoki and K. Suzuki, *J. Mater. Res.* Vol.7 (1992), 888.
116. Y. Ogino, M. Miki, T. Yamasaki and T. Inuma, *Mater. Sic. Forum*, Vols.88-90 (1992), 795.
117. H.Y. Wong, *Mechanical Alloying of Ti-B system*, B. Eng. thesis, Naitonal University of Singapore, (1996).
118. H.D. Hedrich, *New Materials by Mechanical Alloying Techniques*, Proc. of The Confer., Calw-Hirsau, West Germany, 3-5 October, 1988, Ed. E. Arzt and L. Schultz, Deutsche Gesellschaft fur Metall., Oberursel, Germany (1989), 217.
119. G.M. McColvin and M.J. Shaw, *Mater. Sci. Forum*, Vol.88-90 (1992), 235.
120. J.J. Fischer, J.J. deBarbadillo and M.J. Shaw, *Structural Aplications of Mechanical Alloying*, Proc. of an ASM Inter. Confer., Myrtle Beach, South Carolina, 27-29 March 1990, Ed. F.H. Froes and J.J. deBarbadillo, Publ. ASM Inter., Mater. Park, Ohio (1990), 79.
121. G.B. Schaffer and P.G. McCormick, *Appl. Phys. Lett.*, Vol.55 (1898), 45.
122. G.B. Schaffer and P.G. McCormick, *Metall. Trans. A*, Vol.21A (1990), 2789.
123. G.B. Schaffer and P.G. McCormick, *Metall. Trans. A*, Vol.22A (1991), 3019.

# 2

# EXPERIMENTAL SET-UP

## 2.1 Ball Mills

For mechanical alloying, the powder mixture which may be elemental powder particles or alloying particles, is charged in the container (also known as bowl, vial, or jar) of a ball milling machine (or a ball mill), together with some steel or ceramic balls. The powder mixture is subjected to high energy collision by the moving balls as the bowls are rotated. The very basic equipment for mechanical alloying is therefore a ball milling machine.

The alloying process can be carried out using different apparatus, namely, attritor, SPEX shaker mill, planetary mill or a horizontal ball mill. Whichever equipment is employed, however, the principles of operation are the same. Since the powders are cold welded and fractured during mechanical alloying, it is critical to establish a balance between the two processes in order to alloy successfully [1, 2]. The ability of the powders to cold weld and fracture depends upon the alloy system and the milling conditions employed. Soft materials normally have good weldability but high toughness and hence difficult to be fractured. It is believed that the key to successfully produce an alloy is to select the correct impact energy for the milling process.

## 2.2 Planetary ball mill

Planetary ball mill is a very often used machine for mechanical alloying, especially in Europe [3-9]. Because very small amount of powder (for example, as little as a few grammes), is required, the machine is suitable for research purposes in the laboratory [10-12]. A typical planetary ball mill is shown in Figure 2.1 manufactured by Fritsch GmbH. The ball mill consists of one turn disc (sometimes called turn table) and two or four bowls. The turn disc rotates in one direction while the bowls rotate in the opposite direction. The centrifugal forces created by the rotation of the

Figure 2.1. Typical planetary ball mill (Fritsch Puluerisette 5).

bowl around its own axis together with the rotation of the turn disc are applied to the powder mixture and milling balls in the bowl. The powder mixture is fractured and cold welded under high energy impact.

Figure 2.2 shows the motions of the balls and the powder. Since the directions of rotation of the bowl and turn disc are opposing, the centrifugal forces are alternately synchronised. Thus friction resulted from the milling balls and the powder mixture being ground alternately rolling on the inner wall of the bowl, and impact results when they are lifted and thrown across the bowl to strike at the opposite wall. The impact is intensified when the balls strike one another. The impact energy of the milling balls in the normal direction attains a value of up to 40 times higher than that due to gravitational acceleration. Hence, the planetary ball mill can be used for high speed milling.

The impact energy of the milling balls is changeable by altering the rotational speed of the turn disc. The advantage of this type of ball mill is not only that high impact energy could be obtained but also high impact frequency which can shorten the duration of the mechanical alloying process. However, it must be noted that because of high impact frequency, the temperature of the bowl may reach about 393K within a short milling duration of only 30 to 60 min. In cases where relatively high temperature is necessary to promote reaction rate, even this may be an added advantage to the process. In addition, the planetary ball mill may be modified by incorporating temperature control elements.

Two types of bowls are commercially available: steel including hardened chrome steel, stainless CrNi-steel and hardmetal tungsten carbide (WC+Co) and ceramic bowls including sintered corundum ($Al_2O_3$), agate ($SiO_2$) and zirconium oxide ($ZrO_2$). They generally are available in three different sizes of 80, 250 and 500ml. For high energy mechanical alloying, however, steel bowls are recommended since ceramic bowls can cause contamination due to minute chipped off or fractured particles from the brittle surfaces of the milling bowl and balls. Generally, bowls and balls of the same material are employed in the mechanical alloying process to avoid the possibility of cross contamination from different materials.

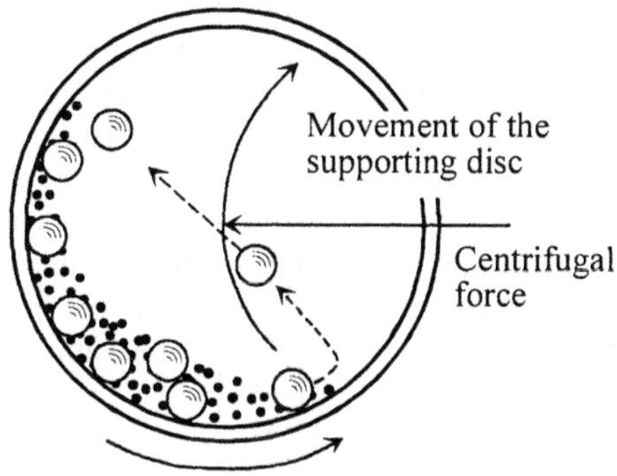


Figure 2.2. Schematic view of motion of the balls and powder mixture.

Based on powder particle size and impact energy required, balls with size of 10 to 30 mm are normally used. If the size of the balls is too small, impact energy may be too low for alloying to take place. In order to increase impact energy without increasing the rotational speed, balls with high density such as tungsten balls may be employed. Table 2.1 gives the recommended number of balls per bowl to be applied.

Table 2.2 gives a summary of abrasion properties and densities for the selection of bowl and ball materials. It can be seen that the oxide materials show the lowest density while tungsten carbide, the highest density. Hence, at the same rotational speed and ball size, the oxide ball with the lowest density will generate the lowest collision energy.

Table 2.1. Recommendation on number of balls per bowl [13].

| Ball size (mm) | Capacity of bowls (ml) | | |
|---|---|---|---|
| | 500 | 250 | 80 |
| 10 | 100 | 50 | 30 |
| 20 | 20 | 15 | 5 |
| 30 | 10 | 6 | |
| 40 | 4 | | |

## 2.3 Conventional horizontal ball mill

Horizontal ball mill (as shown in Figure 2.3) is a conventional apparatus for mechanical alloying. As the name suggests, the mill rotates about a central horizontal axis. Normally, the diameter of the container should be of a size greater than 1 m. The maximum rotational speed should be adjusted just below the critical speed that pins the balls to the internal wall of the container. At this speed, the balls should fall down at maximum height to yield maximum collision energy. Because a large container diameter is required to generate enough impact energy, this type of ball mill is not suitable for laboratory scale research. It is mainly used in industry because large quantity of powder particles can be mechanically alloyed in a single batch.

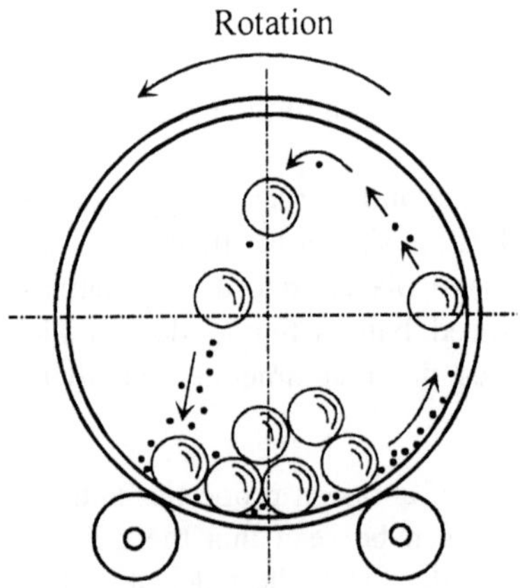


Figure 2.3. Conventional horizontal ball mill.

Table 2.2. Properties of materials used for bowl and ball [13].

| Material | Description | Abrasion behaviour | Density $g/cm^3$ |
|---|---|---|---|
| Agate ($SiO_2$) | microcrystalline felted, weakly transparent chalcedony, high surface quality | Very abrasion proof approximately 200 times more resistant than hard porcelain | 2.65 |
| Zirconium oxide ($ZrO_2$) | light yellow sintered product of zirconium oxide powder, very high sintered density | Very abrasion proof approximately 10 times more resistant than sintered corundum | 5.7 |
| Sintered corundum I ($Al_2O_3$) | almost pure aluminium oxide, very corrosion resistant | good abrasion resistance | 4.0 |
| Sintered corundum II ($Al_2O_3$) | | satisfactorily abrasion proof | 3.5 |
| Hard porcelain | | satisfactorily abrasion proof | 3.1 |
| Hardened steel | material Ck45 No.1191 surface hardened (tennifer) | good abrasion resistance | 7.9 |
| Hardened chromium steel | steel type No.2080 (HRC 60-63) | good abrasion resistance better than CrNi-steel | 7.9 |
| Hardened chromium steel | steel type No.2601 | good abrasion resistance better than CrNi-steel | 7.9 |
| Stainless steel (CrNi-steel) | steel type No.4301 | average abrasion resistance | 7.9 |
| Hardmetal Tungsten carbide WC+Co | Colloidal granulate sintered together with cobalt metal (binding agent); crystallized | extremely abrasion proof approximitely 200 times more resistant than agate | 14.75 |

### 2.4 Horizontal ball mill controlled by magnetic force

It is worthwhile to emphasize that the magnet controlled horizontal ball mill is generally operated in a low energy mode [(14-16)]. However, an important aspect of this device is the possibility of incorporating a magnetic field generated by permanent magnet or magnets to act on the ferromagnetic balls as shown in Figure 2.4. Depending upon the distance between the balls and the magnet, the impact energy can be controlled by changing the spatially adjustable high field magnets. This allows the impact energy to be selected according to the different material systems to be milled. Under an appropriately adjusted magnetic field, the effective mass of the ball can be increased by a factor of about 80 (from 60 g up to about 5 kg). The type of magnetic field can be varied by altering the position of the magnets, which in turn changes the type of energy to be transferred to the powder mixture. Three different energy modes can be obtained: impact mode, shear force mode, and impact and shear force modes.

From Figure 2.4, it can be seen that frictional energy can be altered by changing the intensity of magnet M1, while magnet M2 is mainly used to increase the kinetic energy of the balls. Position M3 controls both the frictional and the kinetic energies of the process. A high energy mode can therefore be obtained by choosing strategic positions for magnets M1, M2 or M1 and M2 together. For low energy mode, the magnet may be placed at position M3. Numerous research works have been carried out based on this innovative design of the ball mill [(17-22)].

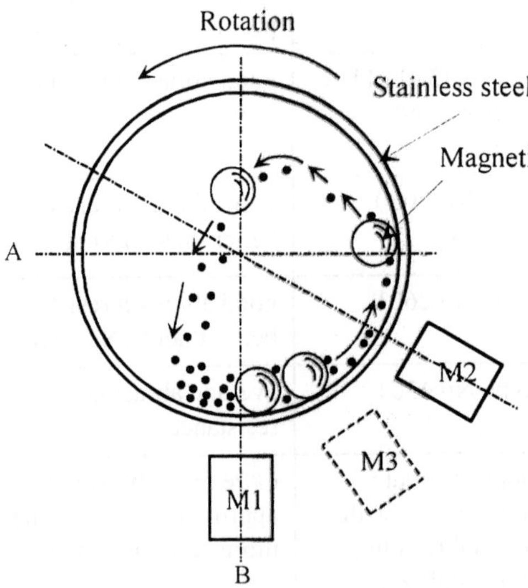


Figure 2.4. Horizontal ball mill incorporated with magnets.

Because of low milling speed, the milling environment in a horizontal ball mill can be made controllable, from vacuum to a pressure of 500kPa overpressure, of gas or liquid The temperature of the milling bowl can also be controlled from room temperature to 473 K by attaching heating elements to the ball mill. It can therefore be seen that the low operating rotational speed of a horizontal ball mill can be made to be advantageous. In addition, less heat is generated and accumulated in the milling process, an important factor to consider in the mechanical alloying of amorphous materials and some high exothermical reaction systems.

## 2.5 Shaker ball mill

Shaker ball mills, as shown in Figure 2.5, are generally used to process small quantity of powder mixture. A typical example of such mill is the SPEX 8000 which was originally developed to pulverize spectrographic samples. This type of mill, which is suitable for research purposes, is widely used in U.S.A. [23-25]. It agitates the charge of powder and balls in three mutually perpendicular directions at approximately 1200 rpm. The container may have a capacity of up to $55x10^{-6}m^3$. Compared to attrition and vibratory ball mills, it is highly energetic. High milling energy can be obtained by using high frequencies and large amplitude of vibration. Table 2.3 shows the number of impact for different number of balls and the resulting kinetic energy occurring over 0.5 or 1.0 second of milling [26]. It can be observed that the majority of impact occurs in the $10^{-3}$ to $10^{-2}$ Joules range.

Table 2.3. Number of impact for different number of balls and the resultant values of kinetic energy [26].

| Number of ball used (2g each) | Milling operating for 0.50 seconds Kinetic energy of impact (J) | | | | |
|---|---|---|---|---|---|
| | $10^{-7}$-$10^{-4}$ | $10^{-4}$-$10^{-3}$ | $10^{-3}$-$10^{-2}$ | $10^{-2}$-$10^{-1}$ | $10^{-1}$-1.0 |
| 5 | 0 | 43 | 297 | 3 | 0 |
| 10 | 0 | 78 | 505 | 13 | 1 |
| 15 | 4 | 124 | 928 | 24 | 0 |
| Mill operating for 1.00 seconds | | | | | |
| 5 | 0 | 78 | 612 | 3 | 0 |
| 10 | 0 | 148 | 1201 | 13 | 2 |
| 15 | 4 | 229 | 1873 | 24 | 0 |

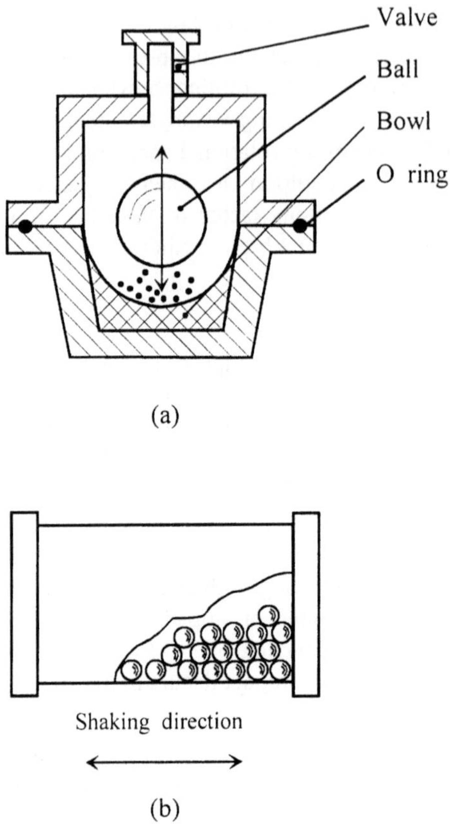


Figure 2.5. Two types of vibrate mills.

## 2.6 Attritor

Mechanical attrition is a versatile process and is one of the earliest mechanical process to synthesize different types of materials in large quantities [29, 30]. This method has been developed since 1970's as an industrial process to successfully produce new alloys. Mechanical attrition has gained considerable attention in material research as a non-equilibrium process resulting in solid state alloying beyond the equilibrium solubility limit [31-33]. A wide range of alloys, intermetallics, ceramics, amorphous and composites have been successfully processed [34].

As shown in Figure 2.6, the central shaft (fitted with paddles) in the attritor rotates at high speed to collide with balls and generate high energy collision between the steel balls and the powder charge to allow mechanical alloying to take place. Small attritors have been utilised for research and development purposes. Normally, 1 kg of powder mixture can be milled in an attritor. The maximum capacity of an attritor for mechanical alloying is about $3.8x10^{-1}m^3$ (100 gallons) with the central shaft rotating at a speed of up to 250 rpm [(2)]. Because of high speed rotation of the shaft, the capacity is somewhat limited. The relatively high frictional motion between the central shaft and the steel balls, and between the container and the steel balls can easily cause contamination to the powder. The rise in temperature during the alloying process is modest and is estimated to be less than 100 to 200°C. Since the milling container is stationary, it can easily be cooled by water. To reduce contamination, milling tools employed can be coated with the same material as the material to be milled.

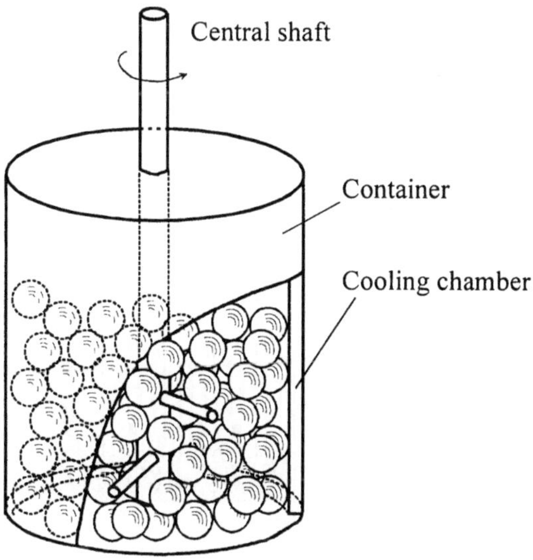


Figure 2.6. High energy attritor-type ball mill used for mechanical alloying.

Another type of attritor known as horizontal attritor has been developed recently in Zoz GmbH, Germany [(35)]. The development of this type of attritor was targeted for applications in the area of mechanical alloying, processing of nanocrystalline materials, amorphous materials, oxide dispersion strengthened alloys, iron based magnetic and other materials.

As shown in Figure 2.7, this type of attritor is a rotary ball mill with a horizontal borne rotor that rotates at high speed to set the balls in motion. Some of the advantages of this attritor are the absence of dead zones due to gravity during milling, possibility of achieving extremely high impact energy and the incorporation of controlled atmosphere.

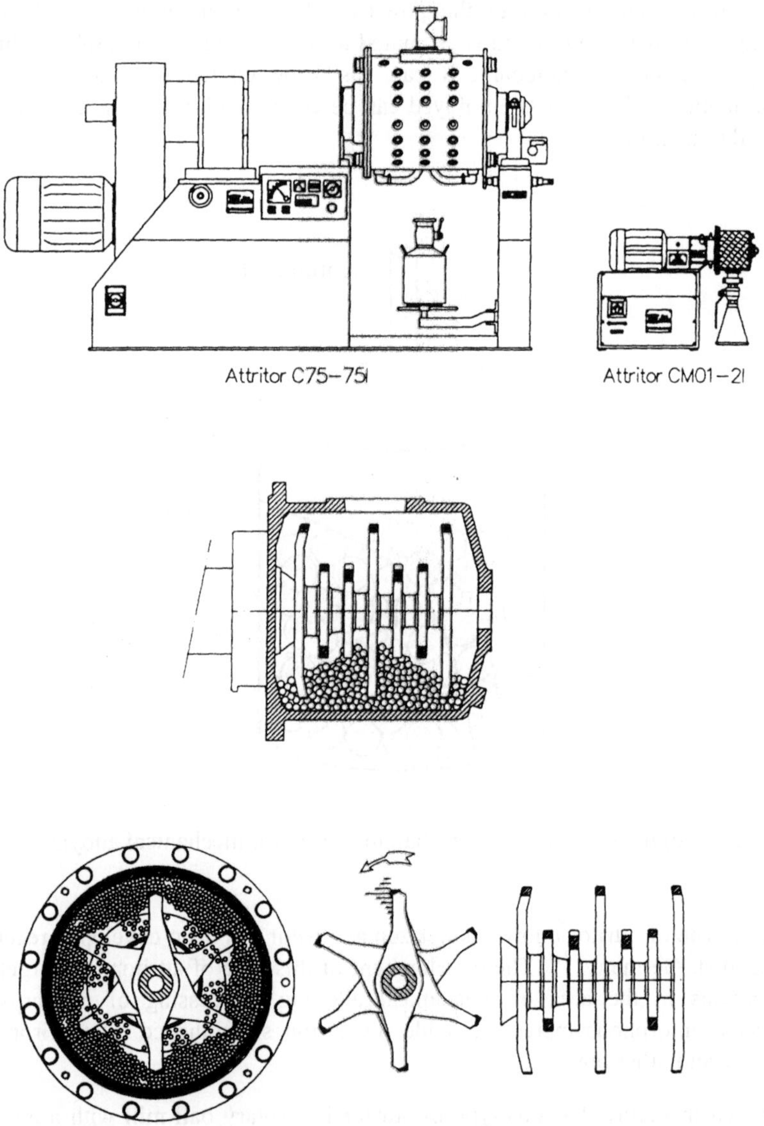


(a)

(b)

(c)

Figure 2.7. (a) Horizontal attritor, (b) grinding chamber, and (c) grinding chamber and rotor [35].

## 2.7 Conclusions

Several types of ball mills are commercially available. Among them, the planetary ball mill, attritor and shake mill are the most widely used in laboratory scale research. These mills generally show high energy behaviour. Because impact energy of these ball mills is dependent upon milling speed, high impact frequency has to be employed to generate high impact energy. However, because of high impact frequency, heat may easily be accumulated. The increase in milling temperature has to be overcome by considering the interval of milling, namely, by following a mechanical alloying sequence of milling and resting alternately. This is especially important for reactive milling, although in some cases, heating of the ball mill is necessary in order to promote chemical reaction.

As SPEX, attritor and planetary ball mills possess high impact energy intensity and high impact frequency, they have been employed to synthesize a large variety of materials. In addition, the high impact energy and frequency enable the mechanical alloying duration to be shortened. Compared to gravitational and centrifugal ball mills, SPEX, attritor and planetary ball mills can achieve a significantly higher collision energy.

Horizontal ball mill possesses high impact energy even though the impact frequency is generally low. The impact energy can be further adjusted by the incorporation of magnets with the appropriate intensity located at strategic positions around the milling chamber. In other words, impact energy can be controlled while the impact frequency remains constant. Since low ball milling frequency and high impact energy can be obtained at the same time, the increase in temperature due to impact and chemical reaction can be minimized. The process, with the disadvantage in efficiency, is nevertheless very useful for mechochemical milling. Horizontal ball mill is generally not suitable for research purposes because container of large diameter is required to achieve successful mechanical alloying.

## 2.8 References


1. J.S. Benjamin, *Mew Materials by Mechanical Alloying Techniques*, DGM Confer., Calw-Hirsau, FRG, 3-5 October, 1988, Ed: E. Arzt and L. Schultz, 3.
2. P.S. Gilman and J.S. Benjamin, *Ann. Rev Mater. Sci.*, Vol.13 (1983), 279.
3. E. Gaffet, *Mater. Sci. Eng.*, A136 (1991), 161.
4. M. Abdelloui and E. Gaffet, *J. Phys IV*, Vol.4 (1994), C3-291.
5. E. Gaffet, C. Louison, M. Harmelin and F. Faudot, *Mater. Sic. Eng.*, A134 (1991), 1380.
6. T. Klassen, M. Oehring and R. Bormann, *J. Mater. Res.*, Vol.9 (1994), 47.
7. M.V. Zdujic and O.B. Milosevic, *Mater. Let.*, Vol.13 (1992), 125.
8. G. Walkowiak, T. Sell and H. Mehrer, *Z. Metallkd.*, 85 (1994), 5.
9. L. Froyen, L. Delaey, X.P. Niu, P. Le Brun and C. Peytour, *JOM*, (1995), 16.
10. S. Zhang, L. Lu, M.O. Lai, *Mater. Sci. Eng.*, A171 (1993), 257.
11. K.Aoki, X.M. Wang, A. Memezawa and T. Masumoto, *Mater. Sci. Eng.*, A179/A180 (1994), *390.*

12. X.P. Niu, A. Mulaba-Bafubiandi, L. Froyen, L. Delaey and C. Peytour, *Scripta Metall. Mater.*, Vol.31 (1994), 1157.
13. Fritsch GmbH, Industriestr. 8, D-55743 Idar-Oberstein.
14. A. Calka and B.W. Ninham, *US Patent*, #5,383,615, Jan. 24, 1995
15. *Anutech Pty Ltd*, ASI., Canberra ACT 0200, Australia.
16. W.A. Kaczmarek, R. Bramley, A. Calka and B.W. Ninham, *IEEE Trans. Mag.*, MAG-26, No.5 (1990), 1840.
17. W.A. Kaczmarek, B.W. Ninham and A. Calka, *J. Appl. Phys.* Vol.70 (1991), 6280.
18. A. Calka and A.P. Radlinski, *Mater. Sic. Eng.*, Vol.134 (1991), 1350.
19. W.A. Kaczmarek, E.Z. Radlinska and B.W. Ninham, *Mater. Chy. Phys*, Vol.35 (1993) 31.
20. A. Calka and W.A. Kaczmarek, *Scripta Metall.*, Vol.26 (1992), 249.
21. W.A. Kaczmarek, A. Calka and B.W. Ninham, *Phys. Stat. Sol.*, Vol.141 (1994), K1.
22. W.A. Kaczmarek, A. Calka and B.W. Ninham, *Phys. Stat. Sol.*, Vol.141 (1994), K123.
23. P.S. Gilman and W.D. Nix, *Metall. Trans.*, Vol.12A (1981), 813.
24. J. Eckert, J.C. Holzer, C.E. Krill and W.L. Johnson, *Mater. Sic. Forum*, Vols.88-90 (1992), 505.
25. E. Ma and M. Atzmon, *Mater. Sci. Forum*, Vols.88-90 (1992), 467.
26. C.C. Koch, *Annu. Rev. Mater. Sci., Vol.*19 (1989), 121.
27. W.E. Kuhn, I.L. Friedman, W. Summers and A. Szegvari, *ASM Metals Handbook*, Vol.7 (1985), 56.
28. C.C. Koch and M.S. Kim, *J. Phys.*, Vol.46 (1985), C8-537
29. W. Mader and K.F. Muller, *Radex Rundschau*, No. 4 (1971), 535.
30. I. Pupke, *Powder Metall. Internat.*, Vol.3 (1971), 94.
31. H.J. Fecht, E. Hellstern, Z. Fu and W.L. Johnso, *Mater. Trans.*, A.21 (1990), 2333.
32. H.J. Fecht, E. Hellstern, Z. Fu and W.L. Johnso, *Adv. Powder Metall.*, Vol.1 (1989), 111.
33. J. Eckert, J.C. Holzer, C.E. Krill III and W.L. Johnson, *J. Mater. Res.* Vol.7 (1992), 1751.
34. H.L. Fecht, *Nanophase Materials, Synthesis-Properties-Applications*, Ed. G.C. Hadjipanayis and R.W. Siegel, Publ: Kluwer Academic Publishers, (1994), 125.
35. H. Zoz, *Mater. Sci. Forum*, Vols.179-181 (1995), 419.

# 3

# THE MECHANICAL ALLOYING PROCESS

## 3.1 Preparation for mechanical alloying

### *3.1.1 Mechanical alloying process*

The mechanical alloying/milling process is a solid state powder process where the powder particles are subjected to high energetic impact by the balls in a vial [1-5]. As the powder particles in the vial are continuously impacted by the balls, cold welding between particles and fracturing of the particles take place repeatedly during the ball milling process. The entire process includes blending of the powder mixture prior to the ball milling, vacuuming and/or filling with protective gases to prevent oxidation and contamination, and the ball milling process itself. Post milling processes after mechanical alloying includes canning, degassing, and finally plastic deformation by extrusion or rolling. The whole process is summarized in Figure 3.1. Before mechanical alloying, it is suggested that the powder particles be blended initially in a blender so that a homogeneous distribution of the different powders can be obtained. For the milling process, several control parameters have to be considered, namely, environment in the vial, ball to powder weight ratio and process control agents.

### *3.1.2 Atmosphere control*

To prevent oxidation and contamination during mechanical alloying, the process is normally performed in an inert atmosphere or vacuum in a ball mill [5-8]. Ball mills are generally not attached with atmosphere control chambers. To prevent oxidation during milling, it is possible to use a glove box filled with inert gas. In most cases, purified argon gas is employed [9-14]. Other inert gases used include helium and nitrogen [15]. The powder mixture blended is firstly loaded in a glove box protected by inert gas. The milling vial should then be sealed with a Teflon or rubber ring before removing from the glove box. Using vacuum vial is another alternative

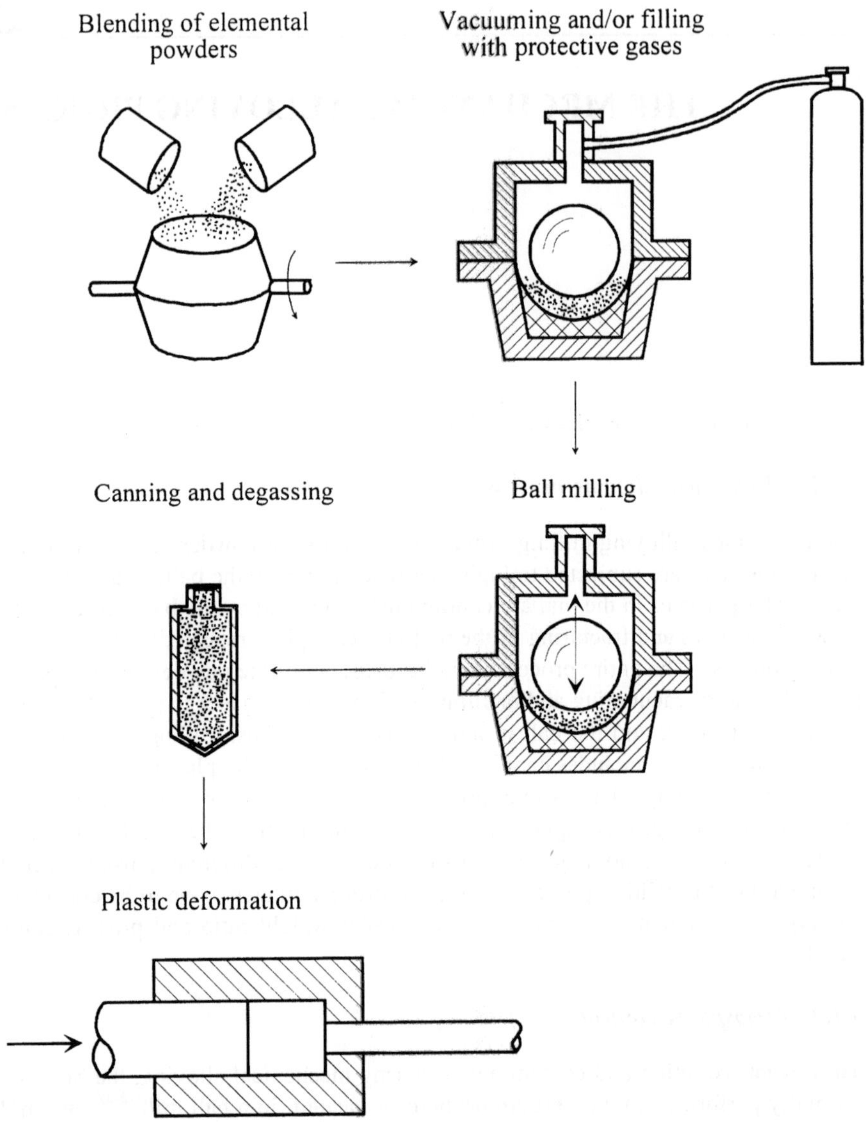


Figure 3.1. The ball milling sequence

solution [16]. In this method, powder mixture can be loaded into a vial in the open air. After sealing, the vial is then vacuumed. The advantage of using a vacuumed system is the reduction of contamination from the atmosphere, a method better than using inert atmosphere.

Nitrogen is a very stable inert gas even when it is heated to very high temperature. In the case of mechanical alloying, however, it does show to be unstable since it reacts with most of elemental materials and compounds resulting in the formation of nitrides [17, 18]. It is therefore not recommended to use nitrogen to provide an inert atmosphere in the mechanical alloying process.

### *3.1.3 Selection of ball to powder weight ratio*

Ball to powder weight ratio is another important parameter in mechanical alloying which has to be properly selected. In general, the higher the weight ratio, the faster is the mechanical alloying process. This is because the number of collision per unit time increases as the number of balls increases. At the same time, increase in collision frequency results in an increase in milling temperature, which in turn leads to a faster diffusion process. The ball to powder weight ratio used is generally in the range between 10:1 and 20:1 [20-22]. Weight ratios of about 5:1 or less are often employed when vibrator type of ball mill or SPEX mill is used [23-25]. In some extreme cases, weight ratio as high as 100:1 to 500:1 [26-28] may be used. Normally, horizontal ball mills tend to use high weight ratio [29, 30], while a ratio of about 20:1 is very often applied in planetary ball mills.

Niu [31] studied the influence of ball to powder weight ratio using a planetary ball mill under a constant milling condition. He found that the microhardness of the powder increased continuously with the increase in weight ratio due to an enhanced and accelerated plastic deformation of the powder particles. The rate of hardening however decreased with the increase in weight ratio as shown in Figure 3.2. The mean particle size on the other hand, exhibited a different trend. It was observed to reduce dramatically initially to its minimum value when the weight ratio was increased from 2:1 until 5:1. The mean particle size started to increase to a maximum value at a weight ratio of 11:1, thereafter, a steady decrease was observed. At each stage of milling, the powder particles showed different morphologies: flaky at a weight ratio of 3:1, flaky plus equiaxed at 5:1 and equiaxed at greater than 10:1 [31].

In practice, balls with the same diameter are often used in a milling process. By using a numerical simulation of a shaker ball mill, however, Gavrilov [32] has predicted that the highest collision energy can be obtained if balls with different diameters are used. This is most likely due most likely to the interference between balls of different diameters. Further experimental research should be performed to verify such prediction.

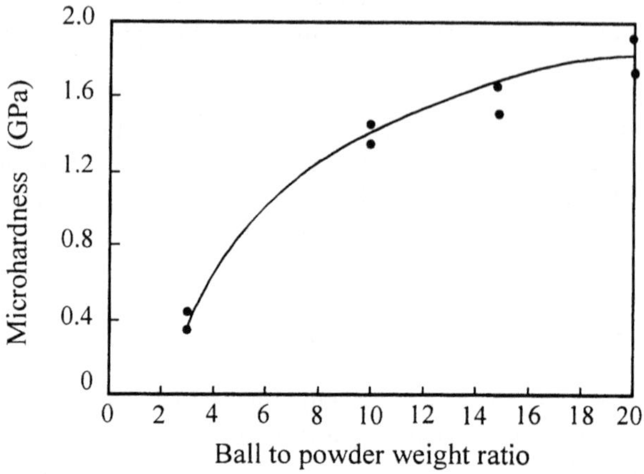


Figure 3.2. Influence of ball to powder weight ratio on powder hardness [31].

### *3.1.4 Selection of process control agents*

Powder particles in the ball mill are subjected to high energy collision [19] which causes the powder particles to be cold welded together and fractured. The cold welding and fracturing process enables powder particles to be always in contact with each other with atomically clean surfaces and with minimized diffusion distance. The essential condition for a successful mechanical alloying process is the balance between cold welding and fracturing. However, this balance in most cases may not be obtainable by the milling process itself, especially if soft materials are used. For such cases, cold welding between powder particles, and between powder particles and milling tools (bowl and balls) becomes a serious problem. The degree of cold welding is dependent on the ductility and the ability to cold welding of the powder to be milled. Depending upon which process is dominant during mechanical alloying, micro-forging or fracturing, the powder particles may grow in size through agglomeration by cold welding, and may change from an equiaxed to a platelet or flake particles by micro-forging or become smaller in size through the fracture process [33]. Figure 3.3 (a) shows the Al powder particles mechanically alloyed at 250 rpm. It can be observed that cold welding leads to an increase in particle size to more than 5 mm while Figure 3.3 (b) gives another example of Al powder being cold welded to a milling ball. Therefore, soft materials such as Al powder cannot simply be mechanical alloyed because the process is impeded by excessive cold welding of the powder particles, preventing them from fracturing. Hence, the critical balance between cold welding and fracturing that is necessary for successful mechanical alloying cannot be achieved [5].

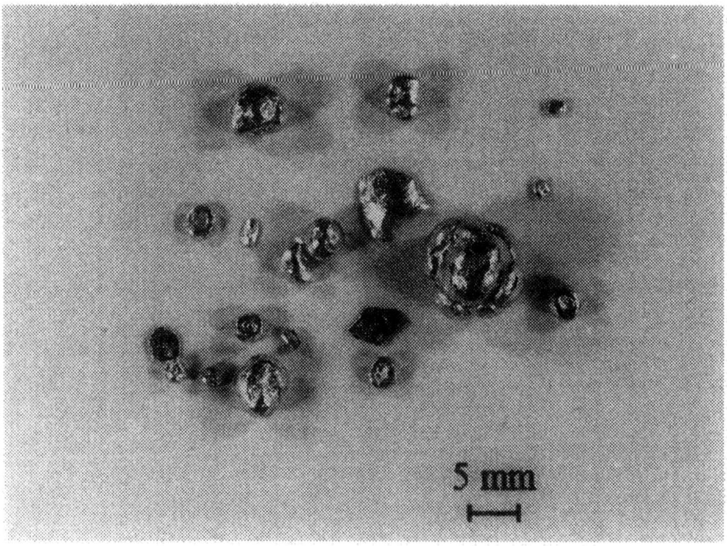


(a)

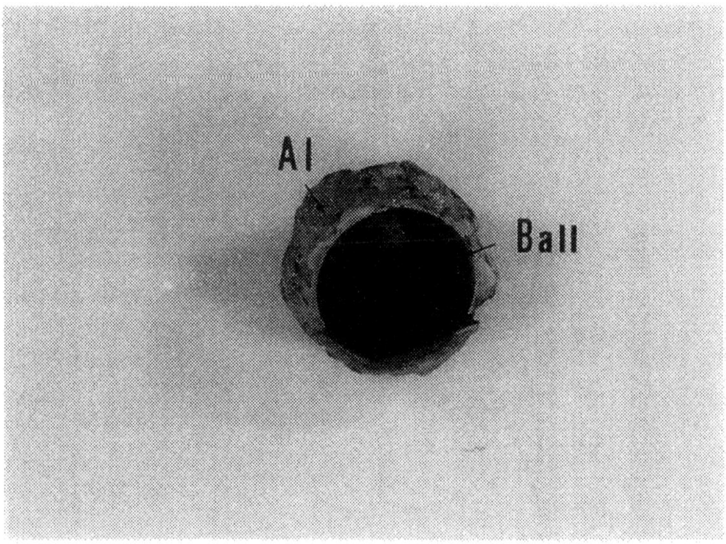


(b)

Figure 3.3. Mechanically milled pure Al powder particles: (a) excessive cold welding leading to an increase in particle size, and (b) serious cold welding of Al particles onto a milling ball.

Two approaches may be taken to reduce excessive cold welding and promote fracturing. The first technique is to modify the surface of the deforming particles by introducing a suitable organic material that impedes the clean metal-to-metal contact necessary for cold welding [34]. The second approach is to modify the deformation mode of the powder particles so that they fracture before they are able to deform to the large compressive strains necessary for flattening and cold welding. This can be accomplished by cryogenic milling. Of the two techniques, the former is more often used [34].

Surface agents and lubricants, generally known as process control agents, are often used to nullify the forces of cold welding during mechanical alloying. The process control agents, being absorbed on the surface of the particles, help inhibit excessive cold welding and therefore agglomeration by lowering the surface tension of the solid materials. Because energy required for milling is a function of plastic deformation of the powder particles and the new surface area generated times the surface tension, a reduction in surface tension results in finer powders [36]. Hence, use of process control agent is necessary. The effectiveness of a given process control agent depends on its reactivity with the metal being milled. Figure 3.4 shows the size of the particles milled with and without process control agent [37]. In the absence of control agent, the mechanical alloying is dominated by cold welding resulting in the fast increase in particle size to about 1 mm. Further growth occurs by the coalescence of these particles. A steady state in size is still not achieved even after 13,320 minutes of milling, although at this stage, the average particle size has already reached 5.2 mm. The rate of increase in particle size is slow after about 120 minutes of milling.

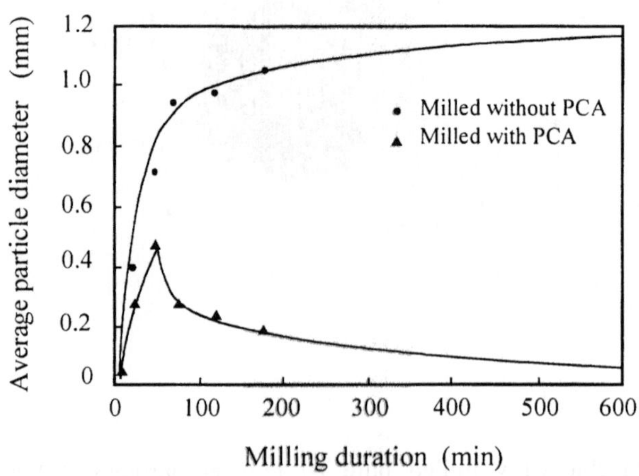


Figure 3.4. Change in average particle diameter as a function of milling time [37].

Although most process control agents are incorporated into the materials to be milled during an intermediate processing stage, cold welding continues to be inhibited until the process reaches a steady state [(38)]. Ovecoglu [(38)] found that the achievement of steady state processing conditions depends on the initial size distribution of the powder, the amount of agent added in and the processing time. At steady state, the particles manifest a uniform size distribution, saturation in hardness, equiaxed shape and a composite structure. For Al alloy, a steady state processing time is between 120 to 240 minutes [(38)].

There exist numerous types of process control agents. Among them, stearic acid is one of the very often used and effective process control agents. The percentage of stearic acid used in a mechanical alloying process is about 1 to 3 wt.% of powder weight, depending upon the properties of materials milled. For example, for ductile materials, more stearic acid is needed and vice versa. Because the melting temperature of stearic acid is 68°C, it exists in a solid state at the beginning of milling. In this form, the stearic acid may not be well homogeneously distributed and hence may result in inhomogeneous distribution of particle size. Niu [(31)] found that the use of ethyl acetate appeared to be more effective in terms of homogenization and particle refinement since it is in a liquid state at room temperature. Table 3.1 gives a list of commonly used process control agents and their typical properties.

Table 3.1. Typical properties of process control agents [(39)].

| Generic name | Chemical formula | Melting point | Boiling point |
|---|---|---|---|
| Stearic acid | $CH_3(CH_2)_{16}CO_2H$ | 67-69°C | 183-184°C |
| Heptane [(33, 40)] | $CH_3(CH_2)_5CH_3$ | -91°C | 98°C |
| Ethyl acetate [(31)] | $CH_3CO_2C_2H_5$ | -84°C | 76.5-77.5°C |
| Ethylenebidi-steramide [(5, 41)] | $C_2H_2$-$2(C_{18}H_{36}ON)$ | 141°C | 259°C |
| Polyethylene glycol [(42)] | $H(OCH_2CH_2)_nOH$ | 59°* | 205°* |
| Dodecane [(37)] | $CH_3(CH_2)_{10}CH_3$ | -12°C | 216.2°C |
| Hexanes [(43, 44)] | $C_6H_{14}$ | | 68-69°C |
| Methyl alcohol [(45)] | $CH_3OH$ | -98°C | 64.6°C |
| Ethyl alcohol [(46)] | $C_2H_5OH$ | -130°C | 78°C |

* Measured from STA.

Figures 3.5 and 3.6 show the morphologies of the Al powder mechanically milled at different percentages of stearic acid of 0.5 and 3wt.%. At a low content of 0.5wt.% stearic acid, large particles with smooth surfaces are formed due to excessive cold welding as shown in Figures 3.5. If the weight percentage of stearic acid is increased to 1%, the particles still increases and shows irregular shape with increase in milling duration. The reason for such large particles is most likely due to mechanical bonding rather than cold welding. Large scatter in distribution of particle size can be found in this case. Thin flakes with curvature may also be formed. Detailed analysis reveals that the curvature of this type of flakes is similar to that of some of the round particles implying that the flakes may have dropped off from the poorly bonded particles. Fine particle size can be obtained if 2wt.% or more stearic acid is used as shown in Figure 3.6. Although less process control agent may promote cold welding, the increase in particle size may lead to insufficient energy to deform the large particles. Surface deformation of the particles may result in an internal tensile stress and finally form a small secondary particle in a large particle. Figure 3.7 shows the cross section of the particle.

The distribution of Al particle size at different percentages of process control agent is given in Figure 3.8 (a). It can be seen that stearic acid has a strong influence on particle size. With the increase in stearic acid or polyethylene glycol, particle size becomes very fine. However, as milling duration is prolonged, the particle size become larger. This is due to the decomposition of process control agent leading to the presence of less process control agent. Another method to measure the amount of cold welding is the weight of powder recovered after milling. More recovered

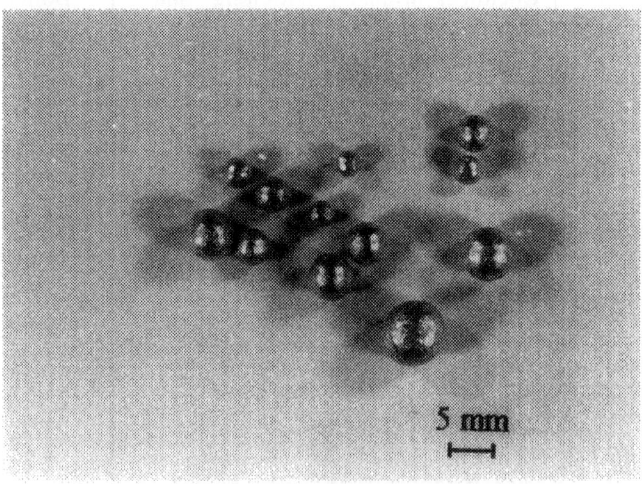


Figure 3.5. Al powder particles mechanically milled with the use of 0.5wt.% of stearic acid.

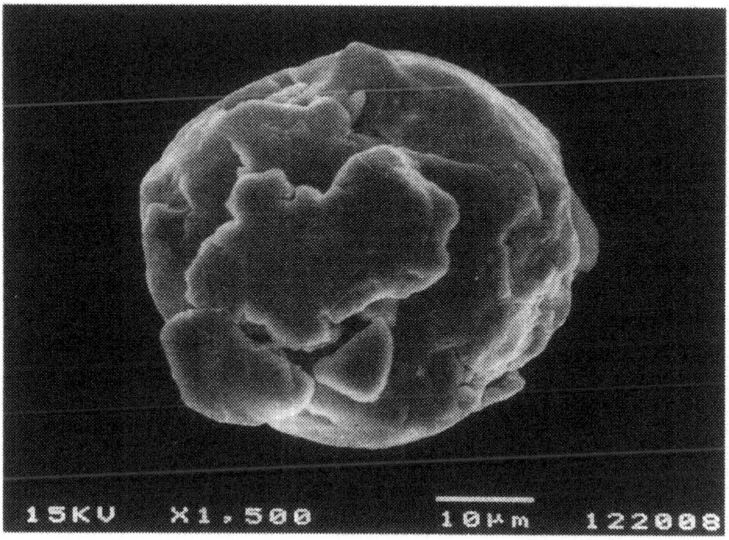


Figure 3.6. Al powder particles mechanically milled with the use of 3wt.% of stearic acid.

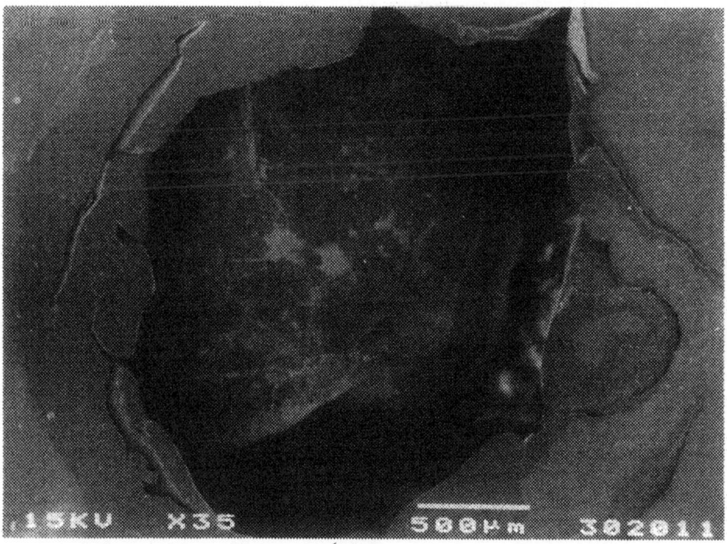


Figure 3.7. Formation of a secondary particle in a large Al particle.

weight means less cold welding. Figure 3.8 (b) shows the weight of powder recovered as a function of stearic acid. The change in the particle size and weight recovered using polyethylene glycol are given in Figures 3.8 (c) and (d). The figures also show that polyethylene glycol is less effective in comparison with

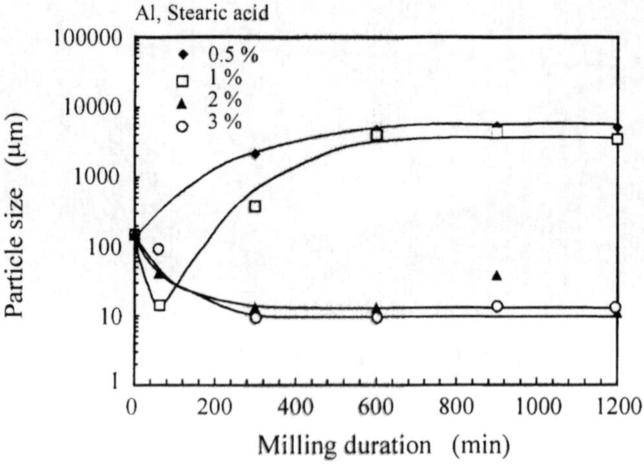


(a)

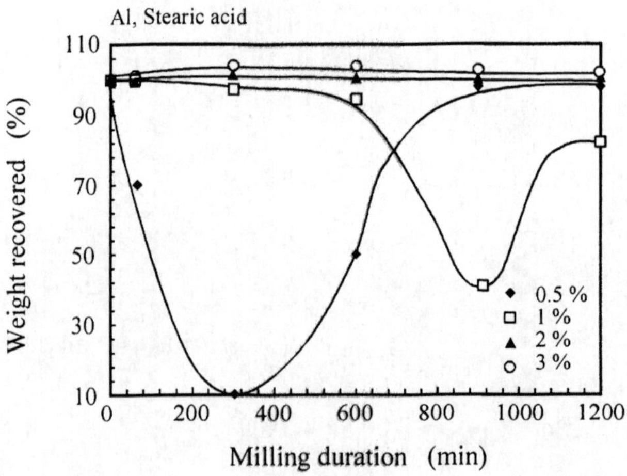


(b)

Figure 3.8. Measurement of average Al particle size and weight of powder recovered as a function of percentage of process control agents and milling duration: (a) and (b) using stearic acid.

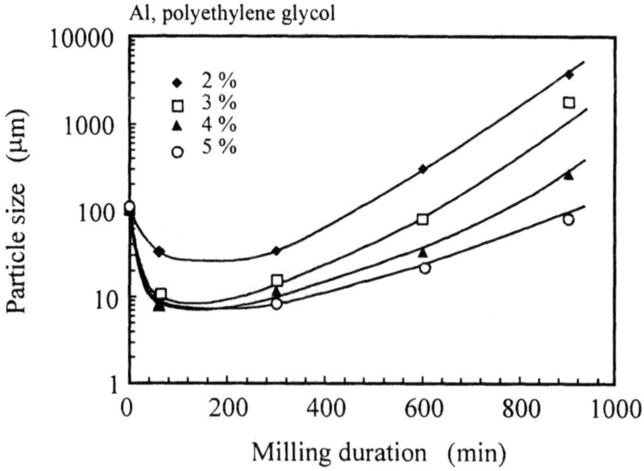


(c)

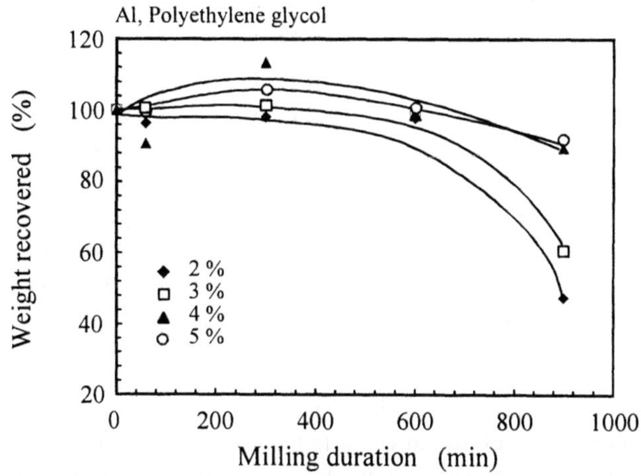


(d)

Figure 3.8. Measurement of average Al particle size and weight of powder recovered as a function of percentage of process control agents and milling duration: (c) and (d) using polyethylene glycol (continued).

stearic acid. After 900 minutes of milling, only 50% of material remains if 2% of polyethylene glycol is used while almost 100% of the material can be recovered if stearic acid is used.

Figure 3.8 also indicates that with the increase in mechanical alloying duration, the change in particle size shows two opposite trends:

(a) Particle size tends to increase if the percentage of process control agent used is below a critical value because of the dominating effect of cold welding.

(b) Particle size tends to decrease if the percentage of process control agent used is above a critical value because of the dominating effect of fracturing.

Figure 3.9 shows the variation of particle distribution as a function of the amount of process control agents and the nature of the process control agent used. It is clear from Figure 3.9 (a) that there is a transition range between 1 and 2 % of stearic acid. Below 2% of process control agent, cold welding is a dominating factor while above 1% towards high percentage of process control agent, fracturing becomes a dominating factor. Below 1% and above 2%, steady process of particle size can be obtained. For polyethylene glycol as shown in Figure 3.9 (b), the particle size decreases continuously as the amount of process control agent is increased. No transition range has been found. When polyethylene glycol is below 2%, mechanical milling cannot progress as a result of excessive cold welding. It should be noted that the size is not only controlled by type and amount of process control agents but also is affected by milling atmosphere.

The effect of the content of the process control agent on microhardness of the mechanically alloyed powders is not significant [31]. The microhardness remains almost constant when the amount of ethyl acetate is increased from 0.5% to 3%. However, an increase in the amount of process control agent significantly reduces the particle size from an initial value of about 1,000 μm at 0.5% of ethyl acetate to 90 μm at 2.0%. At a given mechanical alloying time, very fine powder particle of about 14 μm may be obtainable when the content of ethyl acetate is increased to 3.0% as shown in Figure 3.10.

Some materials like magnesium, which, because of its hcp structure, consists of less slip systems, and are therefore more brittle than materials with fcc crystal structure, such as Al. During mechanical milling, Mg does not form big particles even when low percentage of process control agent is used. Figure 3.11 shows the morphologies of the milled magnesium powder using stearic acid or polyethylene glycol. Flaky morphology at short milling duration (60 minutes) is evident (Figure 3.11 (a) and (c)). The powder particles soon become very fine (Figure 3.11 (b) and (d)). The change in particle size and weight recovered as a function of milling duration for a fixed percentage of process control agents is given in Figure 3.12. It can be seen that particle size increases continuously if less process control agent is used and vice-versa. From measurement of the weight recovered, it is found that Mg encounters less cold welding in comparison with Al powder particles. However, although more Mg powder than Al powder may be recovered after milling, the

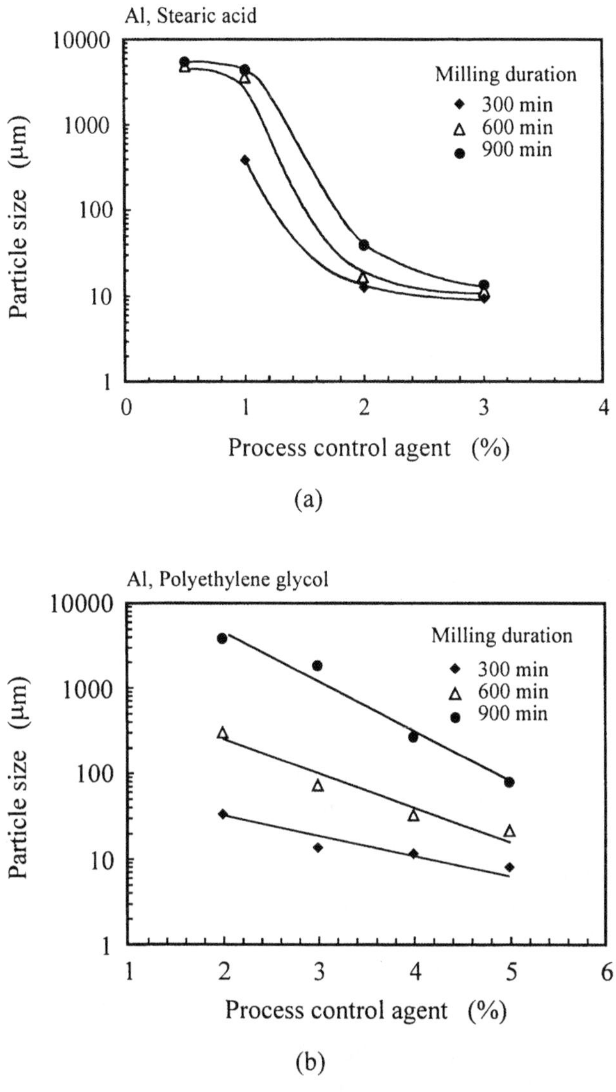


Figure 3.9. Average Al particle size as a function of percentage of process control agent and milling duration using (a) stearic acid and (b) polyethylene glycol.

particle size of Al has been shown to be finer than that of Mg under the same milling condition if stearic acid is above 2wt.%. However, if low content of process control agent is used, Mg shows to be finer than Al. By comparison of the particle size at different percentages of process control agent, it appears that stearic acid is more effective in reducing particle size.

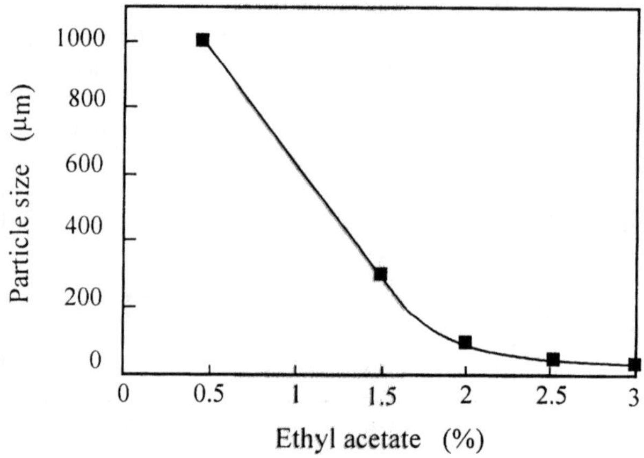


Figure 3.10. Average particle size as a function of percentage of ethyl acetate [31].

Ideally, some mechanical alloying processes should be carried out such that there is no reaction between the mechanically alloyed powders and the process control agents. To achieve this, more stable inert organic materials should be considered. On the other hand, the process control agents should be decomposed easily at the relatively low temperature of post processing.

The amount of process control agents employed is essentially dependent on three parameters, namely, (a) cold weldability of the powder particles, (b) stability of the process control agent and (c) the amount of powder and milling tools used. As can be seen from Figures 3.9 and 3.12, the particle size generally shows a decrease with the increase in the amount of process control agent. Because a balance between cold welding and fracturing has to be achieved during mechanical alloying, use of excessive process control agent may lead to inhibition of cold welding and hence prevents the formation of new materials even though fine particle size may be obtained.

Figure 3.13 shows an example of the influence of the amount of process control agent on the structural change of Al-37wt.%Mg system at mechanical alloying duration of 300 minutes. In the test four different amounts of process control agent were used. X-ray diffraction shows the different spectra although the mechanical alloying duration is the same. Detailed measurement of the lattice parameter indicating the process of interdiffusion is given in Figure 3.14. The change in lattice parameter of mechanical alloyed Al-37wt.%Mg system tends to be slow when more process control agent is used. This is because cold welding is prohibited by a film created by the process control agent. The latter increases the interdiffusion distance between powder particles and hence the difficulty of interdiffusion between two elements increases.

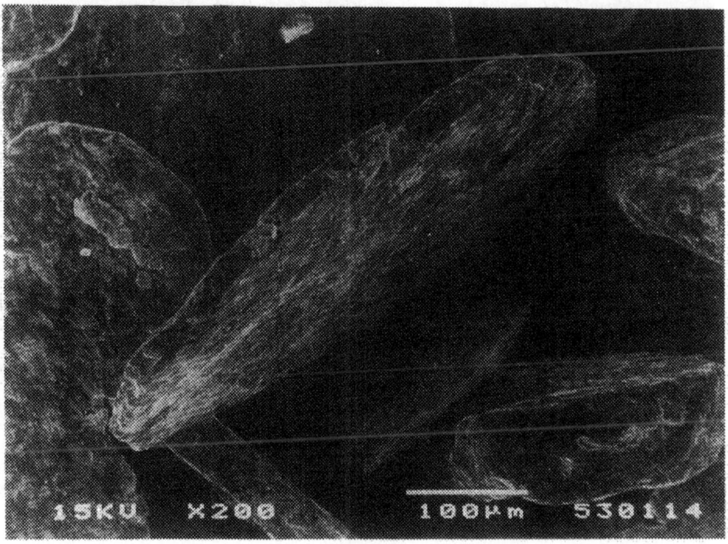


(a)

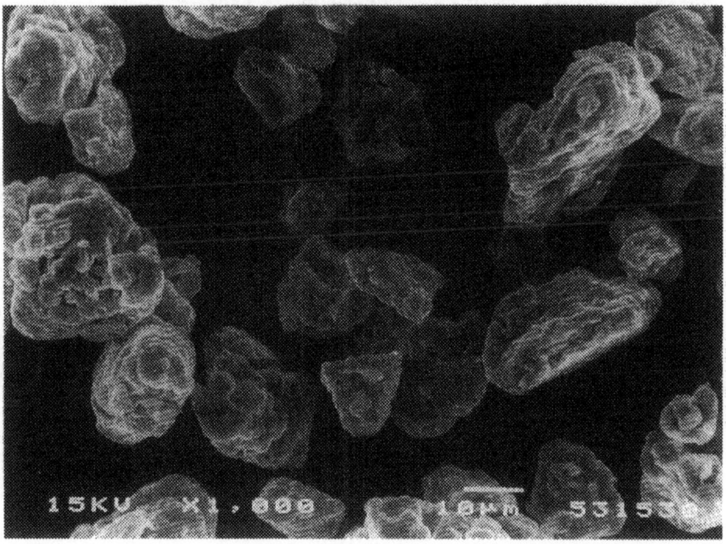


(b)

Figure 3.11. Morphologies of Mg powder particles mechanically milled with the use of 3wt% stearic acid at (a) 60 minutes of milling and (b) 900 minutes of milling.

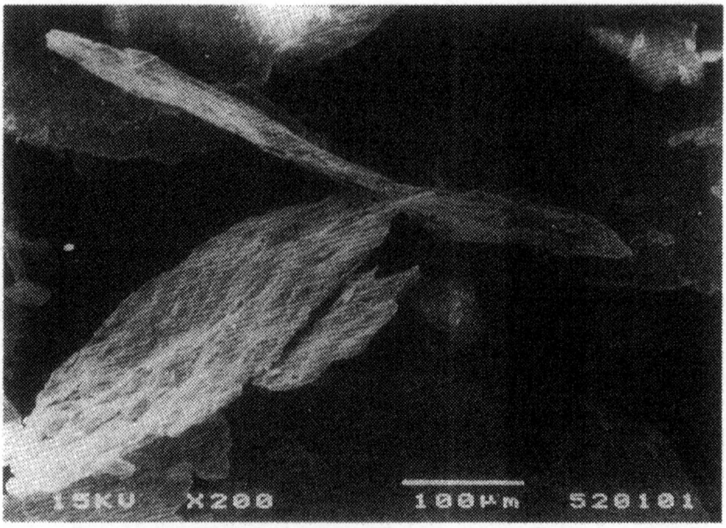


(c)

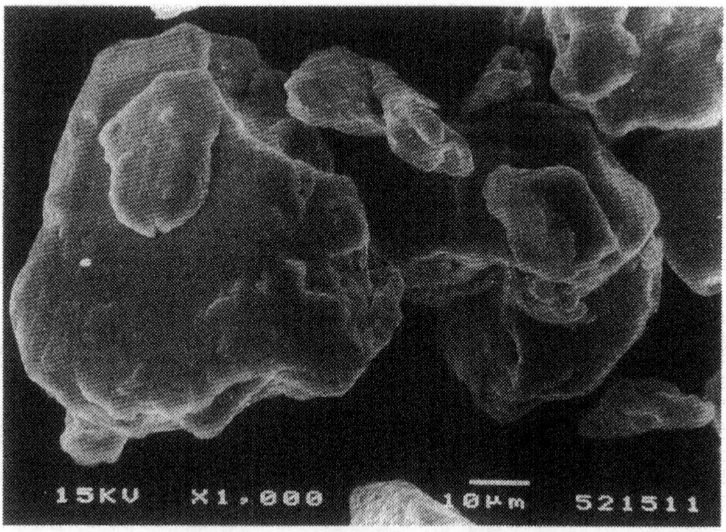


(d)

Figure 3.11. Morphologies of Mg powder particles mechanically milled with the use of 2wt% polyethylene glycol at (c) 60 minutes of milling and (d) 900 minutes of milling.

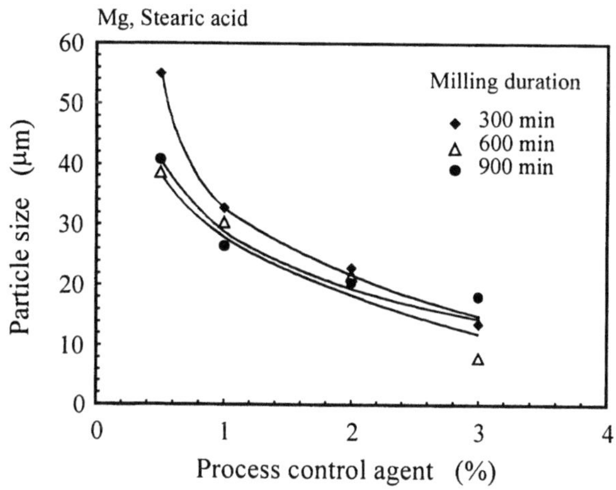


(a)

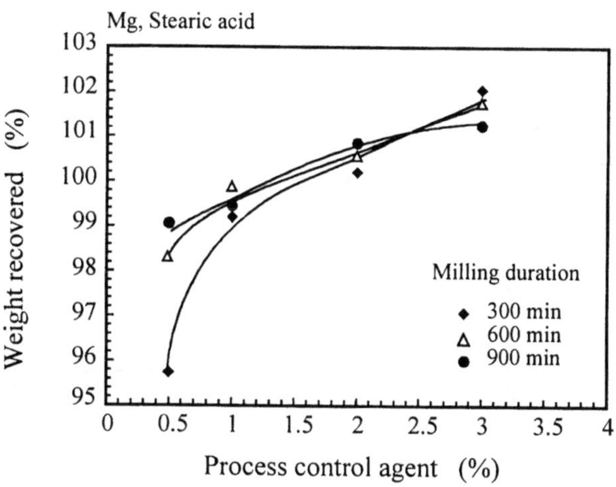


(b)

Figure 3.12. Average Mg particle size and powder weight recovered as a function of percentage of process control agent and milling duration using (a) and (b) stearic acid.

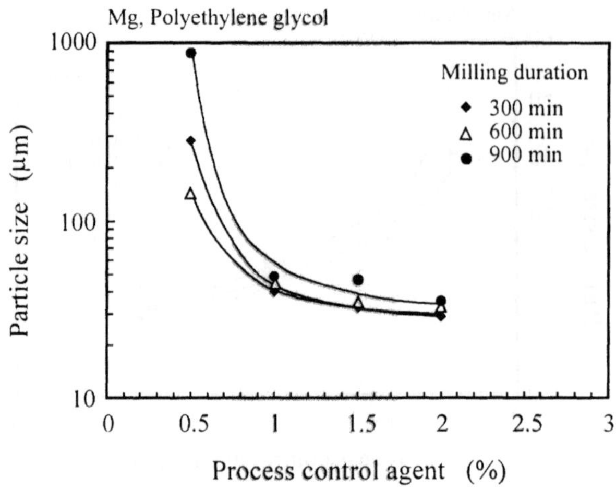


(c)

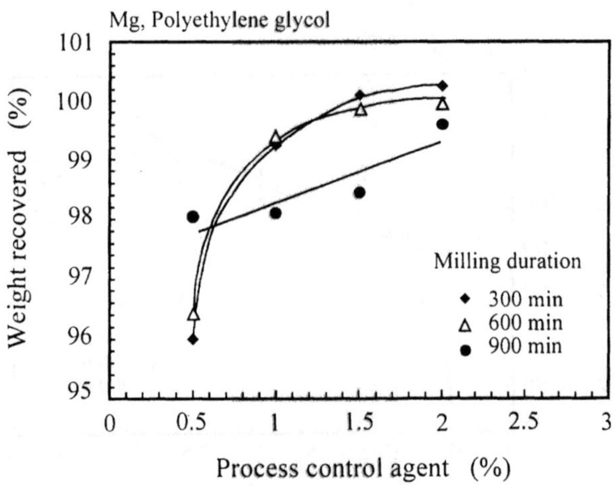


(d)

Figure 3.12. Average Mg particle size and powder weight recovered as a function of percentage of process control agent and milling duration using (c) and (d) polyethylene glycol (continued).

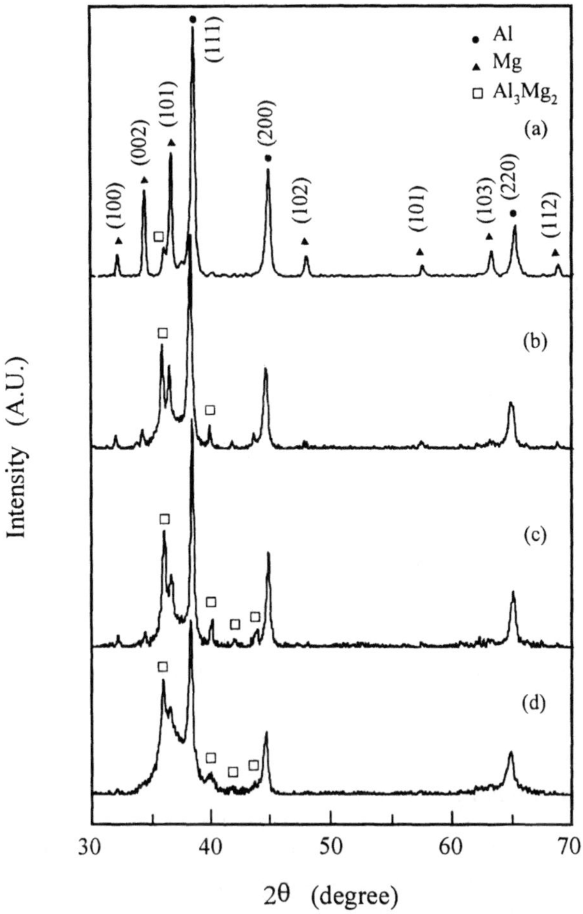


Figure 3.13. X-ray diffraction spectra of mechanically alloyed Al-37wt.%Mg powder using different amount of process control agent of stearic acid: (a) 4wt.%, (b) 3wt.%, (c) 2wt.% and (d) 1wt.%.

Another important consideration is the influence of atmosphere on the particle size. Results show that particle size tends to increase at the same amount of process control agent, if the milling process is well protected by vacuum. Beside the use of organic compounds, cryogenic cooling by liquid nitrogen may also be used to promote fracture and to reduce cold welding [(47)]. Takacs [(48)] proposed to use a combination of large and small balls to avoid excessive cold welding. He studied the milling process of Zn composite. Zn is a malleable metal which tends to collect in a large, compacted chunk on the balls and the wall of the milling vial. However, if a combination of 45.5 g and 3.5 g balls were used at the same time, the above problem could be overcome.

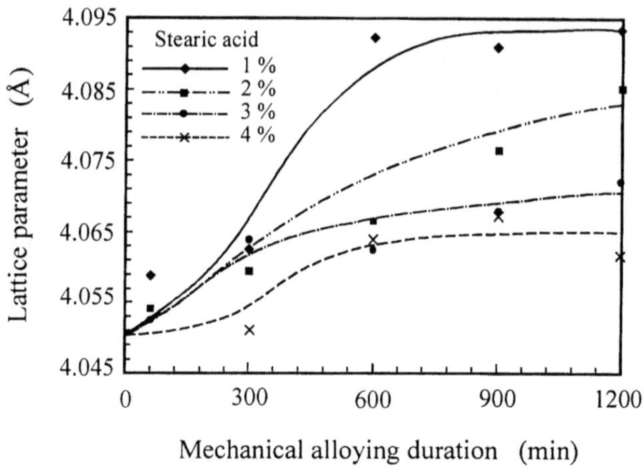


Figure 3.14. Change in lattice parameter of mechanically alloyed Al-Mg powder as a function of different amount of process control agent.

### *3.1.5 Selection of milling intensity*

The intensity of milling, a critical parameter in mechanical alloying [9, 49-51], is a measure of the milling energy which is directly proportional to the power generated in the milling process. The influence of this parameter on the formation of glass materials is first studied by Eckert *et al* [52]. Because the powder particles are subjected to high energy collision, the final products will be influenced by the milling intensity [53-57].

It is obvious that when milling intensity is increased either by increasing the weight of the ball and/or density of the ball materials or by increasing the speed of collision, more energy will be transferred to the powder particles and more energy will be dissipated in the form of heat, which is crucial for the formation of amorphous alloys. When the temperature exceeds the critical temperature of crystallization, partial crystallization can occur during mechanical alloying. Hence, too high a milling intensity may lead to a partial crystallization of the amorphous alloys. If the milling intensity is too low, alloying and amorphozation will take place only after a long time. Because free energy of the powders increases by the introduction of different defects under plastic deformation, the amount of increase in free energy is dependent upon the milling intensity. Milling intensity below a certain value may not be enough to promote the formation of pure amorphous phase. Consequently, the original crystalline phases will remain in the crystalline form [58]. Eckert studied the influence of milling intensity of 3, 5, 7 and 9 on Al65Cu20Mn15 mixture. The intensities investigated corresponded to the calculated ball velocities of 2.5, 3.6, 4.7 or 5.8 m/s respectively ( or to kinetic energies of 14, 29, 49 or 76 mJ) [55]. X-ray diffraction study showed the formation of amorphous phase under an intensity of 5 after 30,600 minutes of milling. In addition, a small irregularity at the amorphous maximum originated from a small amount of unreacted α-Mn appeared. It was seen that the relatively low reaction temperature was not sufficient for a complete solid-state reaction even after 30,600 minutes of milling. An icosahedral phase was formed for milling at an intensity of 7. If an intensity of 9 was used, a nanocrystalline material with CsCl-type structure could be formed. This structure remains unchanged during further milling. It is clear from this study that milling intensity of 5 is too low while high intensity of 9 leads to formation of crystalline material because of increase in milling temperature. The estimated maximum temperatures of the powders under milling intensities of 3, 5, 7 and 9 are 403, 520, 680 and 863 K respectively.

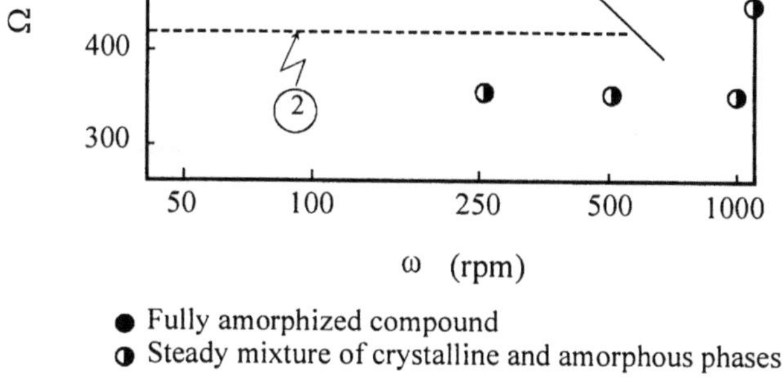


Figure 3.15. Dynamic equilibrium of the amorphous phase in $Ni_{10}Zr_7$ system using a planetary ball mill [59].

Detailed experimental study on planetary ball mills has shown that the end products of $Ni_{10}Zr_7$ alloy are strongly influenced by the rotational speed of the turn table (disk) and bowl although the temperatures during mechanical alloying are in the range of 303 to 327 K. Three phases can be identified in Figure 3.15, namely amorphous, partial crystalline and crystalline phases [59]. A narrow domain for the formation of amorphous is present indicating that amorphization proceeds below a certain power input and above a minimum energy per impact. Transformation from amorphous to crystalline material takes place if the power input is too high and large amount of heat is generated. On the other hand, amorphous will not be formed if the collision energy per impact is below the yield stress of the powder particles.

The same material has been mechanically alloyed using a vibration ball mill with a large variety of vibration amplitudes and ball diameters [60]. The temperature of the bowl was controlled at room temperature or 473 K. The typical evolution of amorphization with milling time and intensity is given in Figure 3.16. Full amorphization can be obtained for the highest milling intensity at room temperature denoted as RT1. Using the same milling intensity but at 473 K, full amorphization can also be obtained but with longer milling time (HT1). If the milling intensity is lowered to RT2 (HT2) and RT3, no full amorphization can be obtained.

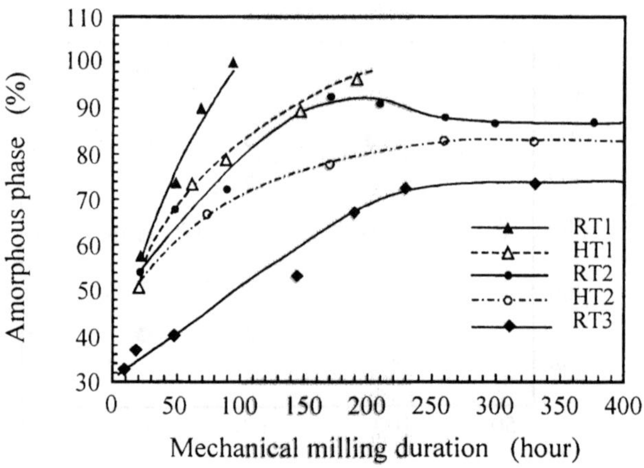


Figure 3.16. Evolution of fraction of amorphous phase as a function of milling time at different milling intensities [60].

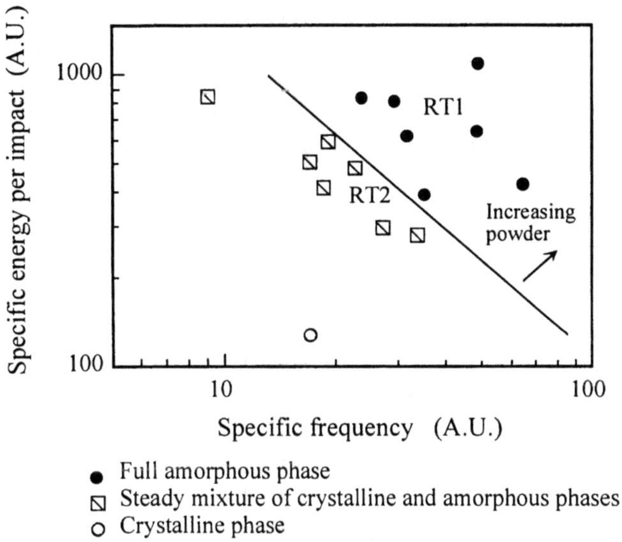


Figure 3.17. Milling map for full amorphization using a vibration ball mill [60].

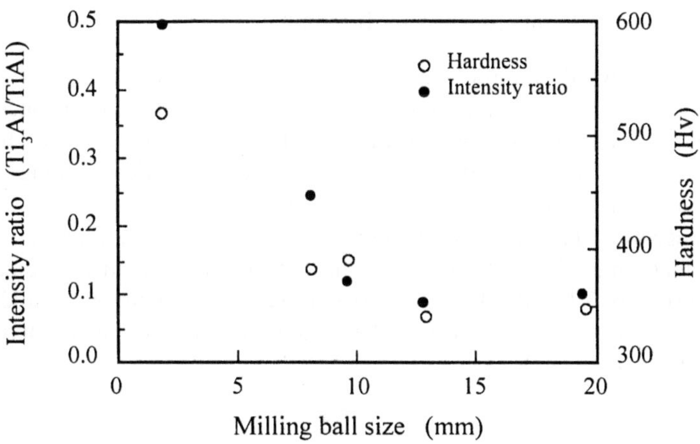


Figure 3.18. Variation of Vickers hardness and diffraction peak intensity ratio of hot isostatic pressed $Ti_3Al$/TiAl compacts [61].

From the summary shown in Figure 3.17, it can be observed that the full amorphization is only obtainable provided that the milling intensity is above a minimum threshold value indicated by a solid line under controlled temperature. Below the threshold value, the steady state amorphous phase fraction decreases with the increase in temperature. If milling temperature is too high, recystallization may occur.

Park *et al.* [61] have demonstrated the influence of milling intensity on the microhardness and diffraction peak intensity ratio of $Ti_3Al$/TiAl by mechanical alloying of Ti-50at.%Al. Figure 3.18 reveals that microhardness decreases when large ball size is used, and that more TiAl compound has been formed if high milling intensity is employed. Optical microscopy shows that the structure of the materialbecomes more refined and homogeneous with increasing size of the ball. In addition, the microstructure of the hot isostatic pressed compacts can be controlled by the milling duration and ball size.

## 3.2 Evolution of structure in mechanical alloying

### *3.2.1 Evolution of particle morphology*

As discussed in section 3.14, cold welding and fracturing are the two essential processes involved in the mechanical alloying process. Microstructurally, the mechanical alloying process can be divided into four stages: (a) initial stage, (b) intermediate stage, (c) final stage, and (d) completion stage.

Figure 3.19 shows the morphologies of the powder particles after different ball milling durations. Scanning electron microscopy micrographs show distinctively the progressive changes in morphology of the ball milled particles. The changes in morphology are the result of the following processes: micro-forging, fracture, agglomeration and de-agglomeration. Depending on the dominant process, each milling stage will display a morphology that best describes the processes taking place at that time. Thus, the particles may (a) become smaller in size due to fracturing (Figure 3.19 (a)), (b) grow in size through agglomeration by cold welding (Figure 3.19 (b)), and (c) become flattened into flakes through micro-forging (Figure 3.19 (c)).

### *3.2.2 Change in microstructure*

Gilman and Benjamin [34] have summarized the mechanical alloying process into four different stages. These stages are the initial, intermediate, final and completion stage.

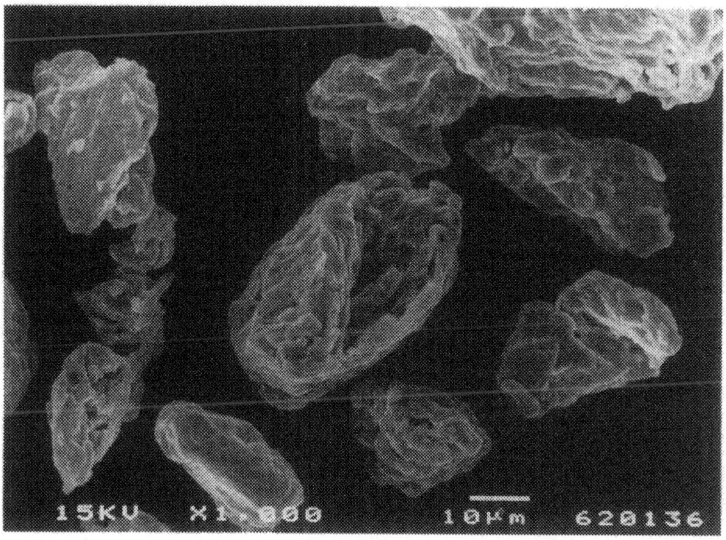


(a)

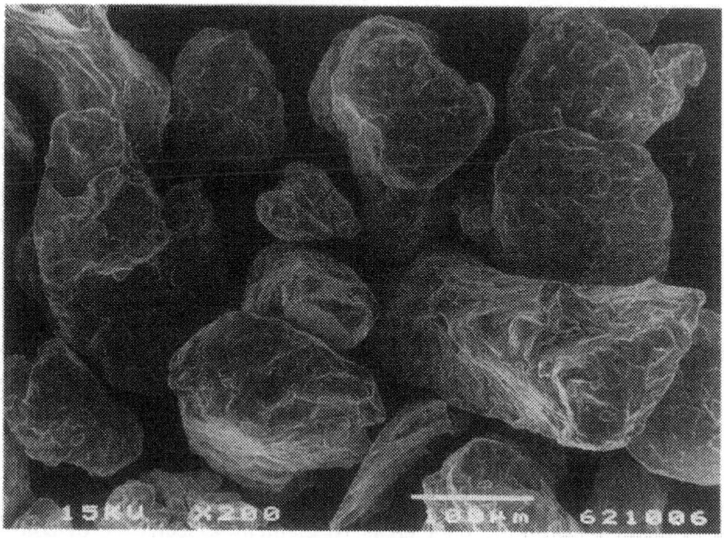


(b)

Figure 3.19. Morphologies of powder particles after different durations of ball milling.

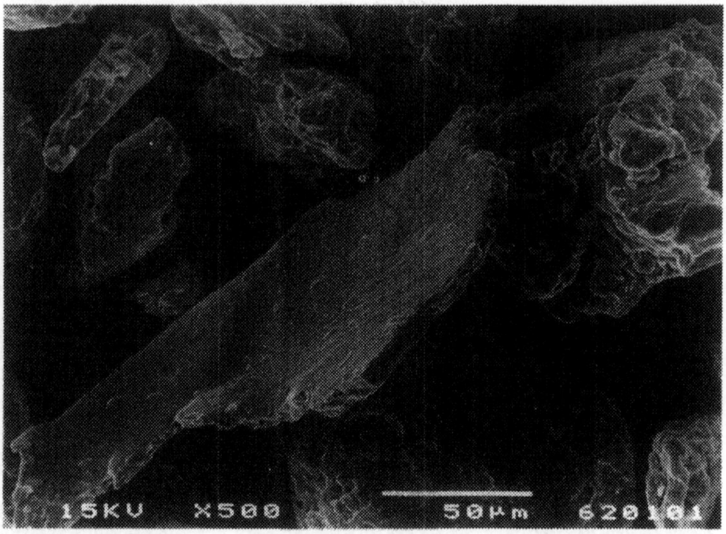


(c)

Figure 3.19. Morphologies of powder particles after different durations of ball milling (continued).

### (a) Initial stage

At the initial stage of ball milling, the powder particles are flattened by the compressive forces due to the collision of the balls since the materials are generally soft at this stage. Micro-forging leads to changes in the shapes of individual particles, or cluster of particles being impacted repeatedly by the milling balls with high kinetic energy. However, such deformation of the powders shows no net change in mass. Experimentally, little or no cold welding of the powder particles can be observed [62].

For ductile-ductile systems, a layered structure with clear separation between different elements is developed during the early stage of milling [8]. The cold welded structure consists of various combinations of the starting ingredients. Figure 3.20 gives an example of mechanically alloyed Ni and Al mixture. The bright areas indicate Ni element while the dark areas Al element. Occurrence of cold welding between the two elements can be seen but there is no formation of layered structure. Prolonged milling will not only refine and homogenize these layers, but also enhance the diffusion process across the composite layers. This is expected because

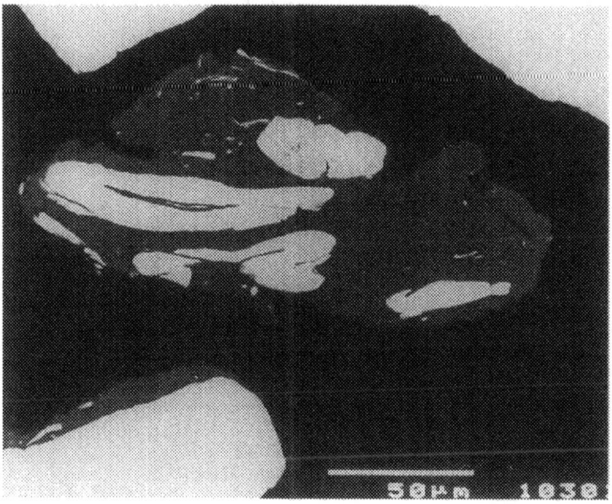


Figure 3.20. Back scattered image of the microstructure of mechanically alloyed Ni-Al mixture.

as the particles become flattened by the compressive forces, the surface area for contact inadvertently increases. Hence, both flattened and un-flattened layers of particles come into intimate contact with each other leading to the building up of layered composite particles consisting of various combinations of the starting ingredients. The chemical composition of the composite material varies significantly within the particles as well as from particle to particle. Figure 3.21 clearly shows the variation of composition as a function of particle size and milling durations [(64)]. Very large scatter in distribution of composition for particles of different sizes can be identified for short milling duration. The big particles are identified as Ni-rich while the small particles, Al-rich.

Scanning electron microscopy and powder size analyses show a wide range of particle sizes present at the early stage of milling. This is partly due to the nature of the particles. Powders of ductile materials can easily be plastic deformed under compressive loading, and hence they become flattened into thin flakes. On the other hand, the relatively hard particles tend to resist the attrition and compressive forces. If the powder mixture contains both ductile and brittle particles, the hard particles may remain less deformed while the ductile particles tend to bind the hard particles together. Fracture is expected to be predominant in bcc and hcp metals as compared to cold welding in fcc metals. Consequently, the difference in ductility of the brittle and ductile powder particles suggests a wide distribution of particle size. The milling time considered at this stage of milling is far too short to achieve homogenous distribution of particle size and shape. A more homogenous distribution is expected when ball milling duration is increased.

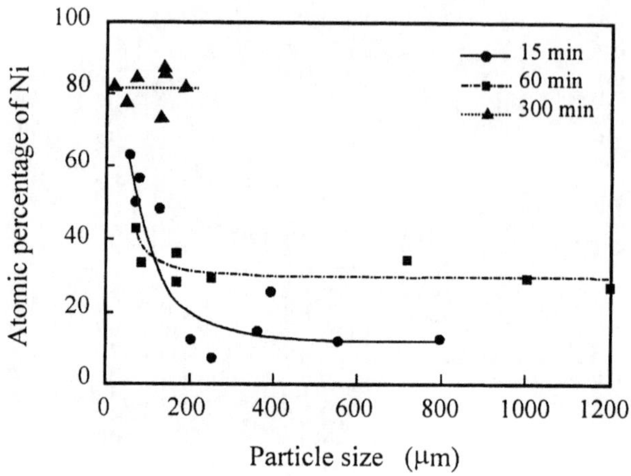


Figure 3.21. Variation in composition as a function of particle size and milling duration [64].

**(b) Intermediate stage**

At the intermediate stage of the mechanical alloying process, significant changes occur as evidenced by the difference in morphologies of the particles in comparison with those in the initial stage. Cold welding is now significant. Greater plastic deformation leads to the formation of layered structures as shown in Figure 3.22. The intimate mixture of the powder constituents decreases the diffusion distance to the micrometer range. Fracturing and cold welding are the dominant milling processes at this stage as indicated by the random orientated lamellae. The laminated structure is further refined as fracture takes place. Thickness of the lamellae is decreased. Although some dissolution may have taken place, chemical composition of the alloyed powder is still not homogeneous.

The change in lamella thickness is a function of the collision energy, mechanical property and the mechanical milling duration. Benjamin and Volin [63] proposed a model to calculate the average rate of change in lamella thickness based on two hypotheses:

(a) The rate at which material is trapped between colliding balls is independent of time.

(b) The energy required to deform the layered structure per unit strain for a constant volume of material is a linear function of the instantaneous Vickers hardness.

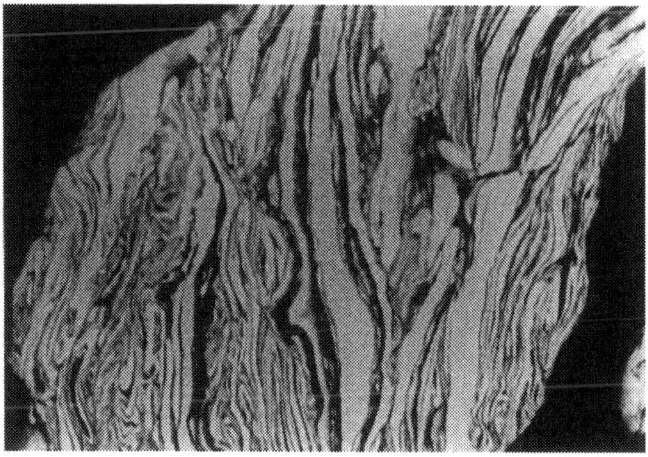

Figure 3.22. Layered structure caused by heave plastic deformation.

Therefore the rate of energy input to the process is constant, given by:

$$\frac{dE_i}{dt} = K_1 \quad [3.1]$$

and

$$\frac{dE_i}{d\varepsilon} = K_2 H \quad [3.2]$$

where $E_i$ is the energy input, $t$, time, $\varepsilon$, true strain (equal to $\ln(L_0/L)$ where $L$ is the lamella thickness), $H$, Vickers hardness of the powder, and $K_1$ and $K_2$, constants. If $H$ can be expressed as a function of $t$, the above equations can be solved. For the case where $H$ is a linear function of $t$, then $H$ may be expressed as:

$$H = a + bt \quad [3.3]$$

where $a$ and $b$ are constants. Combining Eqs.[3.1] and [3.2], it follows that

$$\frac{d\varepsilon}{dt} = \frac{K_1 / K_2}{H} \quad [3.4]$$

By substituting Eq.[3.3] into Eq.[3.4], deformation strain of the powder particles can be obtained:

$$\varepsilon = \frac{K_1 / K_2}{b} \ln(a + bt) - C \qquad [3.5]$$

where $C$ is constant.

When $t = 0$, there is no deformation. Eq. [3.5] can be rewritten as:

$$\varepsilon = \frac{K_1 / K_2}{b} \ln\left(1 + \frac{b}{a} t\right) \qquad [3.6]$$

The ratio of $K_1$ and $K_2$ may be obtained by calculating the deformation strain based on the measurement of thickness of lamella at a given milling time.

**(c) Final stage**

Considerable refinement and reduction in particle size is evident at the final stage of the mechanical alloying process. As shown in Figure 3.23, the microstructure of the particle also appears to be more homogenous in macroscopic scale than those at the initial and intermediate stages. No layered structure can be found as this stage. True alloys may have already been formed

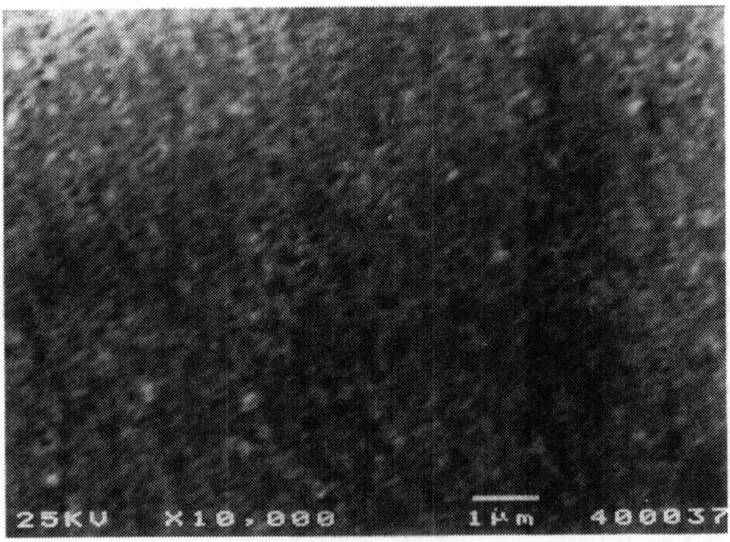


Figure 3.23. Microstructure of mechanically alloyed Ni-Al mixture at the final stage.

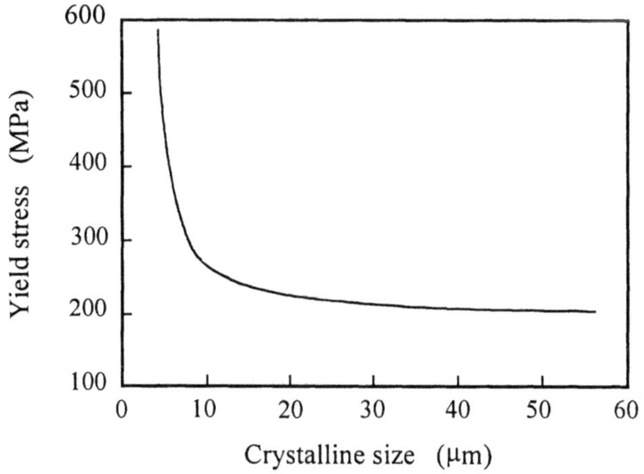


Figure 3.24. Stress required to deform for Al alloy with different crystalline sizes.

Homogeneity of the particles is an outcome of an equilibrium between fracturing and cold welding. The mechanical alloying process for ductile components is a competition between the process of fracturing and cold welding of the particles. Fracture tends to break individual particles into smaller pieces and de-agglomerates particles that have been cold welded. Particles that have become finer in size due to fracturing may alternatively become cold welded together since these two processes are operating in cycle. However, the cold welded particles will require a higher force to fracture them again since the forces that bind them together are stronger when the particle size is finer. On the other hand, some particles that have been cold welded may undergo de-agglomeration because the forces that break them apart are much stronger than those that hold them together. Consequently, depending on the dominant forces, a particle may either become smaller in size through fracturing or may agglomerate by welding as alloying time progresses. The process of cold welding and fracturing will ultimately attain a steady-state equilibrium [(65)]. The average particle size attained is a result of the balance between the strength of the forces that cold weld them and that required to fracture them. The lamellae become finer and eventually disappear. A homogenous chemical composition would have been achieved for all the particles resulting in the formation of a new alloy with composition corresponding to the starting powder mixture. Crystalline size may reach nanometer depending upon the material systems being alloyed. Further deformation is almost not possible since very high deformation stress is needed. The case may be best described by Hall-Petch relation:

$$\sigma = \sigma_0 + kd^{-1/2} \quad [3.7]$$

where $\sigma$ is the stress, $\sigma_0$ and $k$, constants, and $d$, the crystalline size. The relation shows that for materials with very fine crystalline size, high deformation stress has to be applied as shown in Figure 3.24.

**(d) Completion stage**

At the completion of the mechanical alloying process, the powder particles possess an extremely deformed metastable structure. At this stage, the lamellae are no longer resolvable by optical microscopy or scanning electron microscopy. Further mechanical alloying beyond this stage cannot physically improve the dispersoid distribution [(34)]. Real alloy with composition similar to the starting constituents is thus formed.

### *3.2.3 Change in structure*

Change in structures of mechanically alloyed powder mixture can be divided into three stages according to measurement of the lattice spacing. They are grain refinement, solution diffusion and formation of new phase. At the first stage, no solid solution takes place between the powder particles which are only mechanically bonded together. Because of deformation and fracturing of the powder particles, grains may be refined as indicated by the broadening of the x-ray diffraction peaks. Solid solution soon occurs as evidenced by the shift in x-ray diffraction peak patterns indicating the dissolution of different elemental atoms. New phases may be formed at the later stage of mechanical alloying, an occurrence manifested by the formation of new peaks in x-ray measurement.

Figure 3.25 reveals the evolution of the formation of true alloy from $(Cu\text{-}Zn)_xAl_{1-x}$ mixture at different mechanical alloying durations [(66)] whereas the x-ray diffraction of the as-received powder mixture is given in Figure 3.25 (a). At the beginning of the alloying process, very few Al atoms diffuse into the CuZn alloy leading to the shift of x-ray diffraction peaks. The formation of $\alpha$ phase shape memory alloy is followed as shown in Figure 3.25 (b). From Figure 3.26 (a) [(67)], it can be observed that the Al content in the alloy is less than 1%wt. When the Al content in CuZnAl is increased, a lattice structure transition from fcc $\alpha$ phase to bcc $\beta$ phase takes place where concentration of Al is about 2 wt.%. At this stage, an x-ray diffraction peak at 42.94° can be detected after 300 minutes of mechanical alloying. It is understood that this x-ray diffraction peak is an indication of the formation of bcc $\beta$ lattice structure. The diffusivity of $\alpha$ phase is very low at this stage. Even at 973 K, the diffusivity is only about $4.5{*}10^{-11}$ $cm^2$/sec. The diffusivity is, however, increased to

$3*10^{-2}$ at 823 K if diffusion takes place in the β phase. At the final stage, diffusion is enhanced due to the increase in diffusivity in bcc phase. The concentration of Al in CuZnAl increases with mechanical alloying duration. When the Al concentration is above 3%wt, γ phase which also has a bcc lattice structure will be formed (Figure 3.26 (b)). This is indicated by the x-ray diffraction peak at 43.90° after 1,200 minute of mechanical alloying. It can also be noted that another peak occurs at 63.76° corresponding to (200) plane of the γ phase in Figures 3.25 (c) and (d).

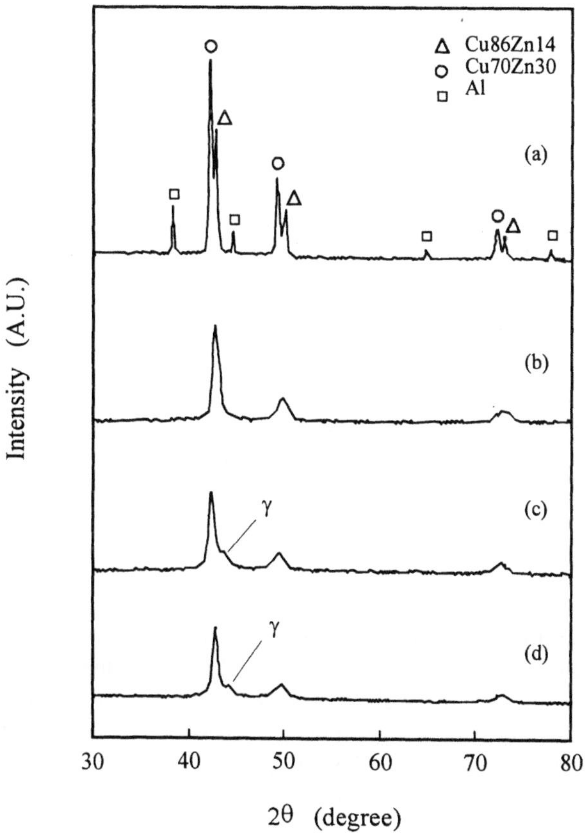


Figure 3.25. X-ray diffraction spectra of mechanically alloyed $(Cu\text{-}Zn)_xAl_{1-x}$ powder mixture at different alloying durations [67].

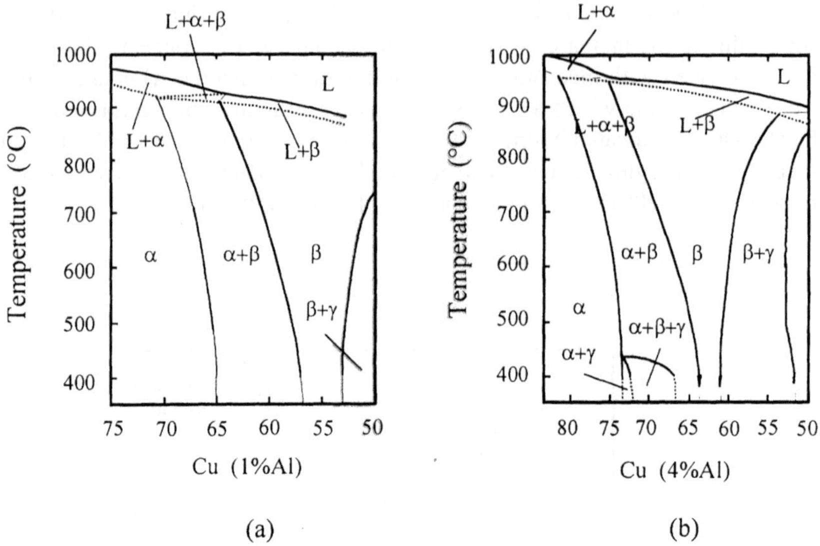


Figure 3.26. Cu-Zn phase diagram at different content of Al.

## 3.3 Contaminations

One crucial problem in mechanical alloying is contamination which has its sources from milling tools, gases and process control agents. Mechanical, physical and chemical properties of the powders may be altered if they are contaminated. Hence, better understanding of contamination may help to reduce its occurrence.

### *3.3.1 Contamination from milling tools*

One of the most common contaminations in mechanically alloyed powders is Fe and Cr elements from the milling vial and balls since most milling tools are made from those types of elements [(68-70)]. During mechanical alloying, the balls impact onto the powder mixture, the vial as well as onto each other. The powder particles are therefore not only cold welded with powder particles themselves but also with the milling tools under high energy collision. The cold welded powder will sooner or later be fractured from the balls and the vial by direct collision and relative friction, leading to transfer of atoms from the milling tool to the powder particles. Measurement of the lattice parameters of the alloyed Ni-Al during mechanical alloying shows a continuous change after formation of $Ni_3Al$ with prolonged mechanical alloying duration indicating a contamination of the elements from the milling tools. The contamination becomes more and more serious after 1200 minutes of milling. The EDX measurement for the mechanically alloyed powder particles normally shows that the contaminations are Fe and Cr which are from the milling tools. It appears that contamination from milling tool is unavoidable especially at high impact intensity and long milling duration. It is a serious problem in the mechanical alloying.

Several methods can be applied to reduce contamination:

(a) Use hardened milling tools;

(b) Employ milling tools with similar composition as the powder mixture so that the contamination will have the same composition as the milled materials (35);

(c) Increase cold welding so that the milling tools may be covered with the powder mixture.

Zoz Company (71) proposed a new design of ball mill using ceramics to prevent contamination from metallic elements. Although contamination from metallic elements can be avoided, the brittleness of the ceramics has to be considered to avoid ceramic contamination.

Figure 3.27 shows the contamination of $Al_2O_3$ to the Mg and Si mixture after 600 minutes of mechanical alloying if the bowl and balls have been made of $Al_2O_3$ running at 200 rpm in a planetary ball mill. However, if stainless steel bowl is used, no contamination could be detected. Therefore, use of high strength ceramic bowl and balls with good impact behaviouris essential. Since ceramic tools generally show poor impact behaviour, applications with relative low milling intensity may be considered.

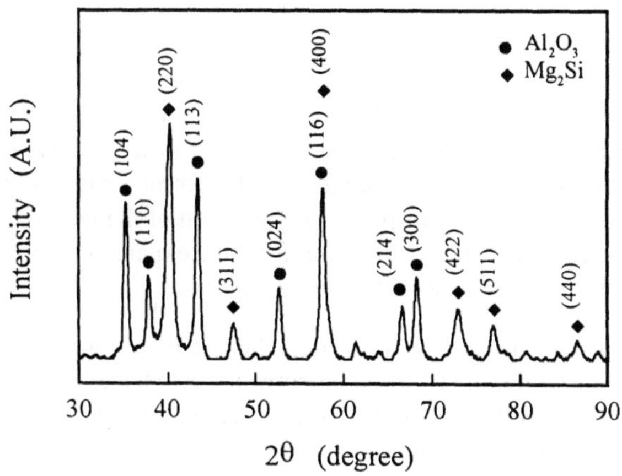


Figure 3.27. $Al_2O_3$ contamination from $Al_2O_3$ bowl.

### *3.3.2 Contamination from atmosphere*

Although inert gases such as argon, nitrogen and helium are usually employed to prevent oxidation, they may themselves react with the powder mixtures being mechanically alloyed. Depending on different material systems, sometime this reaction can be very serious.

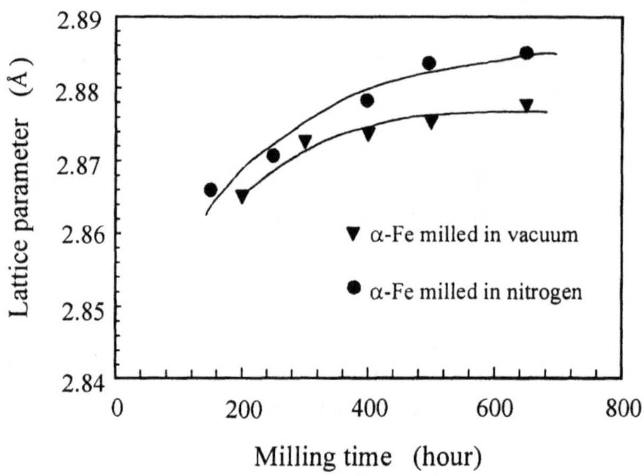


Figure 3.28. Influence of vacuum and nitrogen environments on the change in average lattice parameter of Fe [15].

Figure 3.28 shows the average lattice parameter of Fe powder ball milled under two milling environments, namely, vacuum and nitrogen protection [15]. The lattice parameters change under both conditions: there is a large increment when the Fe powder is milled in nitrogen atmosphere, but only a little increase in the later case. About 0.8% increase in lattice parameter in the former case is attributed to the absorption of nitrogen and a corresponding expansion of the lattice. In the later case, the lack of gases leads to a slower increase in lattice expansion; about 0.4% increase in lattice parameter is associated with the residual gases.

For materials that will readily form nitrides, the use of nitrogen to provide an inert atmosphere should be avoided. Research shows that nitrides may be formed if Ti, Ta, Zr, Mo and Si are milled in molecular nitrogen atmosphere [3]. By milling Ti and Al mixture in nitrogen atmosphere, the nitrogen in the milling vial will rapidly be absorbed by the Ti-Al powder mixture [72, 73]. The amount of nitrogen reaches a saturated value of 50at.%. However, the absorption of nitrogen dramatically decreases with the increase in Al beyond 50at.% as shown in Figure 3.29. The relationship between absorption of nitrogen and concentration of Al in Ti-Al mixture is given in Figure 3.30. From the figure, it can be deduced that the nitrogen molecules must have been dissociated to form free nitrogen atoms by the collision process of the balls. The nitrogen atoms may diffuse into the fresh titanium surfaces. Titanium nitride may be formed in the nitrogen concentrated zone such as grain boundaries and then in the entire powder matrixes (72, 73). If Al is milled together with titanium, nitride in the form of (Ti, Al)N may be formed. Study on mechanical alloying of $Ti_{50}Al_{50}$ shows the formation of ($Ti_{50}$ $Al_{50}$)N with NaCl structure

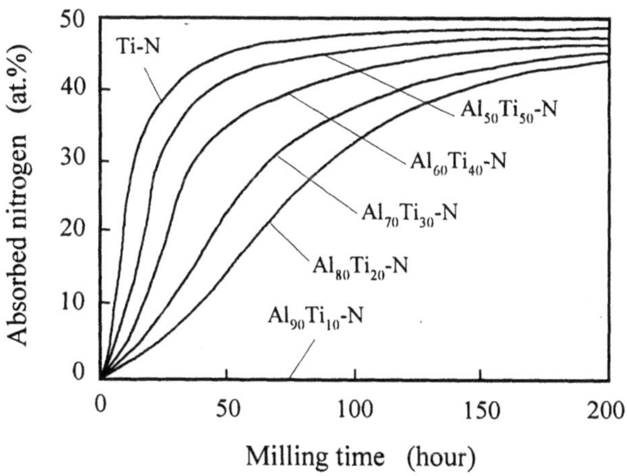


Figure 3.29. Absorption of nitrogen in Ti-Al mixture (73).

having a lattice parameter of 0.4166nm. This structure has been confirmed by transmission electron microscopy. When the nitride is heated to 1,373K, TiN and AlN can be detected (18). Beyond 1,573K annealing, no structural change may be observed. However, if annealed in a vacuum condition, TiN will be stable until 1,173K, but becomes unstable under a high pressure of 100 MPa.

In the case where high percentage of Al is used, the amount of absorption of nitrogen is less. therefore the amount of (Ti, Al)N formed at the same milling duration is less than that in $Ti_{50}Al_{50}$. Miki *et al.* (72, 73) explained this phenomenon of less absorption of nitrogen by considering large Al particle and low free energy of formation of $Al_2O_3$. It is the nature of Al to lead to the formation of large particles by cold welding. The large milled powder particle size will lead to a reduction in nitrogen absorption because the amount of nitrogen atoms absorbed on the fresh surfaces will be decreased by the reaction on the total surface area (72). Since the free energy of formation of $Al_2O_3$ is lower than those of AlN and $TiO_2$, a small amount of oxygen due to contamination from the powder and nitrogen gas will form the oxide layer.

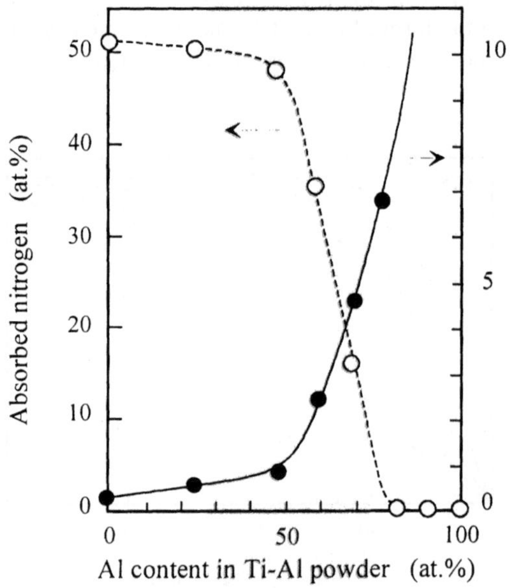


Figure 3.30. Relationship between absorption of nitrogen and Al concentration [72].

Table 3.2. Oxygen and nitrogen contamination of the as-received powder [74].

| Materials | Powder size (μm) | $O_2$ content (ppm) | $N_2$ content (ppm) |
|---|---|---|---|
| Ti-24Al-11Nb | <200 | 900 | 115 |
| Ti | <200 | 2200 | 170 |
| Al | <15 | NA | NA |
| Nb | <45 | 3000 | 150 |
| $Al_3Ti$ | <150 | 4900 | NA |
| Nb/Ti | <45 | 6800 | NA |

Table 3.3. Oxygen and nitrogen contents of powders after 24 hours of mechanical alloying [(74)].

| Materials | Atmosphere | $O_2$ content (wt.%) | $N_2$ content (wt.%) |
|---|---|---|---|
| Ti-24Al-11Nb | AT1 | 3.6 | 6.8 |
| Ti-Al-Nb mixture | AT1 | 4.8 | 7.6 |
| As above | AT1+1% stearic acid | 6.1 | 7.8 |
| Ti-$Al_3$Ti-Nb | AT1 | 5.3 | 7.4 |
| Ti-24Al-11Nb | AT2 | 0.10 | 0.015 |
| Ti-Al-Nb mixture | AT2 | 0.48 | 0.035 |
| Ti-$Al_3$Ti-Nb | AT2 | 0.59 | 0.425 |

To investigate the amount of contamination from different levels of gases, Goodwin *et al.* [(74, 75)] used two levels of purity of argon of 99.995% denoted as AT1 and 99.998% denoted as AT2. The high purity of 99.998% argon was further purified by passing through moisture and oxygen filter columns (<1 ppm $O_2$). No process control agent was used in some of the mechanical alloying processes so as to eliminate contamination from the process control agent. The work showed that $O_2$ and $N_2$ contents are lowered via using purified Ar indicating the importance of atmosphere control. Tables 3.2 and 3.3 show the measurement of oxygen and nitrogen contamination before and after milling.

### *3.3.3 Contamination from process control agents*

Decomposition of process control agents during mechanical alloying cannot be avoided. As these process control agents normally contain carbon, oxygen and hydrogen, their decomposition may cause carbon, oxygen and hydrogen contamination. Use of stearic acid has been found to give contamination of 1.1wt.% carbon and 0.8wt.% oxygen [(76)]. Direct formation of hydride in mechanical alloying of Al-Zr has also been reported [(77)], indicating the decomposition of the process control agent. It has been explained that the process control agent decomposes into hydrogen, oxygen and carbon due to rise in temperature in the mechanical alloying process. Repeated cold welding and fracturing result in the formation of very fine powder particles. Because of large surface area to volume fraction, surface energy of the mechanically alloyed powder particles increase with the increase in milling duration. The fresh surface created by the fracturing process promotes diffusion. In comparison to oxygen and carbon, the diffusion rate of hydrogen is much faster than that of oxygen [78]. Table 3.4 gives the amount of $H_2$, $O_2$ and C contents of different types of process control agent.

Table 3.4. $H_2$, $O_2$ and C contents after decomposition of different process control agent (per gram).

| Generic name | Chemical formula | $H_2$ (%) | $O_2$ (%) | C (%) |
|---|---|---|---|---|
| Stearic acid | $CH_3(CH_2)_{16}CO_2H$ | 13 | 11 | 76 |
| Heptane | $CH_3(CH_2)_5CH_3$ | 16 | 0 | 84 |
| Ethyl acetate | $CH_3CO_2C_2H_5$ | 9 | 36 | 55 |
| Ethylenebidi-steramide | $C_2H_2$-2$(C_{18}H_{36}ON)$ | 13 | 5 | 77 |
| Dodecane | $CH_3(CH_2)_{10}CH_3$ | 15 | 0 | 85 |
| Hexanes | $C_6H_{14}$ | 16 | 0 | 84 |
| Methyl alcohol | $CH_3OH$ | 13 | 50 | 37 |
| Ethyl alcohol | $C_2H_5OH$ | 13 | 35 | 52 |

Depending upon the materials, the degree of absorption of hydrogen is different [79]. Table 3.5 gives the content of free hydrogen in mechanically alloyed powder particles. It can be seen that free hydrogen decreases significantly with the increase in Zr concentration indicating the absorption of hydrogen by Zr.

Table 3.5. Content of free hydrogen in mechanically alloyed powder particles [77]

| Powder | Temperature (K) | Hydrogen (ppm) |
|---|---|---|
| Al-8%Fe | 623 | 992 |
| Al-5%Fe-3%Zr | 623 | 427 |
| Al-53%Zr | 623 | 49 |

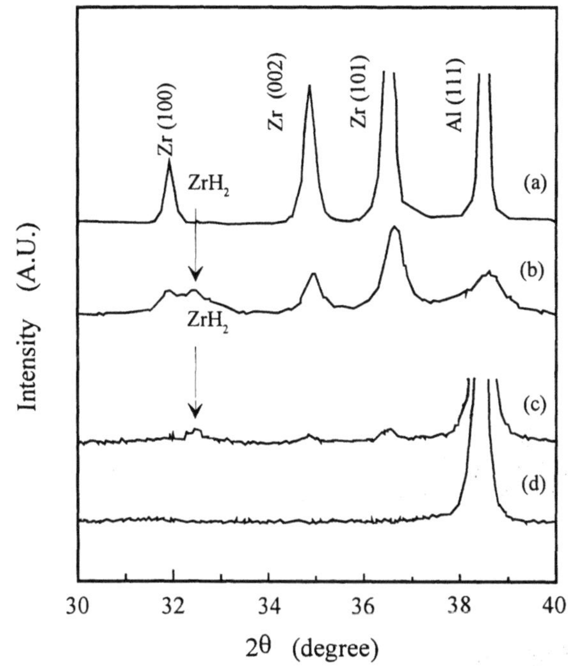


Figure 3.31. X-ray diffraction patterns of: (a) as-received Al-53%Zr mixture, (b) mechanically alloyed Al-53%Zr mixture, (c) mechanically alloyed Al-5%Fe-3%Zr and (d) mechanically alloyed Al-8%Fe [31].

Figure 3.31 shows the x-ray diffraction patterns of different combinations of Al, Fe and Zr [31]. Clear indication of formation of $ZrH_2$ can be seen. If Al-Fe mixture is mechanically alloyed, no hydrides could be detected. The hydride of $ZrH_2$ can be decomposed if the mechanically alloyed powder is heated to 823 K as shown in Figure 3.32 where Figure 3.32 (a) is the spectrum for the as-extruded specimen, Figures 3.32 (b) and (c) show the specimens for powders heat treated at 573 K and 773 K for 300 minutes respectively. It can be seen that decomposition of the hydride of $ZrH_2$ takes place only at 823 K (Figure 3.32 (d)).

Contamination from hydrogen in Mg and Mg alloys is also observed when Mg or Mg-Si mixture is milled with process control agent of stearic acid for 300 minutes using a planetary ball mill. The formation of magnesium hydride implies the decomposition of stearic acid into oxygen, carbon and hydrogen during milling. Table 3.6 gives the decomposition temperature of some hydrides.

Table 3.6 Decomposition temperatures of some hydrides

| Hydride | Temperature T (K)* |
|---|---|
| $MgH_2$ | 633 (79) |
| $TiH_{1.9}$ | 861 (79) |
| $ZrH_{1.6}$ | 1052 (79) |
| $ZrH_2$ | 823 (31, 77) |

* Decomposition temperatures are the endothermic peak temperatures measured by DTA or DSC.

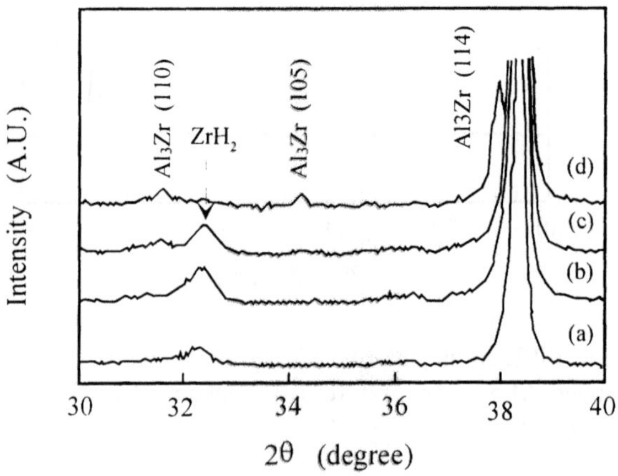


Figure 3.32. X-ray diffraction patterns of heat treated specimens (31).

Decomposed carbon from the process control agents will remain in the mechanically alloyed powder particles forming carbides (80).

## 3.4 Conclusions

Mechanical alloying/milling is carried out in a ball mill. The powder particles are subjected to high energetic impact. The alloying process can be described by the balance between cold welding and fracturing. This balance is generally unobtainable. Therefore, the use of surface modification via organic material called process control agent has to be considered. Domination of either cold welding or fracturing can be controlled by the introduction of process control agent. In general, the percentage of process control agent used in mechanical alloying or milling depends on the cold weldability of the powder particles, ductility of the materials, initial size of the particles, milling duration and stability of the process control agent.

Inert gas or vacuum protection from oxidation or gas contamination should be considered during mechanical milling. Argon gas is very often used in mechanical alloying, however, milling in vacuumed condition shows the best results. For some materials such as B, Mo, Si, Ta, Ti, Zr, nitrogen gas cannot be used because of the possibility of formation of nitrides. Sources of contamination during mechanical alloying come from the milling tools, gases in the vial and decomposition of the process control agent.

Intensity of collision during mechanical alloying is another key factor which controls the final structure of the mechanically alloyed powders. High intensity and large ball to powder weight ratio can promote faster reaction rate but tend to generate high temperature. Intensity of collision is crucial for the formation of amorphous alloys.

## 3.5 References

1. F.H. Froes, *Structural Applications of Mechanical Alloying*, Ed: F.H. Froes and J.J. deBarbadillo, Proceedings of an ASM Intern. Confer., Myrtle Beach, South Carolina, 27-29 March 1990, publ. by: ASM Intern., Mat. Park, Ohio, 1.
2. J.S. Benjamin, *Proc. of the Novel Powder Metall. World Congr.*, San Francisco, CA, USA, 21-26 Jan 1992, 155.
3. A. Calka, J.I. Nikolov and B.W. Ninham, *Proc. of the 2nd Intern. confer. on Structural Appl. of Mechanical Alloying*, Vancouver, British Columbia, Canada, 20-22 Sep. (1993), Ed. J.J. deBarbadillo, F.H. Froes and R. Schwarz, ASM Intern. Mat. Park, OH, USA, 189.
4. H.J. Fecht, *Nanophase Materials, Synthesis-Properties-Applications*, Ed: G.C. Hadjipanayis and R.W. Siegel, Pbl: Kluwer Academic Publishers, (1994), 125.
5. P.S. Gilman and W.D. Nix, *Metall. Trans. A*, Vol.12A (1981), 813.
6. N. Burgio, W. Guo, M. Magini and F. Padella, *Structural Applications of Mechanical Alloying*, Proc. an ASM Intern. Confer., Myrtle Beach, South Carolina, 27-29 March 1990, Ed: F.H.Froes and J.J. deBarbadillo, ASM Intern. Mat. Park, Ohio (1990), 175.
7. M. Otsuka, T. Ishiara, M. Sugamata and J. Kaneko, *ibid*, 221.
8. G.B. Schaffer and P.G. McCormic, *Mater. Forum*, Vol.16 (1992), 91.
9. E. Gaffet, P. Marco, M. Fedoroff and J.C. Rouchaud, *Mater. Sci. Forum*, Vols.88-90 (1992), 383.
10. E. Gaffet, F. Faudot and M. Harmelin, *Mater. Sci. Forum*, Vols.88-90 (1992), 375.
11. Y.H. Park, H. Hashimoto and R. Watanabe, *Mater. Sci. Forum*, Vols.88-90 (1992), 59.
12. T. Tanaka. S. Nasu, K. Nakagawa and K.N. Ishihara and P.H. Shingu, *Mat. Sci. Forum*, Vols. 88-90 (1992), 269.
13. M.D. Baro, S. Surlnach and J. Malagelada, *Proc. of the 2nd Intern. confer. on Structural Appl. of Mechanical Alloying*, Vancouver, British Columbia, Canada, 20-22 Sep. (1993), Ed. J.J. deBarbadillo, F.H. Froes and R. Schwarz, ASM Intern. Mat. Park, OH, USA, 343.
14. P.G. McComick, J. Ding, Y. Liu and R. Street, *ibid*, 349.

15. A. Calka, J. Jing, K.D. Jayasurlya and S.J. Campbell, *ibid*, 27.
16. G.F. Zhou and H. Bakker, *Mater. Sci. Forum*, Vols.179-181 (1995), 79.
17. A. Calka and J.S. Williams, *Mater. Sci. Forum*, Vol.88-90 (1992), 787.
18. Y. Ogino, M. Miki, T. Yamasaki and T. Inuma, *Mater. Sci. Forum*, Vols.88-90 (1992), 795.
19. J.S. Benjamin, *Proc. of the Novel Powder Metall. World Congr.*, San Francisco, CA, USA, 21-26 Jane 1992, Pbl. Metal Powder Industries Federation, Princeton, NJ., Advances in Powder Metallurgy, Vol.7, 155.
20. C.C. Koch, J.S.C. Jang and P.Y. Lee, *New Materials by Mechanical alloying Techniques*, DGM Confer., Calw-Hirsau (FRG), Oct. 1988, Ed. E. Arzt and L. Schultz, Informationsgesellschaft Verlag, 101.
21. F. Petzoldt, B. Scholz and H.D. Kunze, *ibid*, 111.
22. S.E. Lee, H.Y. Ra, T.H. Yim and W.T. Kim, *Mater. Sci. Forum*, Vols.179-181 (1995), 121.
23. U. Herr and K. Samwer, *Mater. Sci. Forum*, Vols.179-181 (1995), 85.
24. M. Burzynska Szyszko, V.I. Fadeeva and H. Matyja, *Mater. Sci. Forum*, Vols.179-181 (1995), 127.
25. R. Schulz, J. Lanteigne, M. Simoneau, P. Tessier, A. Van Neste and J.O. Strom Olsen, *Mater. Sci. Forum*, Vols.179-181 (1995), 141.
26. M. Umemoto, S. Shiga, K. Raviprasad and I. Okane, *Mater. Sci. Forum*, Vols.179-181 (1995), 165.
27. V.M. Lopez Hirata, U. Juarez Martinez and J.G. Cabanas-Moreno, *Mater. Sci. Forum*, Vols.179-181 (1995), 261.
28. J. Secondi, O. Drbohlav and R. Yavari, *Mater. Sci. Forum*, Vols.179-181 (1995), 287.
29. W.A. Kaczmarek, *Mater. Sci. Forum*, Vols.179-181 (1995), 313.
30. P. Millet and A. Calka, *Mater. Sci. Forum*, Vols.179-181 (1995), 321.
31. X.P. Xiu, Ph.D. thesis at KULeuven, Belgium, (1991), 34.
32. D. Gavrilov, O. Vinogradov and W.J.D. Shaw, Proc. of Intern. Confer. on Composite Mat., ICCM-10, Whistler, British Columbia, Canada, 14-18 August (1995), Vol. III, Processing and Manufacturing, Ed. A. Poursartip and K. Street, Woodhead Publishing Ltd., 11.
33. W.E. Kuhn and A.N. Patel, *Modern Developments in Powder Metallurgy, Principles and Process*, Vol.12 (1980), Proc. of the 1980 Intern. Powder Metall. Confer., 22-27 June 1980, Ed. H.H. Hausner, H.W. Ants and G.D. Smith, Metal Powder Industries Federation, American Powder Metall. Inst., Princeton, NJ, 195.
34. P.S. Gilman and J.S. Benjamin, *Ann. Rev. Mater. Sci.*, Vol.13 (1983), 279.
35. W.Y. Lim, M. Hida, A. Sakakibara, Y. Takemoto and S. Yokomizo, *J. Mater. Sci.*, Vol.28 (1993), 3463.
36. *Metals Handbook*, Ninth edition, ASM Intern., Mater. Park, OH (1984), Vol.7, 56.
37. A.M. Harris, G.B. Schaffer and N.W. Page, *Proc. of the 2nd Intern. confer. on Structural Appl. of Mechanical Alloying*, Vancouver, British Columbia, Canada, 20-22 Sep. (1993), Ed. J.J. deBarbadillo, F.H. Froes and R. Schwarz, ASM Intern. Mat. Park, OH, USA, 15.
38. M.L. Ovecoglu and W.D. Nix, *High Strength of Powder Metallurgy Aluminium Alloys II*, Ed. G.J. Hildeman and M.J. Koczak, TMS, Warrendale, PA, (1986), 225.
39. Aldrich, *Handbook of Fine Chemicals*, Aldrich Chemical Company, Inc., WI 53233, USA (1992).
40. T.S. Suzuki and M. Nagumo, *Mater. Sci. Forum*, Vols.179-181 (1955), 189.
41. A. Arias, *Chemical Reactions of Metal Powders with Organic and Inorganic Liquids during Ball Milling*, NASA TN D-8015, (1975).
42. L. Lu, M.O. Lai and S. Zhang, *Key Eng. Mater.*, Vol.104-107 (1995), 111.
43. R.B. Schwarz, P.B. Desch and S.R. Srinlvasan, *Proc. of the 2nd Intern. confer. on Structural Appl. of Mechanical Alloying*, Vancouver, British Columbia, Canada, 20-22 Sep. (1993), Ed. J.J. deBarbadillo, F.H. Froes and R. Schwarz, ASM Intern. Mat. Park, OH, USA, 227.
44. J.R. Groza and M.J.H. Tracy, *ibid*, 327.
45. M. Umemoto, T. Itsukaichi, J. Cabanas-Moreno and I. Okane, *ibid*, 245.
46. A. Malchere and E. Gaffet, *ibid*, 297.
47. P. Nash, S.C. Ur and M. Dollar, *ibid*, 291.
48. L. Takacs, *J. Appl. Phys.*, Vol.75 (1994), 5864.

49. A.N. Streletskii, *Proc. of the 2nd Intern. confer. on Structural Appl. of Mechanical Alloying*, Vancouver, British Columbia, Canada, 20-22 Sep. (1993), Ed. J.J. deBarbadillo, F.H. Froes and R. Schwarz, ASM Intern. Mat. Park, OH, USA, 51.
50. M. Oehring, T. Klassena and R. Bormann, *J. Mater. Res.*, Vol.8 (1993), 2819.
51. U. Mizutani, T. Takeuchi, T. Fukunaga, S. Murasaki and K. Kaneko, *Proc. 1st Intern. Confer. on Proce. Mat. for Properties*, Ed. H. Henein and T.Oki, TMS, Warrendale, PA, (1993), 671.
52. J. Eckert, L. Schultz and E. Hellstern, *J. Appl. Phys.*, Vol.64 (6), 1988.
53. J. Eckert, L. Schultz and K. Urban, *Z. Metallkde.*, Vol.81 (1990), 862.
54. J. Eckert, L. Schultz and K. Urban, *J. Mater. Sci.*, Vol.26 (1991), 441.
55. J. Eckert, *Mater. Sci. Forum*, Vol.88-90 (1992), 679.
56. L. Schultz, *Mater. Sci. Eng.*, Vol.97 (1988), 15.
57. E. Gaffet, *Mater. Sci. Eng.*, Vol.A119 (1989), 185.
58. M. Abdellaoui and E. Gaffet, *Acta Metall. Mater.*, Vol.43 (1995), 1087.
59. G. Martin and E. Gaffet, *Coll. Phys.*, C4 (1990), 71.
60. Y. Chen, R. Le Hazif and G. Martin, *Mater. Sci. Forum*, Vols.88-90 (1992), 35.
61. Y.H. Park, H. Hashimoto, R. Watanabe, J.H. Ahn and H.S. Chung, *Mater. Sci. Forum*, Vols.88-90 (1992), 155.
62. I.A. Ibrahim, F.A. Mohamed and E.J. Lavernia, *J. Mater. Sci.*, Vol.26 (1991), 1137.
63. J.S. Benjamin and T.E. Volin, *Metall. Trans.*, Vol.5 (1974), 1929.
64. L. Lu, M.O. Lai and S. Zhang, Mater. Design, Vol.15 (1994), 79.
65 C.C. Koch, *Annu. Rev. Mater. Sci.*, Vol.19 (1989), 121.
66. L. Lu, M.O. Lai and S. Zhang, *Mater. Sci. Tech.*, Vol.10 (1994), 319.
67. L. Delaey *et al*, "Shape Memory Effect, Super-elasticity and Damping in Cu-Zn-Al Alloys" K.U.Leuven, Metaalkunde, Report 78 R, 1978.
68. J.L. Perez-Rodrigue, L.A. Perez-Maqueda, A. Justo and P.J. Sanchez-Soto, *J. European Ceramic Soc.*, Vol.11 (1993), 335.
69. T.C. Chou, T.G. Nieh and J. Wadsworth, *Scripta Metall. Mater.*, Vol.27 (1992), 881.
70. W.A. Kaczmarek, A. Calka and B.W. Ninham, *Phys. Status Solidi* (A) *Appl. Res.*, Vol.141 (1994), 4.
71. H. Zoz, *Mater. Sci. Forum*, Vols.179-181 (1995), 419.
72. M. Miki, T. Yamasaki and Y. Ogino, *Mater. Trans. JIM*, Vol.34 (1993), 952.
73. M. Miki, T. Yamasaki and Y. Ogino, Mater. Sci. Forum, Vols.179-181 (1995), 307.
74. P.S. Goodwin and C.M. Ward-Close, *Proc. of the 2nd Intern. confer. on Structural Appl. of Mechanical Alloying*, Vancouver, British Columbia, Canada, 20-22 Sep. (1993), Ed. J.J. deBarbadillo, F.H. Froes and R. Schwarz, ASM Intern. Mat. Park, OH, USA, 139
75. P.S. Goodwin and C.M. Ward-Close, *Mater. Sci. Forum*, Vols.179-181 (1995), 411.
76. V. Arnhod and K. Hummert, *New Materials by Mechanical alloying Techniques*, DGM Confer., Calw-Hirsau (FRG), Oct. 1988, Ed. E. Arzt and L. Schultz, Informationsgesellschaft Verlag, 263.
77. X.P. Niu, L. Froyen, L. Delaey and C. Peytour, *Scripta Metall. Mater.*, Vol.30 (1994), 13.
78. B. Lustman and JR. Frank Kerze, *The Metallurgy of Zirconium*, McGraw-Hill (1995), 564.
79. Y. Chen and J.S. Williams, *J. Alloys & Compounds*, Vol.217 (1995), 181.
80. R.F. Singer, W.C. Oliver and W.D. Nix, *Metall. Trans.*, Vol.11A (1980), 1895.

# 4

# FORMATION OF NEW MATERIALS

## 4.1 Oxide Dispersion Strengthened Materials

The first application of mechanical alloying was to develop oxide dispersion strengthened Ni-base alloys (in short, ODS alloys) for use in aircraft engine [1]. Currently, ODS alloys have been developed from different systems [2].

### *4.1.1 Synthesis of aluminium alloys*

Aluminium ranks second only to Fe and steel in the metal industry. Commercial purity of Al varies from about 99.3% to 99.7%. It is one of the lightest metallic materials, with a density of only $2.7 g/cm^3$. Aluminium shows an fcc structure, relatively soft and ductile and has low yield strength. Due to low strength, its usefulness is limited. Fortunately, its mechanical properties can be improved dramatically through the introduction of alloy additions such as copper, magnesium, maganese, zinc, tin and lithium. The alloy additions strengthen the aluminium through solid-solution strengthening and fine dispersion of precipitate mechanisms.

#### *4.1.1.1 Processing*

Beginning with Benjamin who in 1970 successful fabricated IN853 Al alloy using the mechanical alloying technique [1], several industrial laboratories were involved in the development of advanced ODS Al alloys. The attempts to improve the properties of Al by using mechanical alloying technique have shown promising results in providing dispersed oxides and small grain size. The initial application of mechanical alloying to Al matrix materials was to produce an unalloyed Al strengthened by dispersoids, namely, $Al_4C_3$ and $Al_2O_3$ [3-11]. These dispersoids are

Table 4.1. Examples of mechanically alloyed Al alloys

| Material | Composition | dispersoid | References |
|---|---|---|---|
| Al | Al | $Al_4C_3$, $Al_2O_3$ | 3-11 |
| INCOMAP Al-9052<br>INCOMAP Al-9021<br>INCOMAP Al-905XL<br>DISPAL 0<br>DISPAL 3 | Al-4Mg-1.1C-0.8O<br>Al-1.5Mg-4Cu-1.1C-0.8O<br>Al-4Mg-1.5Li-1.2C-0.4O<br>Al-2.5%O<br>Al-3.0%C-0.8%O | $Mg_2Al_3$, $Al_4C_3$, $Al_2O_3$ | 6-13<br><br><br>14<br>14 |
| Al-Cr | Al-Cr | $Al_7Cr$ | 17 |
| Al-Cr-Fe | Al-Cr-Fe | | 18 |
| Al-Cu | Al-Cu | | 19-22 |
| Al-Cu-X<br>(X=Co, Cr, Fe, Mg, Mn, Ni, Zr) | Al-Cu-Co<br>Al-Cu-Cr<br>Al-Cu-Fe<br>Al-Cu-Mg<br>Al-Cu-Mn<br>Al-Cu-Ni<br>Al-Cu-Zr | | 23<br>23<br>23, 24<br>21, 25<br>23, 26-28<br>23, 28<br>30, 21 |
| Al-Fe | Al-Fe | $Al_3Fe$, $Al_{13}Fe_4$, $Al_4C_3$, $Al_2O_3$ | 29-36 |
| Al-Fe-X<br>(X=Ce, Gd, Mn, Mo, Ni, Si) | Al-Fe-Ce<br><br>Al-Fe-Mn | $Al_6Fe$, $Al_{13}Fe_4$, $Al_{10}Fe_2Ce$, $Al_4C_3$, $Al_2O_3$<br>$Al_6Mn$, $Al_6(Fe,Mn)$ $Al_4C_3$, $Al_2O_3$ | 37<br><br>38-40 |
| Al-Li | Al-Li | $Al_4C_3$, $Al_2O_3$ | 41 |
| Al-Li-X<br>(X=Cu, Mg, Si) | Al-Li-Cu<br>Al-Li-Mg | $Al_4C_3$, $Al_2O_3$ | 43<br>43, 44 |
| Al-Mg | Al-Mg | $Mg_2Al_3$ | 19, 42<br>45<br>46 |
| Al-Mg-X<br>(X=Cu, Li, Si) | Al-Mg-Cu-Si<br>Al-Mg-Si | $Mg_2Al_3$, $Mg_2Si$ | 47 |
| Al-Mn | Al-Mn | $Al_6Mn$ | 48 |
| Al-Mn-X<br>(X=Ce, Co, Cr, Si) | Al-Mn-Ce<br>Al-Mn-Co<br>Al-Mn-Cr<br>Al-Mn-Si | $Al_6Mn$ | 49<br>49<br>50<br>51 |
| Al-Mo | Al-Mo | | 52 |
| Al-Ti | Al-Ti | $Al_3Ti$ | 53, 54 |
| Al-Zr | Al-Zr | $Al_3Zr$ | 54 |
| Al-Hf | Al-Hf | $Al_3Hf$ | 54 |

thermally stable with very fine grain size of about 25 nm. Some examples of such dispersoids are DISPAL which is synthesized from Al, carbon and oxygen, and DISPAL Si12 from $AlSi_{12}$, carbon and oxygen.

Because of the nature of its crystal structure, Al is one of the softest materials which can easily be cold welded to each other to form big particles or to be cold welded to milling tools. For this reason, processing control agent must be utilized in the mechanical alloying of Al and it alloys. The process control agent is initially incorporated into the powder particles during the repeated welding, fracturing and rewelding process. It will be automatically decomposed at longer milling duration. Table 4.1 provides some examples of different types of mechanically alloyed Al alloys.

Because of repeated cold welding and fracturing in the milling process, oxide particles are being introduced into the Al matrix as a result of fracturing of the oxide film surrounding the individual powder particles as dispersoids. The dispersoid is further embedded into the matrix by the repeated cold welding and fracturing process. The oxide dispersoids show an amorphous structure after mechanical alloying but a $\gamma$-$Al_2O_3$ structure after being annealed for 4 hours at 923K [6]. Besides oxide dispersoids, carbide dispersoids in the form of $Al_4O_3$ are also commonly found in the Al matrix due to the process control agent. Because of decomposition of the process control agent, hydrated oxides and carbonates may be generated from the presence of hydrogen and carbon.

Table 4.2. Chemical analysis of extruded bars [4].

| Heat | Process control agent (wt.%) | wt.% | | | Vol.% | |
|---|---|---|---|---|---|---|
| | | C | O | $Al_2O_3$* | C | $Al_2O_3$ |
| A | 1.85 stearic acid | 1.53 | 1.85 | 4.05 | 2.07 | 3.26 |
| B | 1.0 stearic acid | 0.79 | 2.40 | 5.20 | 1.08 | 4.20 |
| C | 0.86 methanol | 0.28 | 1.92 | 4.15 | 0.38 | 3.35 |
| D | 0.86 methanol | 0.28 | 1.36 | 2.95 | 0.38 | 2.38 |
| E | 0.65 methanol | 0.29 | 1.44 | 3.11 | 0.39 | 2.51 |

*: Equivalent $Al_2O_3$.

Table 4.2 gives the chemical compositions of some mechanically alloyed and extruded Al [4]. The extruded rods contain significantly large amount of carbon from the process control agents. Generally, the amount of retained carbon is approximately equal to the carbon added via the process control agents. The retained carbon is very stable, unable to be removed by leaching with organic solvents or by vacuum annealing of the processed powder particles. Its content is appreciably lower when heat treated with methanol than with stearic acid. The oxygen contents lie between about 1.5wt.% and 2.5wt.%.

The repeated cold welding and fracturing process also leads to the formation of very fine grain size in the order of 250-500 nm in diameter. Figure 4.1 shows the x-ray diffraction patterns after different mechanical milling durations from which the crystalline size can be estimated from the broadening of the diffraction peaks. The grain size is stable, without much grain coarsening until a temperature of 763K [14]. It is well known that Al has a high stacking fault energy and because of the nature of fcc structure, it exhibits cross slip during deformation. In general, cross slip promotes formation of well defined dislocation cell structure during deformation rather than a high density of dislocation tangles as in case of metals with low

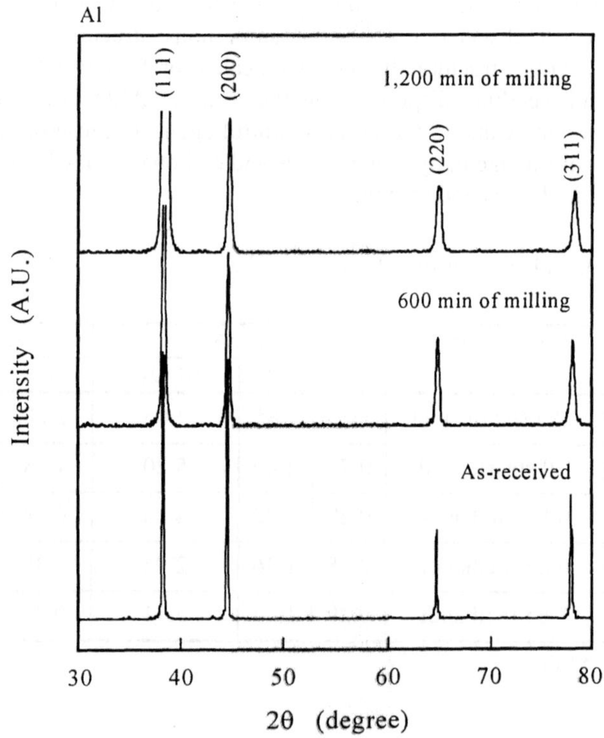


Figure 4.1. X-ray diffraction patterns of mechanically milled pure Al.

stacking fault energies. The formation of cell structure during mechanical alloying leads to a reduction in the stored energy. Hence, the driving force for recrystallization is reduced. Figure 4.2 shows the microstructure of the mechanically alloyed powder after hot consolidation at 773 K. It is clear that no recrystallization has taken place [15]. Prior boundaries are clearly evident. The subgrains in this microstructure is about 0.3 μm.

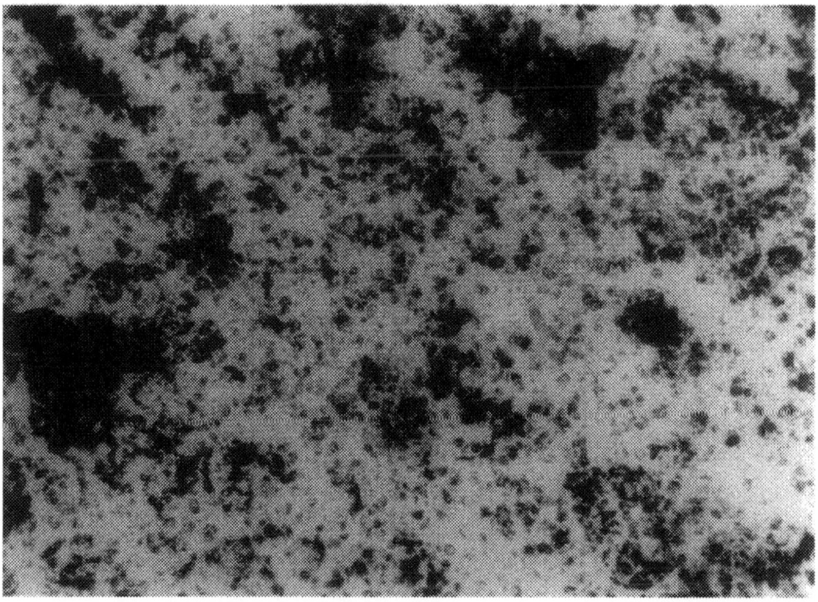

Figure 4.2. Grain structure of mechanically alloyed Al after hot pressing at 773 K and 207 MPa [15].

Mechanical alloying can also be utilized to introduce very fine $Al_nX$ (where X is a transition element) type intermetallic phases within the Al alloys [16]. In general, intermetallic phases have high melting temperature while transition elements typically exhibit low rates of diffusion within Al. Because of their stability and inherent hardness, $Al_nX$ type intermetallics are able to contribute to the elevated temperature strength, stiffness and thermal stability of intermetallic strengthened Al alloys.

### *4.1.1.2 Reaction milling*

Developed by Jangg in 1975 [55-57], chemical reactions between Al, graphite and oxygen during milling are introduced.

In this process, also known as reaction milling, carbon is introduced together with Al or Al alloys. The addition of non-metallic powder particles strongly affects the weldability of the powder particles during milling. The oxygen content can be controlled by adjusting the milling atmosphere of $O_2$, Ar, $N_2$, etc. Reaction between carbon, oxygen and Al takes place during intensive milling leading to partial formation of very fine $Al_4C_3$ and $Al_2O_3$ dispersoids which are stable up to the melting temperature of Al [58]. The chemical reaction between them may not be completed during milling; unreacted materials, such as lampblack, $H_2O$ or hydrocarbons can be observed after milling. The presence of these phases will produce poor mechanical properties. Hence, post heat treatment which enables the reaction to be carried out further has to be followed after mechanical alloying [56].

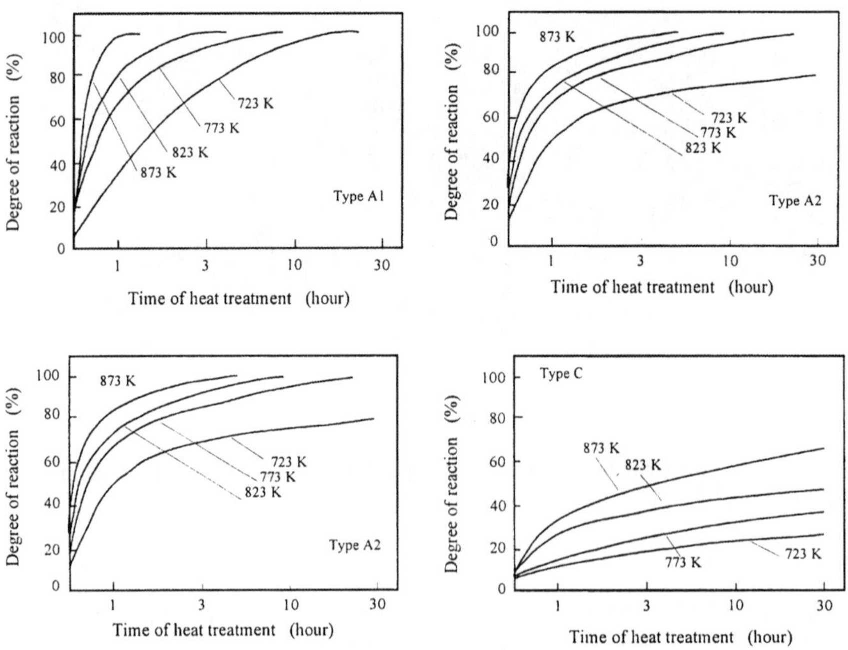


Figure 4.3. Degree of carbon reaction at different heat treatment of granulates containing different types of carbon: type A1-loose lampblack, type A2-furnace type, type B-coke soot, and type C-graphite [57].

Jangg found [57] that the rate of formation of $Al_4C_3$ was strongly influenced by the grindability and crushability of graphite used. Figure 4.3 shows the influence of different types of carbon. It can be seen that lampblack is more suitable for reaction milling. Full reaction may be obtained after 3 hours of heat treatment at 823 K. If low heat treatment temperature is used, time required for completion of chemical reaction will be longer. Graphite is more stable during heat treatment. Even after 30 hours of heat treatment at 873 K, chemical reaction in the mechanically alloyed powder particles still has not completed. The presence of retained carbon tends to lower the mechanical properties of the material. Therefore, selection of type of carbon, milling conditions and heat treatment procedures are important parameters that have to be carefully selected. Such parameters like preparation conditions, analyses of microstructure and the evaluation of stability of mechanical properties at elevated temperature have been investigated in reference (60).

To increase the mechanical properties further, Ma *et al.* [61] employed the in situ $Al_4C_3$ (by introduction of 3wt.% graphite powder) together with 10vol.%SiC particulates into Al. After mechanical alloying, the material was annealed at 873 K and extruded at 693 K. It was found that the yield strength of the alloyed material was 434 MPa which is about 2.5 time higher than that of 20vol.%SiC reinforced Al.

#### *4.1.1.3 Al-Fe systems*

Ordinary Al-Fe alloys can normally be used up to a service temperature of 473 K. The maximum service temperature can be increased to 623 K if the alloys are fabricated by melt spinning or gas atomization. Further increase in service temperature of up to 723-773 K may be achievable by means of mechanical alloying or reaction milling of Al alloys.

Figure 4.4 shows the Al-Fe phase diagram. In equilibrium state, the room temperature phases are Al and $Al_3Fe$ if Fe content is below 26.5at.%. In the case of mechanical alloying, the final phases of the mechanically alloyed powder particles are composition dependent.

Figure 4.5 (a) shows the x-ray diffraction patterns of Al-12.5%Fe powder mixture after 1,200 minutes of mechanical alloying [59]. The diffraction pattern indicates the presence of Al and Fe by the evidence of (111) and (200) Al peaks, and (111) Fe peak although the (200) Al peak overlaps with the (111) Fe peak. However, no evidence of formation of $Al_3Fe$ phase can be detected even after 4,800 minutes of mechanical alloying. There are two possibilities for this lack of evidence: (a) no formation of $Al_3Fe$ phase possible with this composition after 4,800 minutes of mechanical alloying, or (b) formation of $Al_3Fe$ phase possible but in the nanocrystalline form. Crystallographic structural analysis indicates that $Al_3Fe$ phase can only form if the mechanically alloyed powder is annealed. Long range order x-ray diffraction pattern of $Al_3Fe$ can be seen if the powder that has been mechanically alloyed for 4,800 minutes is annealed at 773 K for 30 minutes as shown in Figure 4.5 (b). The intensity of superlattice structure of the powder particles alloyed for 4,800 minutes is stronger than that for 3,200 minutes. Similarly, the degree of long range order appears readily at higher annealing temperature indicating that higher annealing temperature allows a faster relaxation resulting in a more complete restoration of atomic order [62].

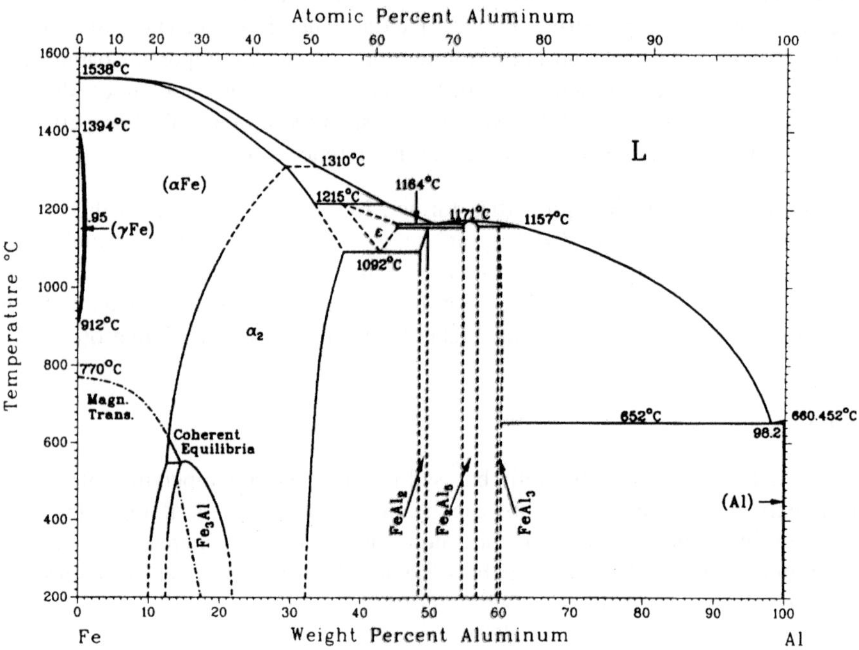


Figure 4.4. Al-Fe phase diagram [63].

When Al-20%Fe powder mixture is mechanically alloyed, it is found that the resultant structure differs from that of Al-15%Fe. Formation of $Al_3Fe$ with distorted structure took place after 3,900 minutes of mechanical alloying (Figure 4.6 (a)). This phase started to form after 2,400 minutes of mechanical alloying although the x-ray diffraction pattern at that time appeared to indicate amorphous phase. The broadening of the peak appeared as a result of the $Al_3Fe$ leading to the appearance of a new peak. Because its intensity was very low at the initial stage of the formation of $Al_3Fe$ phase, no clear peak could be detected. If the Fe content was increased such that the initial composition of the powder mixture was Al-25%Fe, $Al_3Fe$ structure started to form after only 1,200 minutes of mechanical alloying (Figure 4.6 (b)).

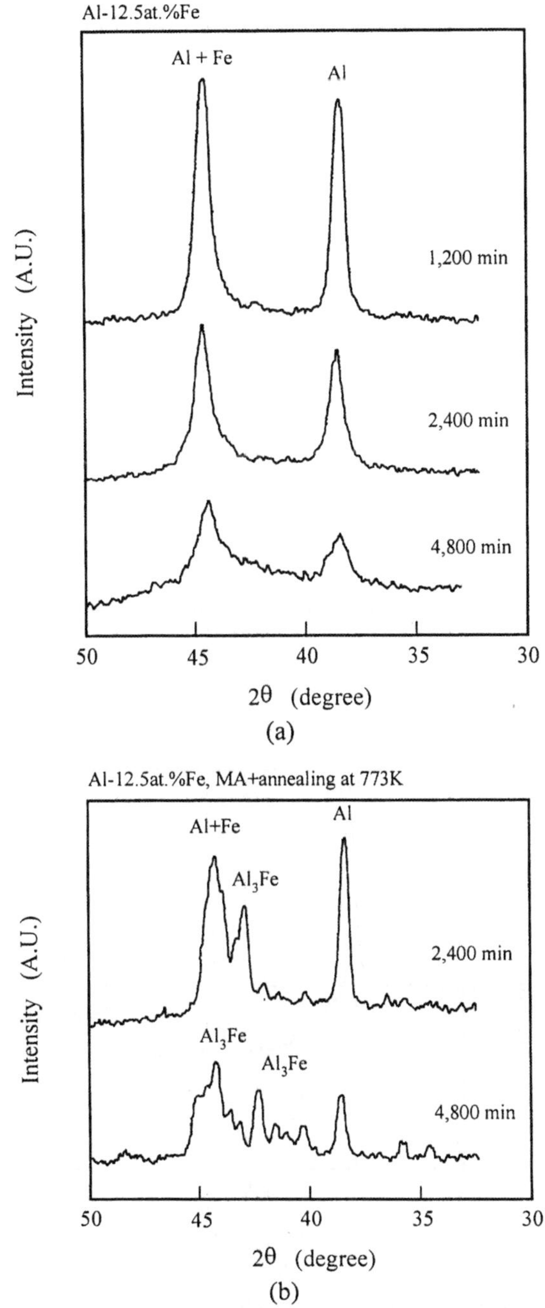


Figure 4.5. X-ray diffraction patterns of Al-12.5%Fe after being (a) mechanically alloyed and (b) mechanically alloyed and annealed [59].

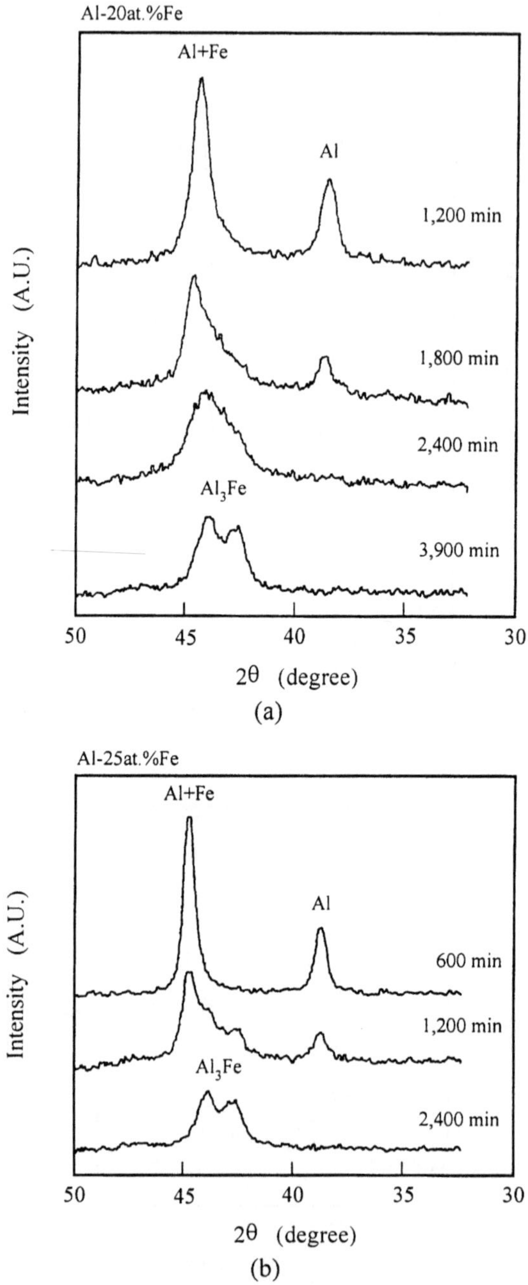


Figure 4.6. X-ray diffraction spectra of mechanically alloyed (a) Al-20%Fe powder and (b) Al-25%Fe powder [59].

It has been found, unfortunately that these alloys have low room temperature strength compared to conventional powder metallurgy alloy [64]. To improve the room temperature properties, Al8%Fe4%Ce powder atomized in a mixture of helium and argon was reaction milled together with 0.5wt.% stearic acid for 90 minutes in air followed by canning, degassing and extrusion. Figure 4.7 shows a comparison between mechanically alloyed Al8Fe4Ce, powder metallurgically prepared Al8Fe4Ce, IM2618 and RM DISPAL 2 Al alloys. Among the four different alloys, mechanically alloyed Al8Fe4Ce shows the highest room temperature yield strength as well as high temperature strength of up to 723 K. Although DISPAL 2 Al alloy shows low strength at lower temperature, the rate of decrease in strength with the increase in temperature is lower compared to specimens prepared by powder metallurgy method or IM2816 material. A comparison of the hardnesses between mechanically alloyed and powder metallurgically prepared Al8Fe4Ce Al alloys held at 723 K is given in Figure 4.8. It can be seen that hardness of the mechanically alloyed material is 33% higher than that prepared by powder metallurgy method. If the materials are held for about 12,000 minutes, the hardness of the former material is almost 100% higher than that of later.

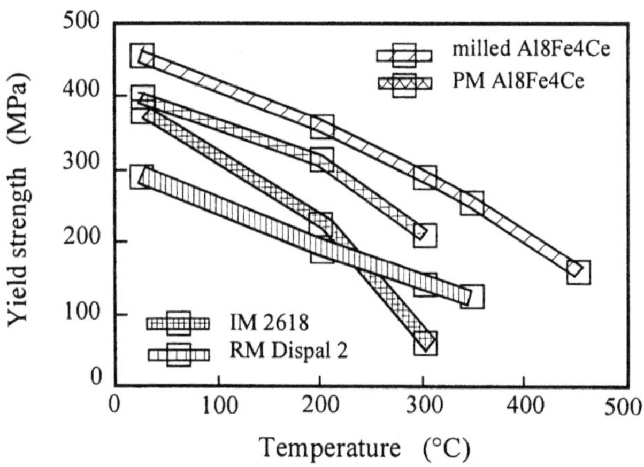


Figure 4.7. Yield strength of mechanically alloyed Al8Fe4Ce compared with other Al alloys [64].

The increase in thermal stability and strength at elevated temperatures is attributed to two effects:

(a) During mechanical alloying or mechanical milling, oxide and carbide are also being synthesised. Because of repeated cold welding and fracturing, oxides and carbides are homogeneously distributed in the Al alloys. These dispersoids act as obstacles to resist dislocation movement when stress is applied to the material. Unlike ordinary precipitation hardened Al alloys, such as $Al_2Cu$, $Mg_2Si$ strengthened Al alloys, strengthening due to precipitates decreases with decoherent of precipitates when the materials are used at elevate temperature, for instance, above 473 K. $Al_2O_3$ and $Al_4C_3$, however, are very stable up to melting temperature of the Al alloys.

(b) These oxide and carbide can stabilise the microstructures by limiting the growth in grain size. Figure 4.9 shows a transmission electron microscopy micrograph of a consolidated reaction milled Al8Fe4Ce annealed at 723 K for 34,200 minutes. No considerable grain growth can be detected after the annealing.

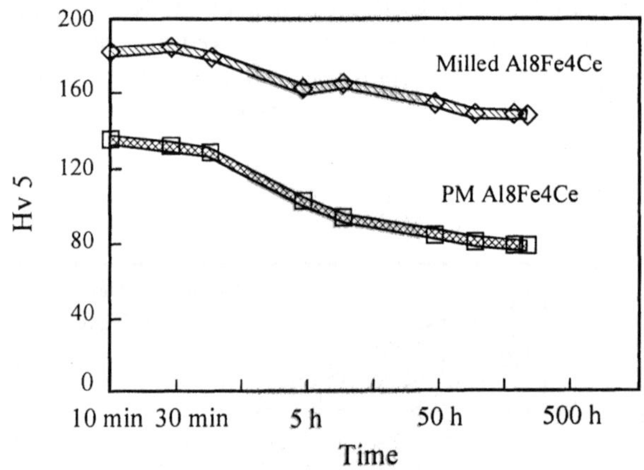


Figure 4.8. Hardness measurement of mechanically alloyed and as-atomized Al8Fe4Ce alloys [64].

The influence of mechanical alloying duration has been studied by Ovecoglu and Nix [37] by mechanical alloying Al-8.4wt.%Fe-3.5wt.%Ce Al alloy for different times. The results show that mechanical properties are influenced by milling duration. At lower temperatures represented by 523 K and 623 K, the values of the flow stress of both mechanically alloyed and non-alloyed materials are close to each other. This observation implies that the deformation mechanisms at low temperatures are similar for both cases. Further refinement of dispersed second phases for the fully mechanically alloyed material is not manifested in an increased contribution to strength at the low temperature. If deformation at low

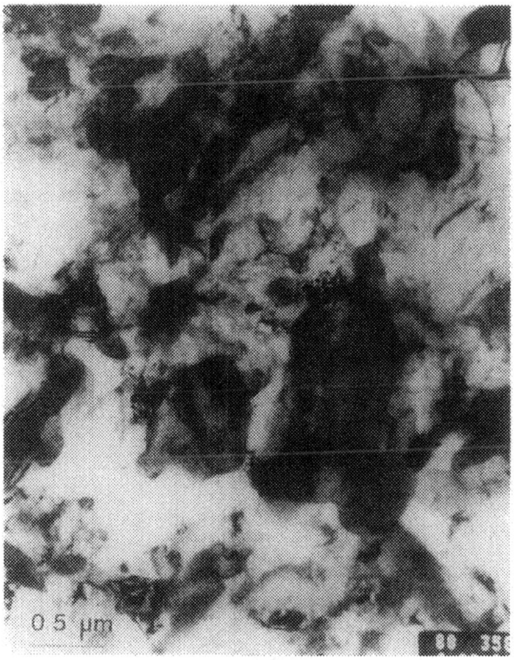


Figure 4.9. Microstructure of reaction milled Al8Fe4Ce Al alloy [64].

temperature is due to the Orowan mechanism, it is expected that due to its more closely spaced oxides and carbides, mechanically alloyed material is much stronger than the non-mechanically alloyed one in which the strengthening might be due to the presence of high volume fraction of intermetallic particles. The same amount of intermetallic particles is also present in the mechanically alloyed material. However, unlike mechanically alloyed materials, recrystallization does not take place in the non-mechanically alloyed materials during consolidation. Hence, low temperature strength of non-mechanically alloyed materials is attributed both to high volume fraction intermetallic phases and heavily cold worked structure. At high temperature, the presence of oxides and carbides become more important because of their stability. These stable and well distributed dispersoids are inherent in the mechanical alloying process. The higher flow stresses are achievable for the material mechanically alloyed for 180 minutes than those for 25 minutes. A strength difference between mechanically alloyed and non-mechanically alloyed Al alloys of 2.5 times is maintained at all temperatures and strain rates. This indicates that the superior mechanical properties of the mechanically alloyed aluminium over the precipitation strengthened aluminium is due to the presence of dispersoids in the Al matrix.

*4.1.1.4 Al-Mn systems*

Mn is a common alloying element for Al alloys. Al alloys are strengthened by the presence of Al-Mn intermetallics. Theoretical and experimental evaluations have shown that Mn can exist in several stable and metastable [65], quasi-crystalline icosahedral [66] and decagonal phases [67] in Al-Mn system. From the Al-Mn phase diagram shown in Figure 4.10, it is known that only $Al_6Mn$ is present if Mn content is below 25wt.%. Although Al may be strengthened by intermetallic $Al_6Mn$, a common drawback is its low ductility.

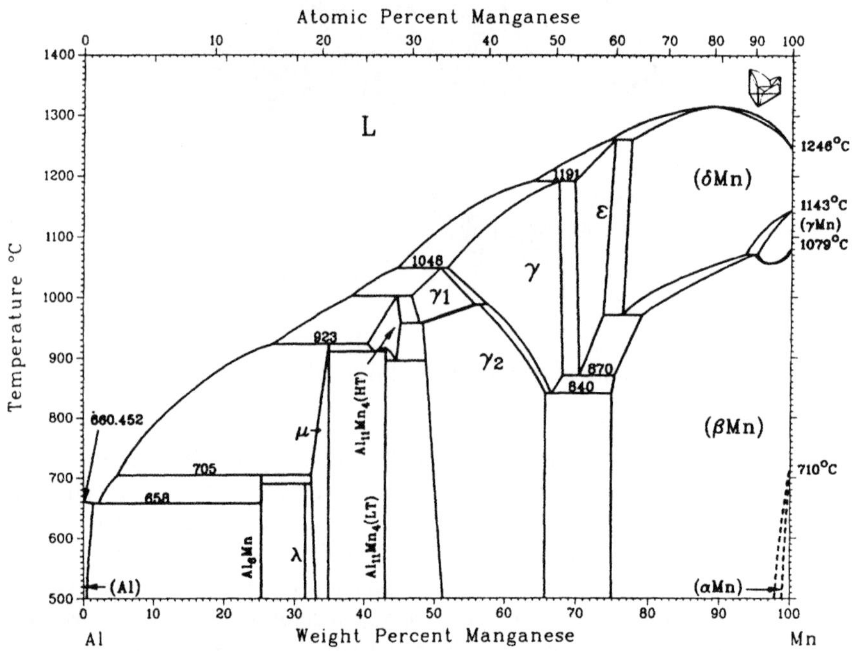


Figure 4.10. Al-Mn phase diagram [63].

Figure 4.11 reveals the structural changes of the mechanically alloyed and extruded Al-8at.%Mn. Although the crystalline size of Al is already refined to between 200 and 500 nm together with $Al_2O_3$ and $Al_4C_3$ dispersoids, no Al-Mn intermetallics can be detected. X-ray diffraction patterns in Figure 4.11 (a) clearly show the presence of fcc Al (111) and (200) diffractions, and Mn diffraction peaks after 2,160 minutes of mechanical alloying by using a SPEX 8000 [68] or a planetary ball mill [48]. This indicates that there is no alloying even to produce an equilibrium intermetallic phase [68]. After mechanical alloying, Mn shows uniform distribution in the Al powder particles. By annealing the mechanically alloyed powders at a temperature above 623 K, an equilibrium intermetallic $Al_6Mn$ may form. The volume fraction of the $Al_6Mn$ intermetallic compound increases from 0.06 at 623 K annealing to 0.21 at 723 K [68]. Annealing at higher temperature will result in further progress in alloying with complete elimination of free Mn. As shown in Figure 4.11 (b), formation of $Al_6Mn$ is evident after hot extrusion. Table 4.3 gives transformation temperature and activation energy for the formation of $Al_6Mn$.

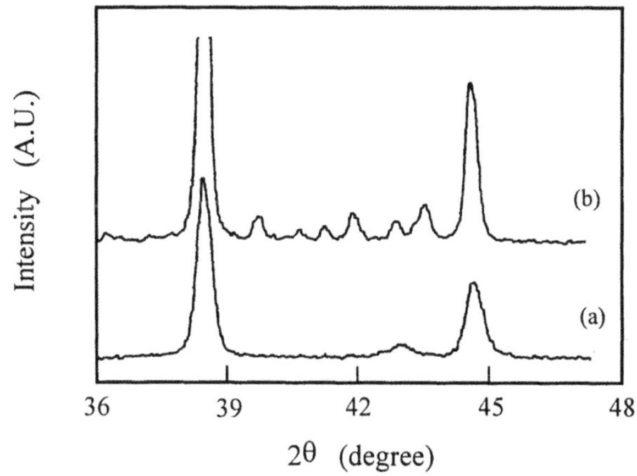


Figure 4.11. X-ray diffraction spectra of mechanically alloyed and extruded Al-Mn alloy [48].

Table 4.3. Temperature and activation energy for transformation of $Al_6Mn$.

| No. | Composition | Peak temperature at 5°C/min, (K) | Activation energy (kJ/mol) | Ref. |
|---|---|---|---|---|
| 1 | Al-8wt.%Mn | 725 | 176 | 48 |
| 2 | Al-33.7wt.%Mn | 705 | 197 | 68 |

If Al-Mn is mechanically alloyed in a $H_2$ atmosphere using a planetary ball mill with a weight ratio of ball to powder of 60:1, structural changes measured by x-ray diffraction can be detected [69]. At the initial stage of mechanical alloying up to 12,000 minutes, the intensities of the x-ray diffraction patterns of Al and Mn decrease and no evidence of formation of new phases can be found. Further mechanical alloying in $H_2$ induces amorphous. Transmission electron microscopy image of the mechanically alloyed powder clearly shows the co-existence of both amorphous and nanocrystalline phases. The crystalline phase is determined as Mn with a crystalline size of about 10 nm. X-ray diffraction and transmission electron microscopy both show no trace of hydrides of Al or Mn although they are milled in $H_2$ atmosphere. It appears that hydrogen does not participate in the solid state reaction but serves as a surfactant impeding the growth of the powder. Crystallization of $Al_6Mn$ takes place if the powder mechanically alloyed for 18,000 minutes is annealed at 493 K. At the same time Al and an unknown phase can also be found. The latter phase disappears after annealed at 673 K.

## 4.2 Intermetallics

### *4.2.1 Ti-Al system*

Intermetallics may be formed between two metallic elements, where an alloy phase is ordered if two or more sublattices are required to describe its atomic structure [(75)]. Ordered aluminides consist of a unique class of metallic materials that are characterized by the formation of long-range ordered crystal structures below their melting points or critical ordering temperatures. The various atomic species in the compounds tend to occupy specific sublattice sites and form superlattice structures [(76)]. The ordered structure exhibits attractive elevated temperature properties of strength, stiffness and environmental resistance because of the long-range ordered superlattice which reduces dislocation mobility and diffusion processes at elevated temperatures [(77)]. However, this reduced dislocation motion also results in extremely low ambient-temperature properties such as ductility and fracture toughness. The lack of both ductility and toughness is the major weakness of aluminides. Many techniques have been used in the synthesis of aluminides to improve their ductility, by macro- and micro-alloying, and by mechanical alloying. Appropriate microstructural modification is to introduce a secondary phase using mechanical alloying.

Since mechanical alloying processing is carried out entirely in the solid state, limitations such as the immiscibility in liquid and solid states do not apply to the process and thus the alloying capabilities are highly rated [(80)]. Mechanical alloying process is characterized by the development of composite layered structures during milling. These layers consist of various combinations of the starting constituent powders which are continually being refined and homogenized by the repeated actions of fracturing, cold welding and rewelding. Formation of lamellaed layers facilitates the dissolution of solute elements. Refinement and gradual reduction of these layers result in an interfacial area large enough for diffusion process to take place across them. For most solid state reactions, diffusion is the rate-limiting step. Hence, reaction rates are directly influenced by the diffusion processes.

This technique has been used in the synthesis of titanium aluminides of $TiAl(\gamma)$, $Ti_3Al(\alpha_2)$ and $Al_3Ti$ [(80-83)]. All three intermetallics have attractive elevated temperature properties of strength and modulus and low density, especially useful in the aerospace industry. The phase diagram of Ti-Al binary system is shown in Figure 4.12. The physical properties of Ti-Al system are given in Table 4.4. Alloys based on TiAl, $Ti_3Al$ and $Al_3Ti$ compositions have high elastic moduli, low densities and enhanced elevated temperature capabilities [84]. Based on these properties, they could be utilized in many elevated temperature aircraft and engine applications, certainly a material for future aircraft designs.

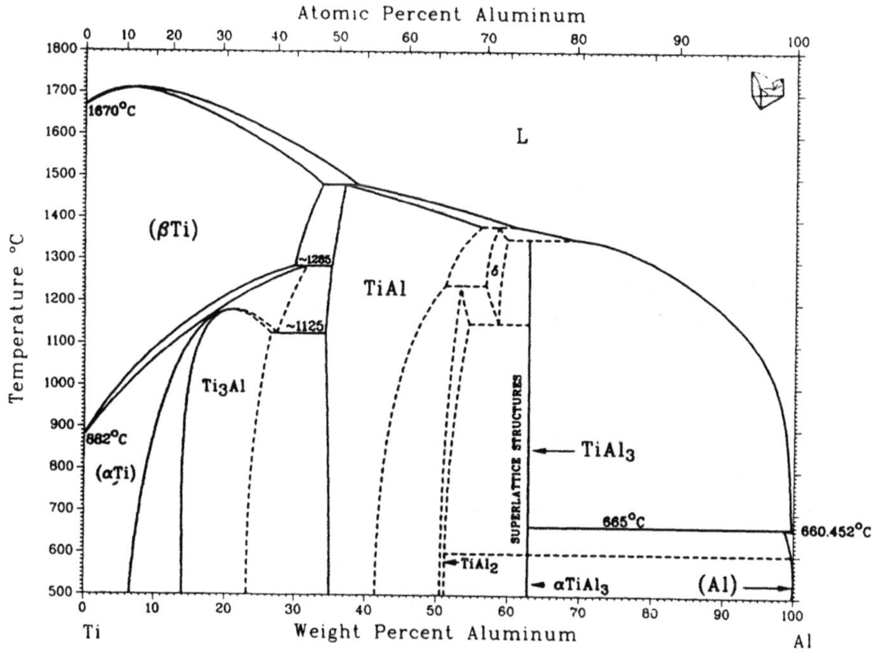


Figure 4.12. Ti-Al phase diagram [63].

Table 4.4. Physical properties of titanium aluminides

| Compounds | Density $g/cm^3$ | Crystal structure | ΔHf, 298 kJ/mol | Tm (K) |
|---|---|---|---|---|
| $Ti_3Al$ | 4.2 | $DO_{19}(\alpha_2)$ | -72.8 | 1873 |
| TiAl | 3.9 | $L1_0(\gamma)$ | -75.3 | 1733 |
| $Al_3Ti$ | 3.4 | $DO_{22}$ | -146.4 | 1613 |

Table 4.5. Some examples of different types of mechanically alloyed Ti-Al intermetallics.

| Composition | Ball mill/Atmosphere | Resultant phases | Ref. |
|---|---|---|---|
| Ti-50at.%Al | Attritor MA1DX/Ar<br>Spex 8000/Ar<br>Horizontal mill/Ar<br><br>Planetary mill/Ar<br>Spex 8000/$N_2$ | Amorphous, fcc TiAl (by over milling)<br>fcc TiAl<br>Amorphous (no formation of Ti aluminide)<br>Amorphous<br>metastable (TiAl)N | 85<br>86<br>87<br><br>88<br>89 |
| Ti-24at.%Al | Spex 8000/Ar | (a) fcc phase<br>(b) $TiAl_3$+fcc phase (annealing at 903K) | 80 |
| Ti-25at.%Al | Planetary mill/Ar<br>Spex 8000/Ar | fcc metastable phase | 90<br>91 |
| $TiH_{1.924}$-3Al | Spex 8000/Ar | $Al_3Ti$ (annealing at 893 K) | 81<br>82 |
| $TiH_{1.924}$-Al | Spex 8000/Ar | TiAl (annealing at 893 K) | 81<br>82 |
| $3TiH_{1.924}$-Al | Spex 8000/Ar | $Ti_3Al$ (annealing at 893 K) | 81<br>82 |
| $TiH_{1.924}$-$Al_3Ti$ | Spex 8000/Ar | TiAl (annealing at 873 K) | 93<br>94 |
| Ti-75at.%Al | Spex 8000<br>Horizontal mill | (a) $L1_2$ phase<br>(b) $DO_{23}$ (annealing at above 663K)<br>(c) $DO_{22}$ (annealing at above 1073K) | 95<br>92 |

*4.2.1.1 Synthesis of fct TiAl*

TiAl is face-centred tetragonal (fct) $L1_0$ structure which has alternate layers of Ti and Al atoms along the (002) plane. The rigid covalent bonds between atoms are the origin of high specific strength and high modulus. However, due to limited number of slip systems, TiAl with single $L1_0$ structure shows very low ductility at room temperature [96, 97]. If the $L1_0$ structure is disordered, the material can be expected to increase its ductility.

Figure 4.13 shows the x-ray diffraction spectra of the mechanically alloyed Ti-58 at%Al as a function of mechanical alloying time. According to the patterns of the x-ray diffraction, three stages of mechanical alloying can be defined.

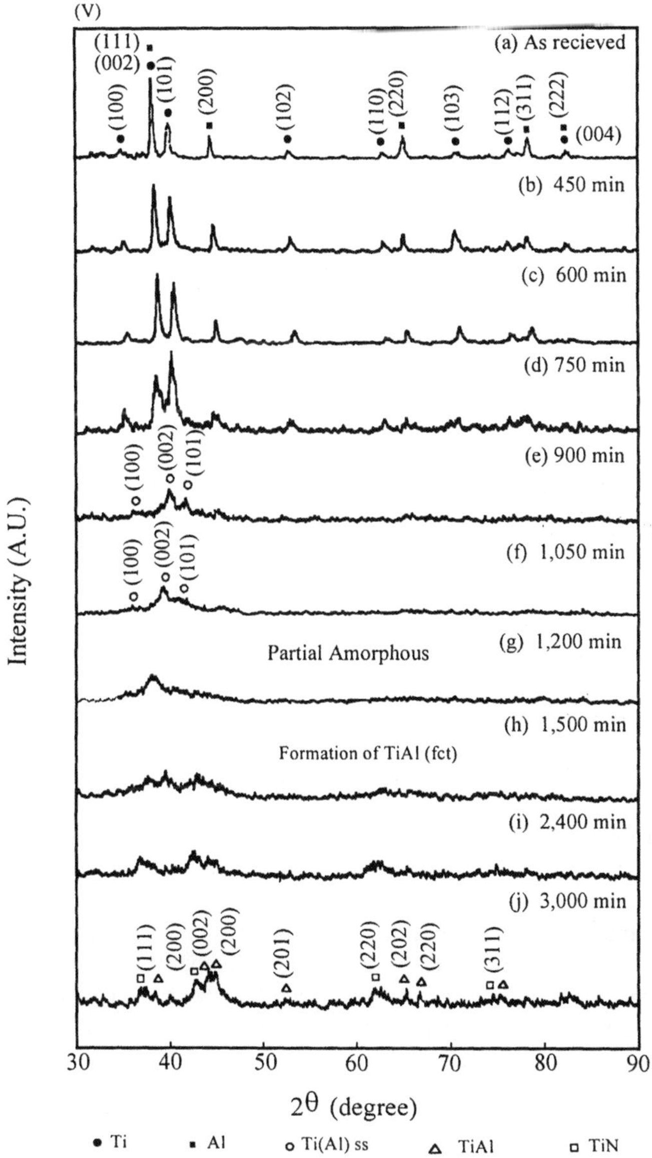


Figure 4.13. X-ray diffraction spectra of Ti-Al powders mechanically alloyed in a vacuum condition.

Stage 1: Solid solution

The x-ray diffraction pattern of the as-mixed elemental Ti and Al is shown in Figure 4.13 (a) as a reference. Qualitatively, the patterns depicted in Figures 4.13 (a) to (c) look similar. However, structural changes had begun to take place at the early stage of mechanical alloying. Observation of shifts in the diffraction angles of Ti to higher angles indicates the formation of a solid solution of Al in Ti. This is valid because the atomic radius of Al is smaller than that of Ti and consequently, dissolution of Al in Ti results in a change in the lattice parameter of Ti. This translates to an increase in the diffraction angles of Ti as deduced from Bragg's equation. Moreover, Al is a faster diffuser in Ti than that of Ti in Al. At room temperature, the diffusivity of Al in Ti is $1.06 * 10^{-21}$ $cm^2s^{-1}$ and that of Ti in Al is $2.9 * 10^{-23}$ $cm^2s^{-1}$, making Al the dominant diffusing element in the Ti-Al system [98]. This can be seen qualitatively in the x-ray diffraction patterns of Figures 4.13 (a) to (d) where it can be observed that the dominant (111) peak of Al decreases in intensity much more quickly than the dominant (101) peak of Ti. The relative weakening of the former peak as compared to that of the latter of Ti supports the conclusion that Al is a much faster diffuser in Ti as compared to Ti in Al. Dissolution of Al into Ti continues with milling duration.

From Figure 4.13 (d), it can be seen that after 750 minutes of milling, the widths of the Ti and Al peaks have begun to broaden due to the reduction in crystallite size and work hardening. During mechanical alloying, heavy plastic deformation of the powder particles induces large amount of dislocations. At the same time, the diffusion of Al into Ti leads to distortion of the Ti structure and hence further increases the density of dislocations. Al crystalline peaks are still present indicating that the dissolution of Al in Ti is far from completion at this stage.

Figure 4.13 (e) shows the diffraction pattern after milling for 900 minutes . The peaks have broadened considerably and are becoming increasingly overlapped as milling progresses. This makes detailed identification of the peaks more difficult than before. Intense peaks such as the (002) and (101) peaks of Ti are however, still distinct. The absence of Al peaks probably indicates that the dissolution of Al in Ti has completed. This is confirmed by differential thermal analysis (Figure 4.14) where the disappearance of the melting endothermic of Al is observed. At this stage of milling, it can be deduced that the existing phase at 900 minutes of milling is a hcp Ti(Al) solid solution. Further milling results in the formation of a supersaturated solid solution of Al in Ti.

Figure 4.13 (f) shows the formation of a non-equilibrium supersaturated solid solution of Ti(Al). This formation represents one of the metastable effects in the mechanical alloying process. It has been well reported in many experiments with mechanical alloying and has been observed even for liquid immiscible systems like Cu-Pb [80]. Since mechanical alloying is carried out entirely in the solid state, limitations such as immiscibility in liquid and solid states as imposed by phase diagrams do not apply to the process.

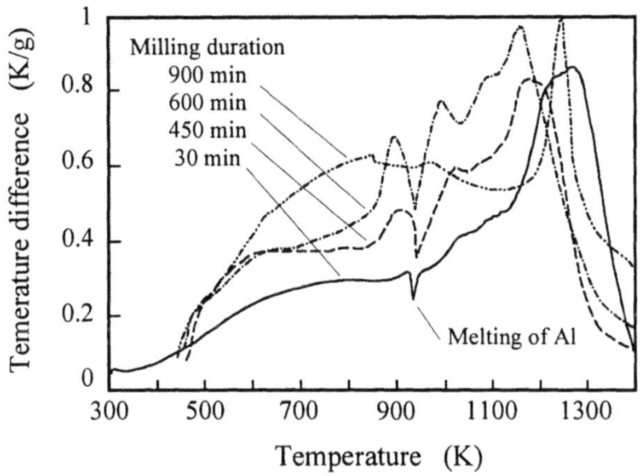


Figure 4.14. Differential thermal analysis of mechanically alloyed powder mixture.

Stage 2: Amorphous phase

Figure 4.13 (g) shows the amorphous stage of the mechanical alloying process. Observation of the 2θ angle in the region of 35-39° shows that the diffraction profile appears to be diffused. This is a typical distinguishing feature between amorphous and crystalline. Continuous deformation severely distorts the unit cell structures, making them less crystalline. Lattice defects brought about by severe deformation during mechanical alloying can destabilize the crystalline structure. More complex theories proposed by Schwarz and Johnson [99] require the elements to have a large negative heat of mixing in the amorphous state and one of the components to be an anomalously fast diffuser in order to form an amorphous phase. Defects introduced by severe deformation during mechanical alloying can sufficiently raise the free energy of the crystalline phases above those of amorphous ones. The diffusivity of Al is at least two orders of magnitude higher than that of Ti. Thus, both the thermodynamic and kinetic requirements of amorphous state have been met and the amorphous phase is reached before the formation of the crystalline intermetallic compound.

More detailed observation of Figure 4.13 (g) shows that the diffraction pattern is not completely amorphous. Some crystalline peaks persist along with the characteristic broad diffused amorphous profile. This may be due to insufficient milling, resulting in some residual crystalline particles residing among the amorphorized ones. Recrystallization of the amorphous phase is possible but unlikely. It is highly improbable that within 150 minutes the Ti-Al system can change from a supersaturated solid solution to the amorphous phase and finally to the onset of crystallization of the amorphous phase. Thus Figure 4.13 (h) represents a partial amorphous stage of the mechanical alloying process. It can also be seen that ambiguity does arise if the studies are based on x-ray diffraction analysis alone. At times, it is difficult to distinguish between broad, overlapping crystalline peaks, amorphous phase and a combination of both. X-ray diffraction patterns by themselves are not adequate and conclusive enough to determine if the phase is completely amorphous containing fine-grained structures or contains small crystallites embedded in an amorphous matrix [(80)].

Stage 3: Formation of fct TiAl

Figures 4.13 (h) and (i) show the x-ray diffraction patterns after milling time has been increased from the partial amorphous phase of Figure 4.13 (g). As mechanical alloying is a high energy process involving the transfer of kinetic energy of highly energetic steel balls to the powder charge, continued milling will generate heat which will ultimately crystallize the amorphous phase. Sharp crystalline peaks are particularly visible at 2θ diffraction angles of 39.6° and 43°. Continued milling results in the formation of more crystalline peaks as can be seen in Figure 4.13 (i). After 3,000 minutes of milling (Figure 4.13 (j)), the identity of the crystalline phase is evaluated as combination of fct TiAl and fcc TiN. This fct phase has a lattice parameter of a= 0.4108 nm; c= 0.4299 nm (c/a = 1.05) with major characteristic peaks at 2θ diffraction angles of 37.1° (111); 42.9° (002); 44.4° (200); and 65° (202). In general, TiAl is characterized by diffraction angles in the 39 ° to 45 ° region and the splitting of the (200) peak, clearly seen in Figure 4.13 (j). TiN has major characteristic peaks at 2θ diffraction angles of 36.7° (111); 42.6° (200) ; 61.8° (220) and 74.1° (311) as indicated in Figure 4.13 (j).

Although the fct TiAl is a stable phase thermodynamically, it is of interest to see if further milling will increase the concentration of defects and cause amorphization to occur again. As far as amorphization is concerned, two possible reaction routes of elements A and B are possible: (a) A+B → $(AB)_{amorphous}$ and (b) A+B → $(AB)_{crystal}$ → $(AB)_{amorphous}$. Amorphization of the mechanically alloyed Ti-Al system as demonstrated here goes through route (a) while further milling results in crystallization of the amorphous phase. Extensive study may be carried out from here to see if amorphization will result from further milling.

The presence of TiN is probably due to the level of atmospheric control during the mechanical alloying process. Although the vial may be back filled with research grade argon evacuated with a vacuum pump and sealed using a leak tight rubber ring, complete protection from the atmosphere is difficult to achieve. Pockets of air may still exist in the vial despite the countermeasures taken. Thus, the level of atmospheric control can at best be classified only as moderate. Since the mechanical alloying process involves the repeated fracturing and welding of milled particles, new surfaces are constantly being produced. Exposure of these new surfaces to the atmosphere will inevitably result in the formation of nitrides and oxides. This is particularly true for highly reactive elements such as Ti. The discovery of formation of nitrides has been well noted by other researchers. Suryanarayana *et al.* [(4)] reported the formation of a fcc phase of (Ti,Al)N despite milling under very carefully controlled atmospheric conditions.

It is interesting to note that the formation of TiN takes place instead of the formation of TiO. Mechanical alloying is a room temperature process and temperature in the milling process is not expected to be high. Based on this fact, it can be deduced that the formation of TiO can be ruled out since TiO forms at temperatures in excess of 1,523 K [(100)].

Nitriding of Al is also ruled out especially in the presence of Ti. Mukhopadhyay *et al.* managed to mechanically alloyed pure Al in air for up to 3,000 minutes without any formation of AlN. From thermodynamic considerations, the Gibbs energy of formation of TiN is -309.155 kJ/mol at room temperature which is much lower than that of -286.997 kJ/mol of AlN at room temperature [(101)]. Thus, the nitriding of Al is deemed not likely in the presence of Ti.

From the above discussion, it can be seen that the structural evolution of mechanically alloyed Ti-58at%Al in Figure 4.13 can be summarized as:

$$\text{Ti+Al} \rightarrow \text{Ti(Al)(ss)} + \text{Al(Ti)ss} \rightarrow \text{Am} \rightarrow \text{TiAl (fct)} + \text{TiN}$$

It must be noted that mechanical alloying time should be reasonably long enough to achieve compositional homogeneity. In the mechanical alloying of Ti-58at%Al 3,000 minutes of mechanical alloying has resulted in producing the fct phase of TiAl and fcc TiN. Thus, the optimum time for mechanical alloying Ti-58at%Al at a milling speed of 200 rpm is between 2,400 and 3,000 minutes .

### *4.2.1.2 Synthesis of fcc TiAl*

Atmospheric control is critical in the mechanical alloying process. Despite milling in purified Ar, Guo *et al.* [(86)] discovered contamination of oxygen and nitrogen contents reaching as high as 20.7 to 32.6at% when Ti - 5at%Al was mechanically alloyed. To understand the effects of milling atmosphere on the mechanical alloying process, the powder particles were mechanically alloyed in Ar atmosphere. It was found that the resultant structure differed from that mechanically alloyed in a vacuum condition. Figure 4.15 shows a series of x-ray diffraction patterns of powder mechanically alloyed over different durations using milling balls of 15 mm diameter at the same rotational speed of 200 rpm. After 600 minutes of processing time, the Al peaks disappeared and a supersaturated hcp solid solution was formed. Further ball milling caused the powders to amorphorize where the sharp peaks could hardly be observed. The Bragg peaks at 3,000 minutes of ball milling were found to be coincidental with that of TiAl having an fcc structure with a lattice parameter $a$ = 0.3998 nm. TiN was found as a side product although it is not obvious in the x-ray diffraction pattern (Figure 4.15) due to its relatively the low intensity.

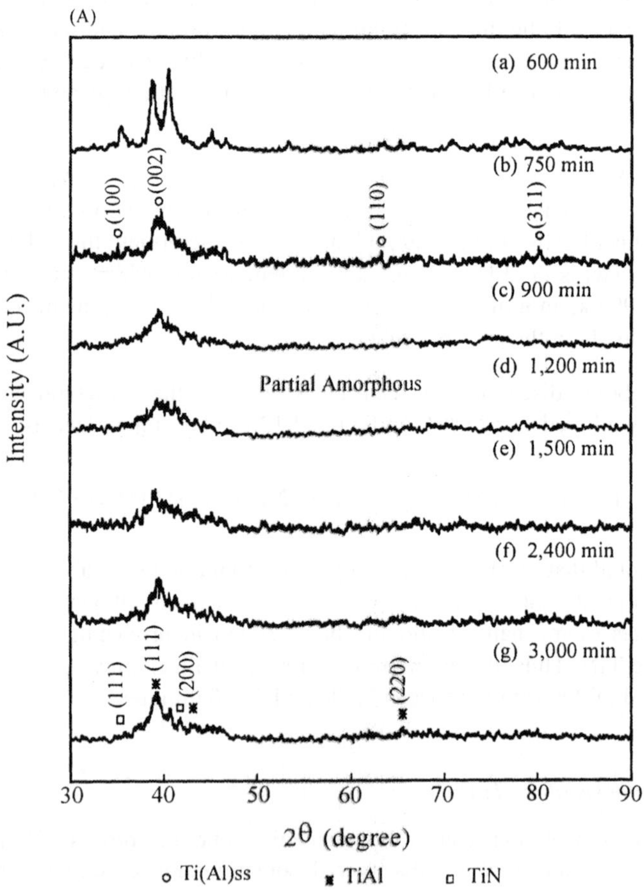


Figure 4.15. X-ray diffraction spectra of Ti-Al powders mechanically alloyed in an argon atmosphere (99.9% pure).

This appears to be a common end product when an fcc structure with $a = 0.4001$ nm is synthesized [102]. The reaction path experienced by the powders can be expressed as follows:

$$\text{Ti+Al} \rightarrow \text{Ti(Al)(ss)} + \text{Al(Ti)ss} \rightarrow \text{Am} \rightarrow \text{(TiAl)N} + \text{TiN}$$

According to Goodwin *et al.* [102], limiting the total contamination level will suppress the formation of the fcc phase usually found to occur in the mechanically alloyed Ti alloys. In the present case, air was first pushed out by passing Ar gas into the milling vial which was then evacuated to decrease the amount of $N_2$ content. Without evacuation of the vial, there would be excessive air in the vial. Therefore, formation of the fct phase of TiAl may be due to low contamination level during the milling operation.

Based on analysis of the diffraction patterns, it was found that there was little significant difference between the two levels of atmospheric control in the early stages of milling. As mentioned before, the mechanical alloying process will produce new surfaces for reaction due to repeated fracturing and cold welding. Exposure of these surfaces to the atmosphere will result in contamination. Sufficient milling must be allowed for the creation of these reaction surfaces and hence the effects of atmosphere on the powders will be more significant in the later stages of milling.

From Figure 4.15, it can be seen that the diffraction profile of the powder milled in Ar appears to be broader and more diffused than that milled in a vacuum condition shown in Figure 4.13. It also appears to be in the more advanced stage in the mechanical alloying process. Due to the considerable degree of peak broadening and overlapping of peaks, it is difficult to identify the peaks conclusively. This observation has also been confirmed by Suryanarayana *et al.* [103] whose investigation showed that the time taken to obtain amorphous phase was much shorter when the powders were milled in air than in argon or helium. The explanation for such a phenomenon is as follows: Nitrogen atoms may segregate preferentially to the grain boundaries of the lamellae structure during the initial stages of milling [104]. This solid solution of nitrogen in the Ti-Al system will result in an accumulation of strain by hindering defect mobility. Subsequently, the rate of work hardening is higher for powder samples milled in air than in pure argon or helium, leading to the formation of amorphous phase being accelerated. It can therefore be deduced at this stage that as milling progresses under a less protective environment, it is likely that the transformation stages of the Ti-Al system are accelerated considerably.

For mechanically induced reactions, the reaction rate can increase with a decrease in temperature and a change of phase [117]. This is due to a number of reasons. Firstly, the interfacial area of the reaction increases as milling continues due to the refinement of the crystallite size [118]. This increases the chemical reactivity during milling. Secondly, the repeated fracture and cold welding processes during milling create dynamically new surfaces for the reactions to proceed. Unreacted materials are continually brought into intimate contact with each other. Thirdly, the large number of defects introduced through plastic deformation allows significant diffusion to take place at room temperature [119]. In addition, the slight heating effect due to the transfer of kinetic energy from the highly energetic ball charge enhances and sustains the diffusion process [84].

Suryanarayana *et al.* discussed three different possible mechanisms for the formation of fcc TiAl phase [80]. The first possible mechanism is due to the introduction of stacking faults when Al diffuses into hcp Ti structure. If this is true, the lattice parameter of the new fcc phase should satisfy the relation $a_{fcc} = 2^{1/2} a_{hcp}$ (where $a$ is lattice parameter). Analysis has shown that $a_{fcc} = 0.42$ nm which is very close to the observed fcc phase. However, this mechanism is unsatisfactory because both $Ti_3Al$ and TiAl phases should be observed in the material. The second mechanism proposed is due to crystallization of the amorphous phase [91, 120-122]. The third possible mechanism is the stabilization of fcc Ti-Al phase by interstitial impurities of nitrogen and oxygen from the atmosphere and carbon from the process control agent. Because TiN has a fcc structure, it will stabilize the fcc structure of the mechanically alloyed Ti-Al mixture. Another possibility is the formation of AlN which has a fcc structure with a lattice parameter of $a_{fcc} = 0.4120$. This value is very close to the TiN lattice parameter. However, no formation of AlN has been found if Al is milled in nitrogen [80]. It has been suggested that this fcc phase is (Ti, Al)N with a lattice parameter 0.42 nm.

#### *4.2.1.3 Synthesis of $Ti_3Al$*

A similar structural evolution can be observed if the powder mixture containing Ti-25at.%Al is mechanically alloyed. Three stages of structural changes have been recorded with x-ray diffraction spectra as shown in Figure 4.16.

Stage 1: Solid solution

At the first stage, solid solution takes place leading to a shift of peaks of hcp Ti due to the nature of diffusivity of Ti and Al. According to Ti-Al phase diagram, the maximum dissolution of Al in Ti is about 13at.% while according to Oehring's study, the maximum value is 10at.%Al. Above this value, hexagonal $DO_{19}$ $Ti_3Al$ phase will form. The relative decrease in intensity of the spectra of fcc Al indicates a reduction in pure Al while the peak broadening both from Ti and Al indicates a refinement of the crystalline sizes and lattice disorder. The presence of unreacted crystalline elements rules out the possibility of melting and suggests ascribing the formation of the observed macroscopic grains to the action of milling [91].

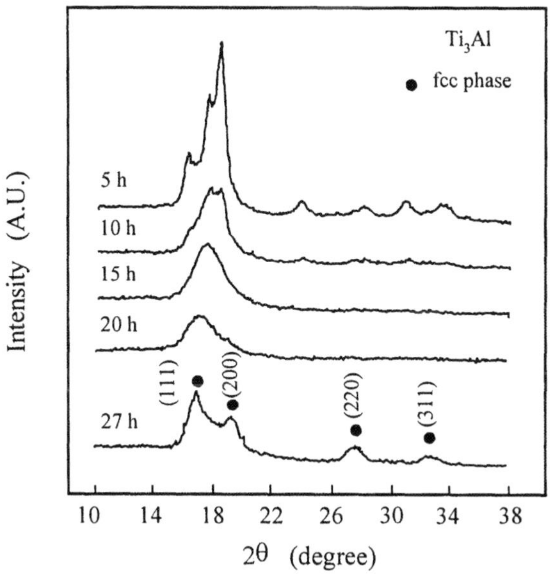


Figure 4. 16. X-ray diffraction spectra of mechanically alloyed Ti-25at.%Al [122].

Stage 2: Amorphous phase

With broadening of the x-ray diffraction peaks, some of the x-ray diffraction spectra become more clearly depicted with increasing mechanical alloying duration after the first stage of solid solution. As shown in Figure 4.16, after about 600 to 1,200 minutes of mechanical alloying, only one very broad and halo peak, which is a typical diffraction of amorphous phase, is present. In comparison with mechanical alloying of Ti-58at.%Al system, the solid solution period of Ti-25at.%Al system is very short. In addition, amouphization starts earlier than the Ti-58at.%Al system.

Stage 3: Formation of fcc $Ti_3Al$

From the x-ray diffraction peak of 1,620 minutes of mechanically alloyed powder shown in Figure 4.16, it is clear that a new material with a metastable fcc crystalline structure having a lattice parameter of 0.4220 nm is formed. The same phenomenon has been noted by detailed analyses from Cocco [91] as shown in Figure 4.17 and Guo [90]. The fcc crystalline structures with lattice parameters of 0.4220 [91] and 0.4200 nm [90] have also been reported

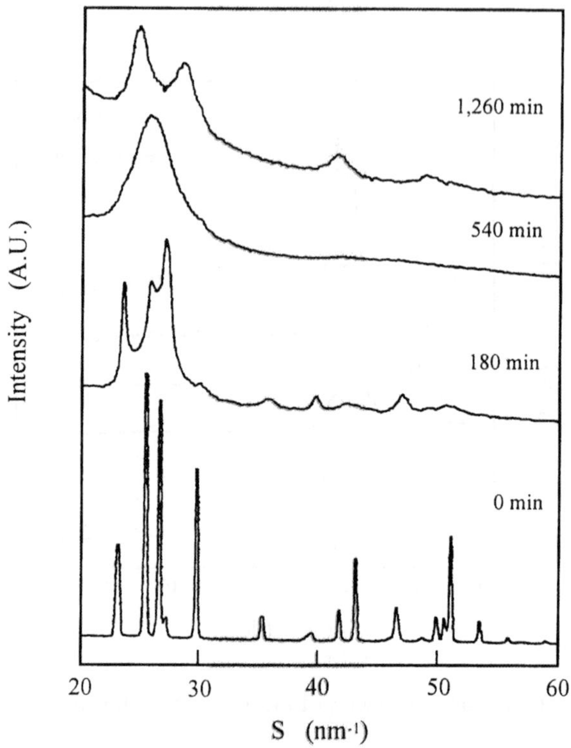


Figure 4.17. X-ray diffraction spectra of mechanically alloyed Ti-25at.%Al [91].

*4.2.1.4 Synthesis of $Al_3Ti$*

Schwarz *et al.* [95] claimed that the mechanically alloyed Al-25at.%Ti powder mixture possesses a metastable $L1_2$ structure. However, it must be realized that the as mechanically alloyed Al-25at.%Ti powder mixture is not fully ordered. Its long-range order parameter, *S*, is significantly lower than 1. When this metastable $L1_2$ phase is heated above 663 K (measured using differential scanning calorimeter), an irreversible transformation from $L1_2$ phase to another metastable $DO_{23}$ phase with a transformation enthalpy of 130 J $g^{-1}$ takes place. The metastable $DO_{23}$ phase again transforms to a stable $DO_{22}$ phase with a transformation enthalpy of 13.4 J $g^{-1}$ at about 1073 K. Schwarz *et al.* [95] suggested the following sequence for the formation of $Al_3Ti$ structure:

$$L1_2\ (<663\ K) \rightarrow DO_{23}\ (663 \sim 1073\ K) \rightarrow DO_{22}\ (1073\ K)$$

Since the solubility of Ti in Al is as low as 1.15wt.% at 938 K and zero below 673 K, Ti in Al shares other merits of transition metal additions [123]. Recently, Kim *et al.* [124] reported that the maximum solubility of Ti in Al was less than 2.46wt.% when Al-20wt.%Ti powder mixture was mechanically alloyed. Since only 2.46wt.% of Ti can be dissolved in Al, the remained Ti can only exist in two forms, namely, in the formation of $Al_3Ti$ intermetallic compound and in the form of dispersoids in the Al matrix. Based on solubility and x-ray analysis, however, it is not possible to form $Al_3Ti$ directly from mechanical alloying because Al shows a super saturated structure with homogeneous distribution of nano-sized Ti particles within the grains and around the grain boundaries after mechanical alloying. Therefore, as shown in Figure 4.18 [(123)], there is no manifestation of Ti diffraction pattern after 900 minutes of mechanical alloying. If the mechanically alloyed Al-X wt.% (X <15wt.%) powders are annealed at different temperatures, $Al_3Ti$ with $DO_{22}$ crystal structure can be formed from precipitation at 673 K annealing. However, if the mechanically alloyed Al-20wt.%Ti is annealed, $Al_3Ti$ with $L1_2$ crystal structure may be formed at about 573 K and is stable up to about 873 K. This $L1_2$ structure will transform to $DO_{22}$ structure only when it is annealed at a temperature above 873 K.

$L1_2$ structure in Al-Ti system can be formed either by rapid solidification or by introducing $L1_2$ structure stabilizer, such as Cu [(125)], Fe [(126)], Ni [(127)], Mn [(128)] or Cr [(129)]. In mechanical alloying of Al-Ti system, Kim [(123)] has suggested two routes for the formation of $L1_2$ $Al_3Ti$ structure:

(a) Precipitation may take place from the super-saturated Al to form $Al_3Ti$ dispersoids.

(b) Because of the existence of a large amount of Ti in the form of nano-particles within the Al powder particles, the nano-sized Ti particles may react with Al via interdiffusion at elevated temperature.

The second suggestion seems more reasonable since no shift of x-ray diffraction peak for Al has been observed during formation of $Al_3Ti$. If the formation of $Al_3Ti$ is through a reaction related to the precipitation of dissolved Ti from Al, a shift of Al diffraction peaks should be observed [(123)]. The absence of such a shift indicates that the formation of $Al_3Ti$ may be formed through the reaction between Al and fine Ti particles. Transmission electron microscopy observation clearly shows the presence of metastable fcc phase ring of the mechanically alloyed Al-25at.%Ti. Some ring patterns reveal cubic $L1_2$ structure, but the intensity of the pattern is too low to be clearly detected. The $L1_2$ structure may be due to the presence of nitrogen and oxygen which modify the $Al_3Ti$ structure [(80, 86)]. Similar to the conclusions of Schwarz *et al.* [(95)], Kim *et al* [(124)] and Suryanarayana *et al.* [(80)], Ahn *et al.* [(92)] suggested the following sequence for the structural evolution of mechanically alloyed and annealed Al-25at.%Ti:

$$\text{fcc} + L1_2\ (<573\ \text{K}) \rightarrow DO_{23}\ (673 \sim 873\ \text{K}) \rightarrow Al_{24}Ti_8\ (1073\ \text{K}) \rightarrow DO_{22}\ (1273\ \text{K})$$

Kim [(123)] reported to have observed the formation of $Al_{24}Ti_8$ ordered superstructure phase, but not the occurrence of $L1_2$ structure via mechanical alloying.

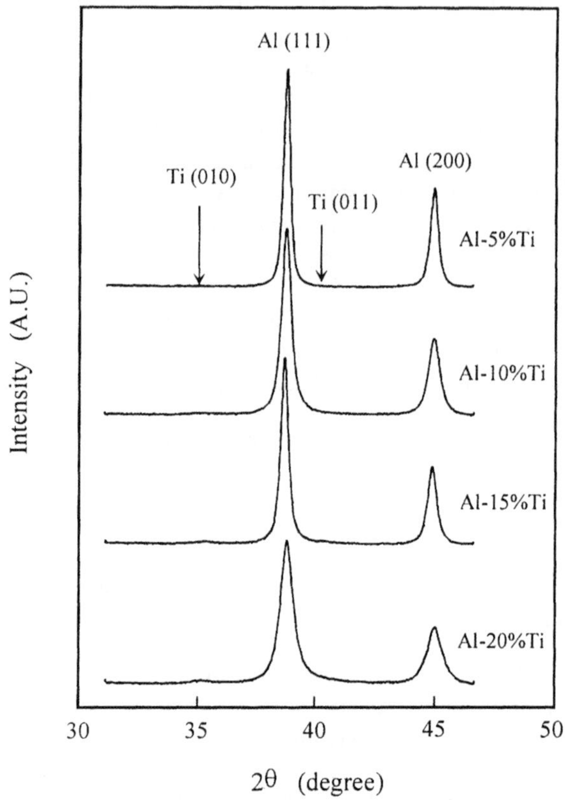


Figure 4.18. X-ray spectra of mechanically alloyed Al-X wt.%Ti (for X=5, 10, 15 and 20) after 900 minutes of mechanical alloying [123].

*4.2.1.5 Novel technique for the synthesis of TiAl, $Ti_3Al$ and $Al_3Ti$*

Introduced by Suryanarayana *et al* [94], TiAl intermetallic compound can be synthesized from $Al_3Ti$ and $TiH_2$ via mechanical alloying followed by thermochemical processing. This is based on the principle that the reaction of hydrogen with Ti is reversible and hydrogen may be easily removed from the hydrides by a vacuum thermal treatment [94, 130].

The three major phases that appear after 1,680 minutes of mechanical alloying as shown in Figure 4.19 (a) are tetragonal TiAl with indices a = 0.4005 nm and c = 0.4070 nm, tetragonal $Al_{24}Ti_8$ and fcc $TiH_{1.924}$. At this stage, the formation of TiAl is evident even though the reaction is not complete. The volume of $Al_{24}Ti_8$ is 41%. After 3,120 minutes of mechanical alloying (Figure 4.19 (b)), only TiAl phase with a volume fraction of 55% and $TiH_{1.924}$ phase with volume fraction of 45% appears to be present. The chemical analysis of the mechanically alloyed powder shows that the Al content corresponds almost exactly to the equiatomic TiAl phase. However, x-ray spectra only show the presence of TiAl and $TiH_{1.924}$ phases without any Al phase present. The absence is most probably due to dissolution of Al in TiAl phase and to the nanocrystalline structure of Al.

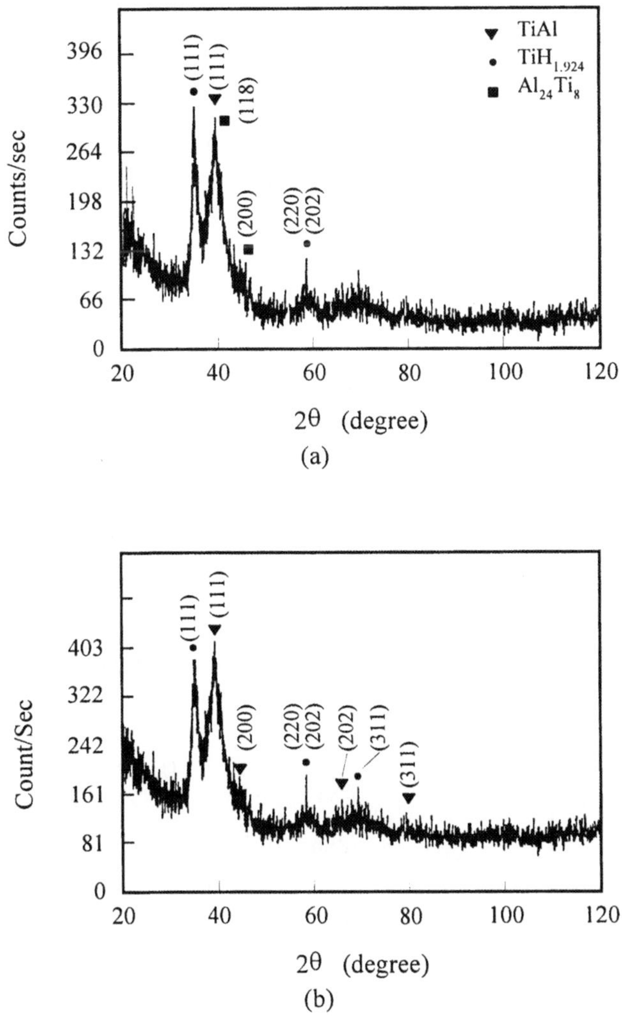


Figure 4.19. X-ray diffraction spectra of mechanically alloyed $Al_3Ti$ and $TiH_2$ mixture [94].

The increase in the content of TiAl phase can be promoted by annealing the mechanically alloyed TiAl and $TiH_{1.924}$ phases at high temperature. Under high annealing temperature, $TiH_{1.924}$ phase decomposes by the removal of hydrogen from the compound. The reduced Ti reacts with Al to form TiAl upon annealing. The small amount of hydride and $Al_{24}Ti_8$ phases can be reduced by controlling the proportion of the starting powders and/or by hot isostatic pressing. 95vol.% of TiAl can be obtained after hot isostatic pressing at 1023 K for 300 minutes. The formation of TiAl via $Al_3Ti$ and $TiH_{1.924}$ phases may be written as [94]:

$$Al_3Ti + 2TiH_2 \rightarrow 3TiAl + 2H_2 \uparrow$$

Titanium aluminides of TiAl, $Ti_3Al$ and $Al_3Ti$ can also be synthesized from titanium hydride of $TiH_{1.924}$ and element of Al with $TiH_{1.924}$ to Al ratios of 1:1, 3:1 and 1:3 respectively [81, 82].

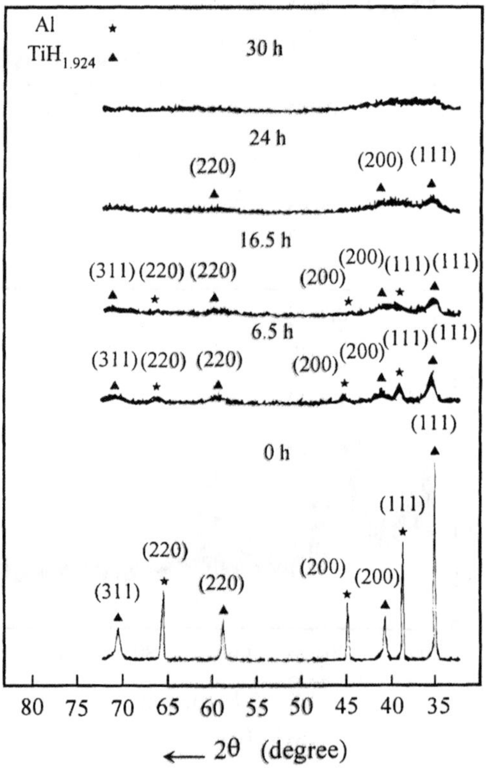


Figure 4.20. X-ray diffraction spectra of mechanically alloyed Al and $TiH_{1.924}$ mixture [81].

For the powder mixture with Al to $TiH_{1.924}$ ratio of 1:1 [81], Al x-ray diffraction peaks of (111), (200), (220) and (311) planes as shown in Figure 4.20 are still evidence after 990 minutes of mechanical alloying. No Al diffraction can be detected while the $TiH_{1.924}$ diffraction is still detectable after 1,440 minutes of mechanical alloying indicating that full dissociation of hydride is not completed at that time. Al appears probably as a solid solution within the hydride. Further mechanical alloying to 1,800 minutes leads to the formation of amorphous phase. 100% TiAl can only be achieved after a thermal treatment at 893 K for 6 days. The same structural evolution has been found in the powder mixture with ratio of 3:1. Full $Ti_3Al$ structure may be obtained after a thermal treatment at 893 K for 10 days. In the case of the 1:3 powder mixture, $TiH_{1.924}$ and Al phases co-exist up to 2100 minutes of mechanical alloying. Formation of $Al_3Ti$ structure requires a thermal treatment at 893 K for 2 days.

Because $TiH_{1.924}$ is brittle, problem due to sticking can be avoided even in the case of 75at.%Al-25at.%$TiH_{1.924}$ [81]. No process control agents are required during mechanical alloying and hence, contamination is minimized.

### *4.2.2 Ni-Al system*

In recent years, extensive research on Ni-Al system using mechanical alloying and other technique has been carried out [131-135]. Some investigations made use of shock-induced chemical reaction to produce Ni-Al intermetallic compounds. Horie *et al.* [136, 137] attempted to form Ni-Al by shocking Ni and Al powders together. They studied nickel aluminides using mechanically mixed powders of the respective elemental constituents in appropriate ratios as starting materials to form stoichiometric $Ni_3Al$. In addition, Ni-Al composite particles in which Ni has been deposited onto a spherical Al core to yield a nominal composition of 80 wt.% Ni in the composite particle were used. It was observed that in the case of mechanically blended powder mixtures, the nickel aluminide products were readily synthesized and controlled by the shock condition. Large quantities of $Ni_3Al$ were produced. The extent of the reaction and nature of the reaction products were dependent upon the shock pressure. In contrast to the mechanically blended powder mixture, the Ni-Al composite powders (Ni deposited on Al core) showed no large scale reaction region as the latter was confined only at the Ni-Al interface. Some research work on the formation of Ni-Al intermetallic compounds using mechanical alloying technique has also been carried out. Koch *et al.* [138] reported the formation of $Ni_3Al$ compound after milling Ni and Al powders for 300 minutes. Ivanov *et al.* [131] studied the synthesis of $Ni_2Al_3$ using mechanical alloying of mixtures of Ni and Al powders at a composition of $Ni_{40}Al_{60}$. It was found that mechanical alloying of Ni-Al led to the formation of metastable β' NiAl phase which reverted to the rhombohedral $Ni_2Al_3$ phase after annealing. Because mechanical alloying is a process in which powders with different compositions are impacted by high energetic balls, this process naturally involves two simultaneous actions: cold-welding between powders and fracturing of the powders. Hence, it can be deduced that under certain condition, a Ni-Al intermetallic compound with a nanocrystalline microstructure can be formed [(139)].

### *4.2.2.1 Synthesis of NiAl*

NiAl intermetallic compound can be directly formed using mechanical alloying of elemental Ni and Al powders. Due to very high negative mix enthalpy of Ni and Al (-71 kJ/mol at 298 K), formation of NiAl via mechanical alloying is accompanied by an abrupt increase in temperature after a very short milling duration of about 90 to 180 minutes, indicating the occurrence of a mechanochemical reaction between the two elemental powders [(140-142)] where the chemical reaction is driven by thermodynamic force of $\Delta H<0$. After this exothermic reaction, almost all the elemental powders are transformed into NiAl. Further milling will lead to a complete chemical reaction. X-ray diffraction spectra show an almost completely ordered B2 structure. Before the main reaction takes place, layered Ni and Al composite particles are formed. Low interdiffusion between Ni and Al layers occur because of the low diffusion coefficient of the diffusion couple. No significant formation of the compound or shifts of the Ni and Al peaks, which would indicate a gradual phase formation by solid solution between the two elements, can be observed at short milling duration [(143)]. The final steady grain size measured by transmission electron microscopy and x-ray diffraction is about 6-15 nm [(140, 143)]. This nanostructure phase is stable until about 810 K at which the energy release due to grain growth and stored energy is about 1.04 kJ/mol [(140)]. The lattice parameter of the mechanically alloyed NiAl is 0.2877 nm smaller than that of cast NiAl with a lattice parameter of 0.2887 nm.

If Ni content is above 33at.%, no formation of $Al_3Ni$ intermetallic compound has been detected as a result of exothermic reaction at the early stage of mechanical alloying. This phenomenon is in contrast with the fact of formation of $Al_3Ni$ during annealing of Ni and Al composites [(141)]. Thermogram of the mechanically alloyed equimolar Ni-Al powder mixture during heating show that there are three exothermic reactions at about 503, 613 and 773 K [(141)]. If the mechanically alloyed powder is annealed at the first two peak temperatures, formation of $Al_3Ni$ compound may be observed. Further annealing at the third peak temperature leads to the formation of a mixture of phases of $Al_3Ni_2$, NiAl and $Ni_3Al$. By increasing the annealing temperature to 973 K, $Al_3Ni_2$ phase disappears.

One of the apparent difference between mechanical alloying of Ni and Al mixture and thermal treatment is the process of diffusion. Significant interdiffusion takes place before nucleation of the intermediate phase. Ni concentration in Al during mechanical alloying may increase so rapidly as to be able to bypass the nucleation of $Al_3Ni$ phase [(141)]. Another reason why nucleation of $Al_3Ni$ phase is impeded is due probably to the large reaction enthalpy of NiAl phase than that of $Al_3Ni$ since NiAl has the largest enthalpy of reaction in the Ni-Al system.

*4.2.2.2 Synthesis of $Ni_3Al$*

The evolution and formation of $Ni_3Al$ has been examined by means of x-ray diffraction patterns after each stage of mechanical alloying. These x-ray diffraction spectra are shown in Figure 4.21. The change in lattice parameter of Ni and Al elements estimated from the peaks corresponding to the (220) planes of Ni and Al as a function of milling time is shown in Figure 4.22. Figure 4.21 (a) shows the x-ray diffraction pattern of the as-received Ni and Al powder after blending for 420

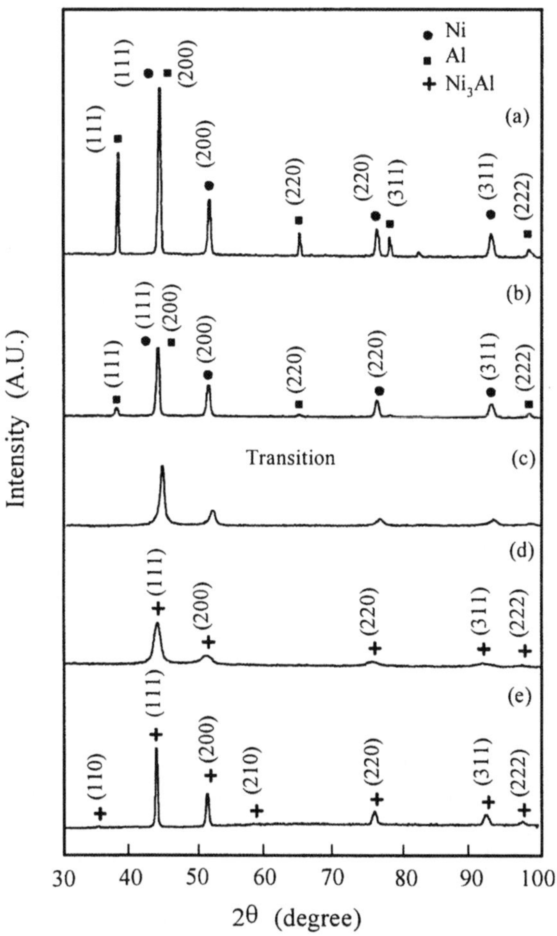


Figure 4.21. X-ray diffraction spectra of Ni-Al powder mixture at different mechanical alloying durations: (a) 0 minute, (b) 120 minutes, (c) 300 minutes, (d) 600 minutes, and (e) 600 minutes and annealed at 873 K for 20 minutes.

minutes. It can be seen that with the increase in milling time, the patterns for Ni are gradually shifted to a lower θ-angle while for Al, to a high θ-angle indicating a change in lattice parameter which may be the result of mutual dissolution between Ni and Al. The intensities of Ni and Al peaks are lowered while the full widths at half maximum of the diffraction peaks are broadened indicating a decrease in effective crystalline size with mechanical alloying time. All the peaks for Ni and Al contents in Figure 4.21 (b) are still visible at 120 min of mechanical alloying. It is noted that although the intensity of the x-ray diffraction pattern is greatly lowered, there is only a slight change in the lattice parameters of Ni and Al before about 120 minutes of mechanical alloying. During this period, the laminate structure is fine but with high densities of defects. The latter provides short diffusion circuit. Figure 4.21 (c) at 300 minutes of mechanical alloying is referred to as the transition period where Ni, Al, $Ni_3Al$ and the other phases co-exist. Formation of $Ni_3Al$ with disordered fcc structure occurs after 600 minutes of mechanical alloying.

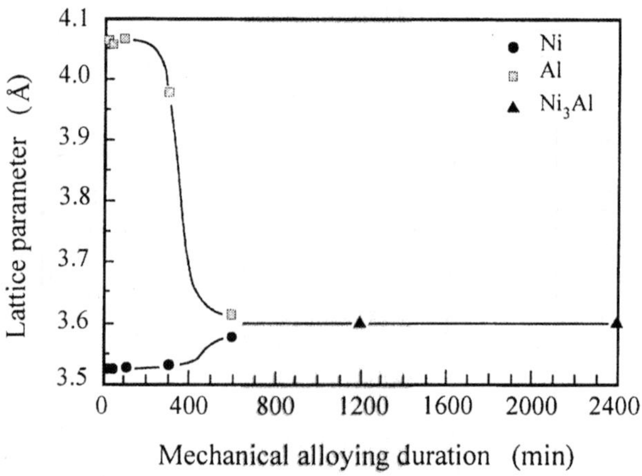


Figure 4.22. Change in lattice parameter of mechanically alloyed Ni-Al mixture.

Disordered fcc $Ni_3Al$ is a metastable phase which will transform to a stable phase upon heating. This disordered to an ordered transformation releases heat during heating. Therefore, this ordering process can be monitored by means of thermal analysis instrument. Figure 4.23 shows the thermogrames of the mechanically alloyed powder at 300 and 600 minutes of mechanical alloying at a heating rate of 20 K/min. Thermal analysis of the powder after 300 minutes of mechanical alloying, however, reveals only one exothermic peak at about 703 K although three peaks have been found for short milling duration. This peak is associated with the ordering transition. After 600 minutes of mechanical alloying, it can be seen that the exothermic peak becomes larger due to the increase in volume percent of $Ni_3Al$

intermetallic compound and also due to the increase in stored energy after longer mechanical alloying duration.

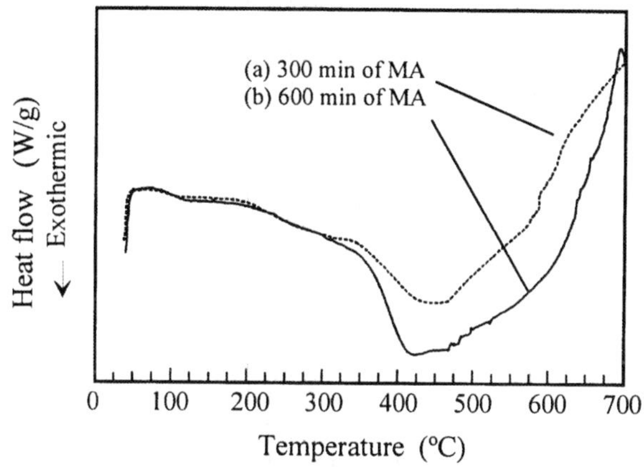


Figure 4.23. Differential scanning calorimeter measurement of mechanically alloyed Ni-Al powder mixture after: (a) 300 minutes and (b) 600 minutes.

This mechanical disordering from an ordered γ' phase has been observed in the mechanical milling of an ordered $Ni_3Al$ [145-147]. The disordering measured increases dramatically at the beginning of the mechanical milling process by a long range order parameter, *S*, which can be calculated from the relative integrated intensities of the superlattice/fundamental intensity ratio of $I_{100}/I_{111}$. Differential scanning calorimeter measurement shows an increase in enthalpy with the increase in mechanical milling duration indicating an increase in disordering. No saturation value is reached after 900 minutes of mechanical milling [138]. Similar to the mechanical alloying of Ni-Al powder mixture, peak temperatures for ordering decrease with prolonged mechanical milling duration. After 900 minutes, the peak ordering temperature is about 650 K. Surinach *et al.* [147] milled ordered $Ni_3Al$ powder particles in an attritor, and found that the powder particles underwent a transition from an ordered γ' to a disordered γ' during the milling process. Reordering took place over a wide range of temperatures when the powder particles were heated. According to their conclusions, the exothermic peak at about 703 K was an indication of the ordering transition from the disordered γ' to the ordered γ'. To confirm this transition process, the powder particles milled for 600 minutes were annealed at 873 K. X-ray diffraction spectrum is given in Figure 4.21 (e). It can be seen that the peaks correspond to an ordered superlattice structure. The enthalpy released is about 6 kJ/g-atom. The activation energy for the ordering is about 1.78eV [144].

Hida *et al.* [148] also found variation in reaction onset temperature as a function of milling time when $SiO_2$-Al mixtures were milled in a planetary ball mill. It was observed that unmilled powder mixtures reacted at approximately 873 K. The reaction onset temperature of the milled powder, however, decreased with increase in milling duration. This enhancement of reactivity is a mechano-chemical activation phenomenon which strongly influences the reaction mechanisms, kinetic parameters, as well as the composition and structure of the end products.

### *4.2.3 Synthesis of magnesium intermetallic systems*

#### *4.2.3.1 Synthesis of $Mg_2Si$*

$Mg_2Si$ has a relatively high melting temperature of 1358 K, low density of 1.99 g/cm$^3$, high hardness of 4,500 MN/m$^2$ and low thermal expansion coefficient of 7.5 x $10^{-6}$ K [149]. It is one of the candidate materials for high temperature applications. Due to the large difference in melting temperatures for the two elements, it is relatively difficult to synthesize $Mg_2Si$ via casting. Because mechanical alloying is a solid state process, it offers a possible means to synthesize $Mg_2Si$ at room temperature.

Figure 4.24 shows the x-ray diffraction spectra of the as-received Mg-Si powder mixture with the stoichiometric composition of $Mg_2Si$ (Figure 4.24(a)) and the powders mechanically alloyed for 360 (Figure 4.24(b)), 600 (Figure 4.24(c)), 1,200 (Figure 4.24(d)) and 1,800 minutes (Figure 4.24(e)). The phases consist mainly of pure Mg and Si after 360 minutes of milling. After 600 minutes, the formation of magnesium silicide, $Mg_2Si$, is evident by the presence of the (111) and (220) diffraction peaks which are far from the original Mg and Si peaks. The relative intensity of the peaks of $Mg_2Si$ phase increase with increasing milling duration, indicating that further milling would lead to the formation of more magnesium silicide. The formation of $Mg_2Si$, however, even after 1,800 minutes of milling was not complete with $Mg_2Si$ coexisting with some Si element. The reason for the remaining Si is due to the lack of Mg as a result of either the oxidation of Mg or the cold welding of Mg with the milling tools.

Differential scanning calorimetric study shows the reaction behaviours occurred in mechanically alloyed Mg-Si powders. Figure 4.25 shows the analysis of the 300 minutes of mechanically alloyed powder measured at a heating rate of 20 K/min. As shown in Figure 4.25, an endothermic reaction occurs at about 626 K. This reaction is probably due to the decomposition of the magnesium hydride, $MgH_2$, formed during milling by the reaction of Mg with hydrogen decomposed from the organic additives. Because the content of $MgH_2$ is low and its crystalline size is very small, its presence could not be revealed by x-ray diffraction analysis. Formation of hydrides has been reported during mechanical alloying of Mg [150] as well as Al-Zr [151]. These hydrides are metastable and will decompose at certain temperatures depending upon the type of hydrides. The decomposition temperature for magnesium hydride observed has also been noted by Chen and Williams [150].

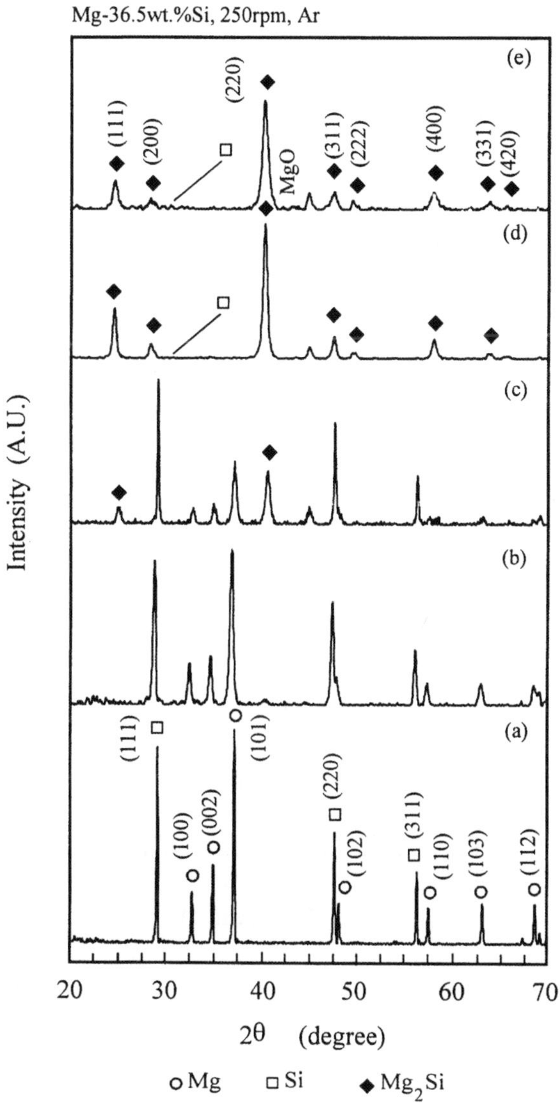


Figure 4.24. X-ray spectra of the as-received and mechanically alloyed Mg and Si mixture using a planetary ball mill (Fritsch 5).

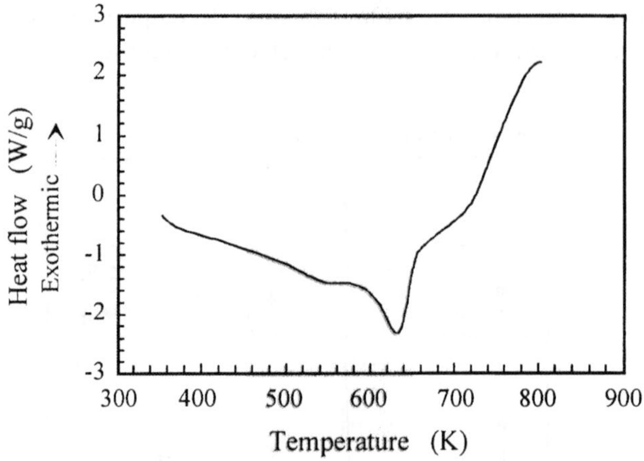


Figure 4.25. Differential scanning calorimetry measurement of Mg-Si powder mixture after 300 minutes of mechanical alloying.

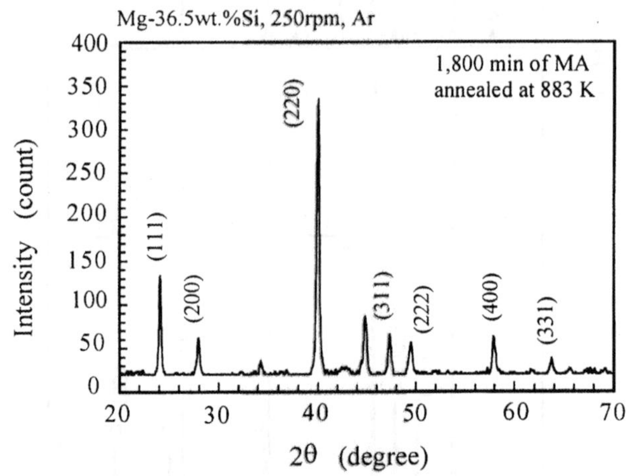


Figure 4.26. X-ray diffraction spectrum of the Mg and Si powder mixture mechanically alloyed for 1,800 minutes followed by thermal annealing at 883K for 120 minutes.

Although there is direct evidence of the formation of $Mg_2Si$ via mechanical alloying, its full formation from elements of Mg and Si seems to need long milling duration which may lead to some contamination. To avoid such contamination, the partially alloyed Mg and Si mixture can be thermally activated to form $Mg_2Si$. The result shows the reduction in Mg and Si content as given by x-ray diffraction peaks in Figure 4.26. Obviously, the amount of $Mg_2Si$ has increased with either milling duration or by the thermal annealing process. This relation between Mg and Si may be represented by a milling map which is calculated based on the intensities of Mg, Si and $Mg_2Si$ peaks. Using this map as given in Figure 4.27, the milling duration and annealing duration can be selected.

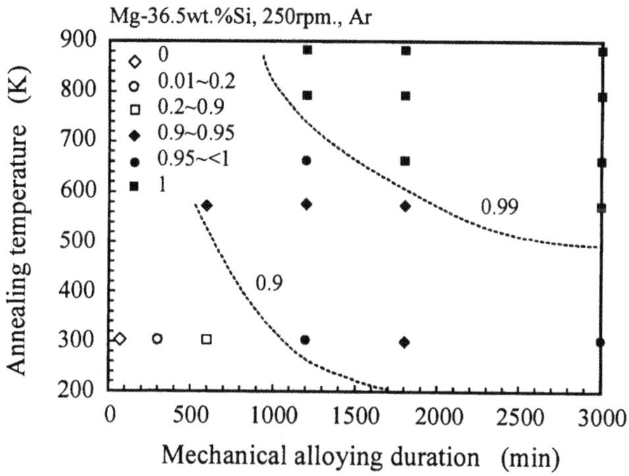


Figure 4.27. Milling map of formation of $Mg_2Si$.

## 4.3 Ceramics

Formation of carbides, borides and nitrides is usually accompanied by the release of reaction heat. In the case of Ti, Zr and Hf based carbides and borides, the adiabatic reaction temperature $T_{ad}$ either exceeds or is close to the melting point, $T_m$, of the resulting products. Large amount of research work on reactive milling of ceramics from elemental powders has been carried out. Table 4.6 shows a list of ceramics successfully synthesized via reactive milling.

### *4.3.1 Synthesis of borides*

#### *4.3.1.1 Synthesis of $TiB_2$*

Figure 4.28 shows the x-ray diffraction spectra of Ti-B as a function of milling duration. The spectrum of the as-received mixture of Ti and B is shown in Figure 4.28 (a) as a reference. All the major peaks were found to correspond to that of Ti as the B used was amorphous and did not contribute to any significant peaks in the diffraction pattern.

Table 4.6 List of some ceramics produced via reactive milling.

| Compound | Synthesis | ΔH298 (kJ/mol) | Ref. |
|---|---|---|---|
| $TiB_2$ | Ti+B | -323.8 | 152, 154 |
| TiC | Ti+C | -184.6 | 155-157 |
| SiC | Si+C | -73.2 | 158 |
| VC<br>$V_2C$ | V+C | <br>-117.2 | 159 |
| BN | $B+N_2$ | -254.4 | 159, 161 |
| CrN | $Cr+N_2$ | | 162 |
| $Cu_3N$ | $Cu+N_2$ | | 161 |
| $Mo_2N$<br>MoN | $Mo+N_2$<br>$Mo+NH_3$ | -81.6 | 159<br>162 |
| NbN | $Nb+N_2$ | -235.1 | 162 |
| $Si_3N_4$ | $Si+N_2$ | -744.8 | 159, 161 |
| TaN | $Ta+N_2$ (annealed at 973K)<br>$Ta+NH_3$ (annealed above 953K) | -252.3 | 159, 162 |
| TiN | $Ti+N_2$ | -337.9 | 161-163 |
| VN | $V+N_2$ | -217.2 | 161, 162 |
| WN, $W_2N$ | $W+NH_3$ | | 161, 162 |
| ZrN | $Zr+N_2$, $Zr+NH_3$ | -365.3 | 159-162 |

Figure 4.28 (b) shows the x-ray diffraction pattern after 30 minutes of ball milling. No shift in peaks could be detected. The diffraction pattern appears to be similar to that of Figure 4.28 (a) suggesting that no new phase had formed. However, significant changes took place after 60 minutes of ball milling as evidenced in Figure 4.28 (c). Characteristic peaks of $TiB_2$, (100), (101), and (110) occurring at diffraction angles of 34.1°, 44.4° and 61.1° respectively can be observed. A comparison is made with the x-ray diffraction pattern of commercial $TiB_2$ shown in Figure 4.28 (d). The two patterns match very closely with each other. Based on this observation, the formation of $TiB_2$ from mechanical alloying of Ti-66.7at%B powder mixture is confirmed. Figure 4.28 (e) shows the diffraction patterns of Ti-66.7at%B mechanically alloyed for 60 minutes and annealed 1,273 K. No structural change took place after annealing; from which it can be deduced that the

mechanically alloyed $TiB_2$ is in an equilibrium state and is very stable thermodynamically.

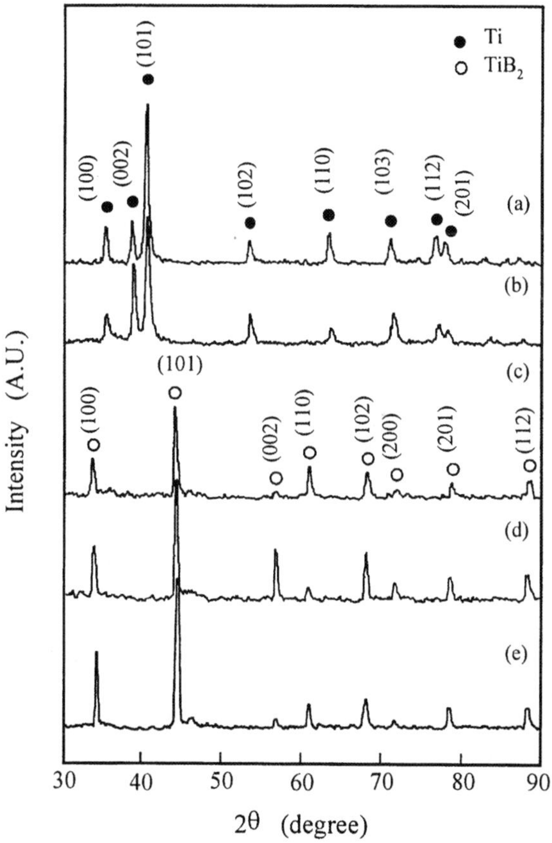


Figure 4.28. X-ray diffraction patterns of mechanically alloyed Ti and B mixture, commercial $TiB_2$ and annealed powder mixture.

The rapid formation of $TiB_2$ from Ti-66.7at%B powder mixture can be explained by the highly exothermic reactions taking place during milling. Formation of borides, carbides, nitrides, ceramic composites and other refractory compounds are usually accompanied by strong exothermic reactions. Such reactions are capable of being self-sustained and are thus defined as self-propagating high temperature synthesis [176] or combustion synthesis.

Combustion synthesis reactions during mechanical alloying have been reported in the literature. Park *et al.* [154] reported the formation of $TiB_2$ from combustion synthesis while milling Ti-B powder mixture in a vibrator ball mill. An exothermic temperature spike was seen to occur during milling. Formation of $TiB_2$ was detected after the occurrence of the temperature spike. Combustion synthesis was also reported during mechanical alloying of metal oxide systems [100]. In both the cases, abrupt change in temperature was observed. This sudden rise in temperature had been perceived as the occurrence of combustion. The rate of combustion and hence the formation of $TiB_2$ is collision energy dependent. Park *et al.* [154] measured the combustion process of Ti and B mixture mechanically alloyed using a vibrator mill operated at a frequency of 25 Hz with a small gyratory motion of 2.5 mm. Balls of six different sizes were used in six different mechanical alloying processes with ball to powder weight ratios of 50:1 and 100:1. The temperature rise due to the exothermic reaction was different when different ball sizes were used. Exothermic temperature spikes were noted at about 5,400, 6,300 minutes, and at about 7,800 minutes when 25.4, 20.6 and 19.1 mm size balls were used respectively. No spike could be observed when ball size of 17.5 mm or below was used. This indicates that the exothermic temperature spike depends not only on the impact energy given to the powder but also on the magnitude of impact energy per collision given to the powder [154]. Amorphization of Ti-B just before the temperature spike has been observed [171].

The rate of formation of $TiB_2$ from Ti-66.7at.%B clearly shows that mechanical alloying is a technique that can be used to increase reaction rate in the solid state. The formation of $TiB_2$ does not seem to follow the conventional diffusion theory because conventional diffusion in mechanical alloying normally takes a long time by gradual diffusion through formation of lamellaed layers of the constituent elements. As a result, there is a progressive change in the structure of the milled material before the formation of the final product.

Mechanical alloying is a room temperature process. The temperature during milling is not expected to increase significantly. It may be high if the reaction enthalpy of the material being milled is considered [100]. From thermodynamics data available, the enthalpy of formation of $TiB_2$ at room temperature is -323.8 kJ/mol [101] where the negative sign indicates that it is a highly exothermic reaction. Davis *et al.* [177] and Schwartz *et al.* [178] successfully modeled the temperature rise due to collision events during milling. However, these models did not take into account the reaction enthalpy of the materials being milled. This can be significant for materials whose formation results in large amounts of energy release as in the case of $TiB_2$. In highly exothermic systems, heat generated by the reaction enthalpy is faster than it is dissipated [180]. In addition, the creation of new surfaces for reaction and very finely distributed B in Ti during the mechanical alloying process can increase the possibility of fast reactivity between the Ti and B elements. This enhancement of reactivity creates better conditions for self-propagating high temperature synthesis to take place. Once the reaction has taken place in a small local volume, the highly exothermic reaction will provide the energy for initiating reactions in the neighbouring regions [(157)]. The reaction becomes self-sustaining and thus greatly increases the reaction rate to form $TiB_2$. Hence, the formation of $TiB_2$ is attributed to the combustion synthesis caused by the high formation enthalpy between Ti and B elements under high impact energy. The condition for combustion synthesis is induced by mechanical alloying and thus the formation of $TiB_2$ is described as mechanical alloy induced combustion synthesis.

Radev *et al.* [(153)] has also reported the formation of $TiB_2$ phase by explosion kinetics after 80 minutes of mechanical alloying in a planetary ball mill. The boron content of the resultant $TiB_2$ was determined by alkalimetric titration of mannitol-boric acid with a NaOH solution to be 28.84wt.%.

### *4.3.2 Synthesis of carbides*

#### *4.3.2.1 Synthesis of TiC*

Direct formation of TiC has been made possible using mechanical alloying of elemental powders of Ti and C in a planetary ball mill at an angular velocity of 250 rpm [(155)] or a Spex mill [(157)]. Similar to the formation of $TiB_2$, at a short milling duration (240 minutes), no formation of TiC could be found and x-ray diffraction peaks of Ti are broadened. There is no significant shift in the Ti peaks. At this stage, Ti and graphite powder particles are well mechanically alloyed together. The two very fine and well distributed constituents provide for the possibilities of fast reaction to occur. Self-propagating high temperature synthesis shows easy formation of new compounds if the particle size is small. The reaction becomes difficult or even impossible if large particles are used [(157, 179)]. In-situ temperature measurement also shows an abrupt increase after an additional 60 minutes of milling. X-ray diffraction shows full formation of TiC after this abrupt change in temperature. After the abrupt increase in milling temperature, a decrease in a temperature followed. From kinetic point of view, self-propagating high temperature synthesis takes place more easily than the interfacial, gradual diffusional reaction. If the TiC compound is formed by the diffusion of Ti and C atoms through the interface, a thin film of TiC will form first. Further reaction required the Ti and C atoms to diffuse through the TiC diffusion layer. The activation energied of Ti and C diffusing in TiC is 737 and 400 kJ $mol^{-1}$ respectively. On the other hand, the activation energy of self-propagating high temperature synthesis is less than 117 kJ $mol^{-1}$ [(157)].

The duration of the exothermic reaction is composition dependent. When the initial content of carbon decreases, the duration for the formation of TiC also decreases. The time for the appearance of the abrupt increase in temperature decreases from 186 minutes to 178 or 148 minutes if the carbon content is lowered from 50at.% to 43 or 35at.%. In the later case, some residual Ti remains after milling. Since the increase in milling temperature is an indication of the formation of TiC compound, further milling leads to reduction in grain size. The crystalline size measured by x-ray diffraction is between 10 to 25 nm. The crystalline size of the TiC is very stable until about 873 K [155]. Teresiak [155] has shown that below this temperature, there is only a slight increase in the crystalline size. Beyond this temperature, however, the increase in crystalline size becomes fast. The thermal stability of the materials is also milling duration dependent. The rate of crystalline growth is faster in short rather than those of long milling duration. For example, the crystalline sizes are about 56, 30 and 24 nm for the powders milled for 480, 1,440 and 2,880 minutes respectively after annealing at 873 K while they grow to about 150, 56 and 48 nm when annealed at 1,123 K. The rates of crystalline growth of the powder after 480 minutes of milling are 0.05 nm/K below 873 K but 0.2 nm/K between 873 and 1,223 K while for the powder milled for 1,440 and 2,880 minutes, the rates are 0.01 nm/K below 873 K but 0.1 nm/K between 873 and 1,223 K.

*4.3.2.2 Synthesis of other carbides*

Malchere and Gaffet [158] have confirmed the possibility of formation of SiC via the process of mechanical milling of elemental powders of Si with 25, 50 and 75at.% carbon using a Fritsh G7 planetary ball mill. Although SiC can be directly synthesized from powders of all compositions, a large amount of Si and C still remains after 5,760 minutes of mechanical alloying if the following milling conditions are employed: $\Omega = 438$ rpm and $\omega = 765$ rpm. If $\Omega = 484$ rpm and $\omega = 250$ rpm, however, the x-ray diffraction intensities of Si and C are greatly reduced, indicating that less Si and C remaining after the same duration of mechanical alloying. Figure 4.29 shows the x-ray diffraction patterns of mechanically alloyed Si and C mixtures under two different milling conditions. It is clear that the strongest chemical reaction between Si and C can be induced when stoichiometric composition is used. In the case when $\Omega = 484$ rpm and $\omega = 250$ rpm, SiC can almost be fully formed from elemental powders. The crystalline size measured from the broadening of x-ray diffraction peaks is in the range of about 3 to 20 nm. Secondary electron image of the powders shows agglomeration of small particles ranging about 50 nm in size.

Other carbides synthesized via elemental powders include VC and $V_2C$. Calka *et al.* [159] found that the milling intensity is very crucial. Under high intensity milling condition, 50at.%V and 50at.%C can directly form VC compound. Medium milling energy leads firstly to the formation of nanocrystalline structure which transforms into VC compound upon heating. When low milling energy is used, however, the V and C powder mixture initially yields forms nanocrystalline structure with a size of 65 nm. Following that, the nanocrystalline structure transforms to $V_2C$ compound and not to VC phase. The presence of unreacted amorphous carbon cannot be excluded. Slight increase in milling energy may firstly produce nanocrystalline structure with size of 150 nm, followed by the formation of VC and V.

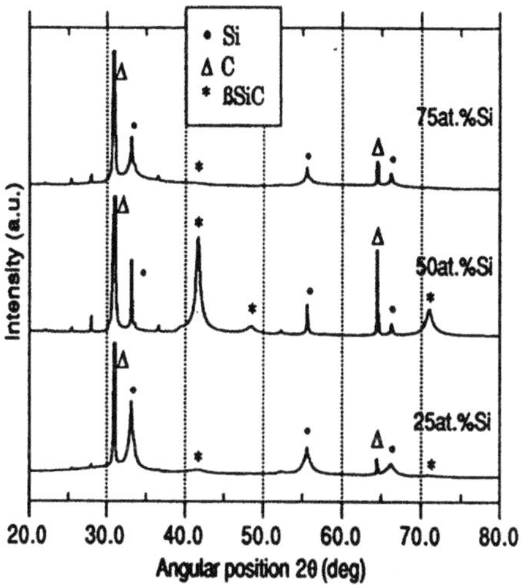


(a)

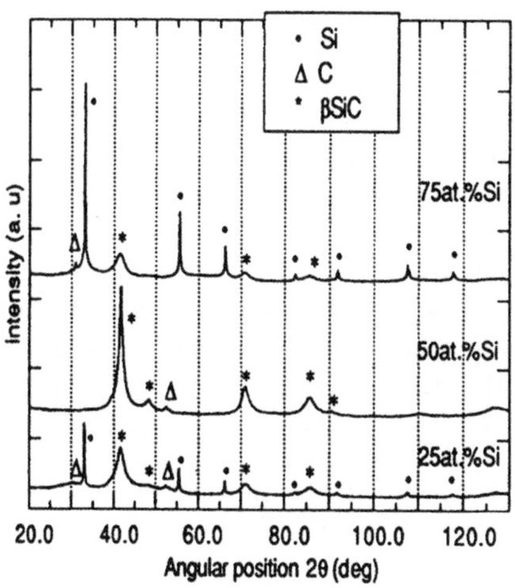


(b)

Figure 4.29. X-ray diffraction spectra of mechanically alloyed Si and C mixtures: (a) $\Omega = 438$ rpm and $\omega = 765$ rpm and (b) $\Omega = 484$ rpm and $\omega = 250$ rpm [(158)].

### *4.3.3 Synthesis of nitrides*

Nitrides can be formed using conventional methods such as carbothermal reaction of oxides, liquid-phase reduction, vapour-phase reduction and gas-phase reaction through organometallic precursors, combustion nitridation and synthesis [(159)]. Direct nitridation method is a very often used process in the industry. However, this process involves a series of heating cycles. Mechanical milling process on the other hand, has demonstrated to be a good and cheap means to synthesize nitrides at room temperature.

Formation of nitrides from elemental powder can be divided into two categories: (a) milling in $N_2$ gas and (b) milling with $NH_3$. The nitrides obtained via mechanical milling include such as TiN, ZrN, VN, BN, $Mo_2N$, $Si_3N_4$, $Cu_3N$, $Mg_3N_2$, and $W_2N$ [(161)].

In the first category, elemental powders may be milled in $N_2$ gas. During milling, the powder particles are fractured resulting in large area of irregular surfaces. It is known that absorption of nitrogen into rough metal surfaces is many times higher than that of perfectly smooth surfaces. With fresh and clean surfaces of the powder particles after fracturing, $N_2$ gas may be dissociated under impact pressure [(164)]. As shown in Table 4.6, formation of nitrides may generate large negative heat of mixing which provides the driving force.

Calka *et al.* [(161)] found that TiN with crystalline size of 9 nm was formed after 3,600 minutes of milling. The TiN structure was very stable even after 60 minutes of annealing at 1,073 K. However, Eskandarany *et al.* [(192)] synthesized a fcc $Ti_2N$ phase which was thermally stable only until about 600 K. Similarly, VN and ZrN with crystalline sizes of 11 and 14 nm respectively have also been directly synthesized. Compared with the formation of TiN and VN, formation of BN follows a different route. The polycrystalline B is firstly transformed to amorphous during milling. There is no indication of the formation of BN structure after milling. However, x-ray diffraction spectra clearly show the formation of BN if the milled B powder is annealed at 1,073 K. Depends upon the materials, nitrides can be formed via milling in $N_2$. Special examples of such materials are Mo, Si, W and Cr. If these elemental powders are individually milled in $N_2$ filled vials, multi-phase structures of $Mo+Mo_2N$, $Si+Si_3N_4$, or $Cr+Cr_2N+CrN$ may coexist even after annealing.

Ogino *et al.* [(162)] shows that the range of absorption of nitrogen is different for different materials. For materials like Cr, Mo, Nb, Ta, Ti, and Zr, about 40 to 50at.% nitrogen may eventually be absorbed by the powders. The rate of absorption of nitrogen estimated from the slope of the absorption at the half values of the saturated nitrogen concentration versus milling duration show that Zr has the fastest nitrogen absorption rate while Ti ranks the second. The nitrogen absorption rates for Nb, Ta, Mo and Cr are lower. Theoretical calculation using Miedema's theory [(193)] shows that the rate of nitriding increases with the increase in the enthalpy of nitride formation. The rate of nitriding is proportional to the cube of the milling cycle. This indicates that the nitrogen absorption in the powder particles occurs at the instance of ball collision and the quantity of nitrogen absorbed by one collision event is nearly proportional to the mechanical energy supplied to the powder particles.

In the second category, elemental powder may be milled in ammonia ($NH_3$). Calka *et al.* [(159)] studied the different nitride systems and showed the different rates of nitridation when $N_2$ or $NH_3$ is used. For milling of Zr, ZrN can be formed in $N_2$ or $NH_3$. Milling time for the completion of nitridation reaction is much shorter when $NH_3$ is used. If Zr is milled in $N_2$ gas, the time of completion is about 5 times longer than that in $NH_3$. However, 10at.% contamination has been found in the case of using $NH_3$. Although full completion of reaction can be found in both cases, x-ray diffraction spectra of the milled powder after annealing at 1273 K showed the presence of $ZrO_2$ if Zr is milled in $N_2$ atmosphere, meaning that oxidation has taken place during milling.

Milling of Ta in $NH_3$ showed the formation of nanophase which could be transformed into nanostructure and TaN upon annealing at 853 K, and into βTa and TaN upon annealing at 1073 K. Although $Mo_2N$ can be formed, no completion of nitridation for Mo has been found when Mo is milled even for 16,800 minutes. Milling W in $NH_3$ followed by annealing at 1273K showed the formation of a mixture of $W+W_2N+WN$ phases, but no evidence of formation of tungsten nitrides has been detected when W is milled in $N_2$ atmosphere.

### 4.4 Composites

Attractive physical and mechanical properties can be obtained with metal matrix composites which combine metallic properties with those of ceramics [(194)]. Reinforcement materials in metal matrix composites may be carbides, nitrides, oxides and other elemental materials. One of the major problems encourtered in the production of metal matrix composites using powder metallurgy is the agglomeration of reinforcement particles due to size difference between powders for the matrix and the particles, or the development of static charge on the particles. To achieve homogeneity of particle distribution, several methods may be employed, such as a proper choice of the particle size and the use of polar solvents which neutralize the charges on the surface of the particles [(195)]. Another possible method to improve particles distribution is to use mechanical alloying technique [(196, 197)] or ball mill process [(198)]. Mechanical alloying or ball mill enables metallic materials to be coated onto ceramic powder particles [(185)]. The constituent powder particles are repeatedly fractured and cold welded by continuous collisions so that powder particles with very fine structure can be obtained [(84)]. Investigation on the microstructural evolution of 7010-SiC composite during mechanical alloying using an attritor mill by Bhaduri [(108)] showed that the formation of equiaxed composite particles was dependent on several milling conditions. The higher speed of rotation and larger ball to powder ratio hastened the process while the addition of SiC particles retarded the process due possibly to their inhibiting effect on the formation and welding of lamellae in the initial stages of mechanical alloying. It was suggested that it may perhaps be advantageous to introduce the reinforcement at some intermediate stage of milling [(200)].

### *4.4.1 Synthesis of aluminium-based composites*

#### *4.4.1.1 Al-SiC system*

An early investigation on the possibility of the formation of SiC with Al matrix in three different compositions [(158)]: Al-7.5at.%Si-7.5at.%C, Al-11at.%Si-11at.%C and Al-14.2at.%Si-14.2at.%C corresponding to 10, 15 and 20vol.%SiC as reinforcement particulates respectively was carried out using a planetary ball mill. Although SiC could be formed via milling Si and C for all three different compositions, there was no evidence of the formation of SiC even after 6,240 minutes of milling. No SiC could be detected even after annealing at 793 K. This indicates that Si and C may be present as solid solution during mechanical alloying. Since it is not possible to form SiC via mechanical alloying of Al-Si-C powder mixture, the formation of composite powder particles reinforced by SiC must be through mechanical milling of Al and SiC particulates.

Distribution of reinforcement in the matrix metal is a critical consideration for metal matrix composites. Therefore, distribution and size of reinforcement are important issues to be considered during mechanical milling. Figure 4.30 shows the typical morphologies of the powder milled at 100 rpm without using process control agent after 30 and 2,100 minutes of milling (Figure 4.30 (a) and (b) respectively). It can be noted that the size of the particles increased with milling duration. After about 600 minutes of mechanical alloying, flattened particles could be observed. After 2,100 minutes of alloying, all the particles became flattened as shown in Figure 4.30 (b).

The typical morphologies of the particles mechanically alloyed with process control agent at 250 rpm are shown in Figure 4.31. The difference in morphology for powder mechanically alloyed at 100 and 250 rpm with process control agent may be compared. The powder alloyed at 250 rpm showed very fine size and rounded form. No flattening of the particles could be observed. Three stages of change in particle size can here be identified: (a) Particle size decreased at the first 240 minutes (Figure 4.31 (a)). This mainly reflects the dominating effect of fracturing of the powder particles. This observation differs from the mechanical alloying of other materials in that normally, an increase in particle size is observed at the beginning of mechanical alloying due to the powder particles being relatively soft at the start of the mechanical alloying process and are easily cold welded together. As the process control agent gradually becomes exhausted an increase in cold welding results: (b) the size of the powder particle slightly increased due to decreasing effect of the remaining process control agent (Figure 4.31 (b)), and (c) the unchanging size of the particle after 840 minutes of mechanical alloying due to the balance between cold-welding and fracturing (Figure 4.31 (c)).

(a)

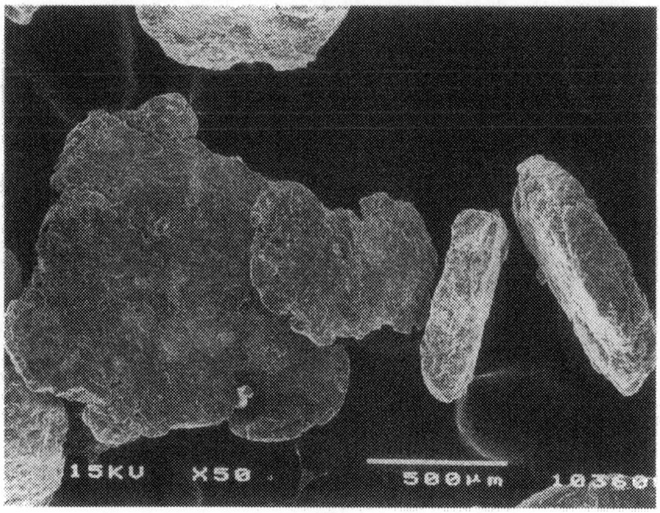


(b)

Figure 4.30. Morphologies of powder particles mechanically milled at 100 rpm.

(a)

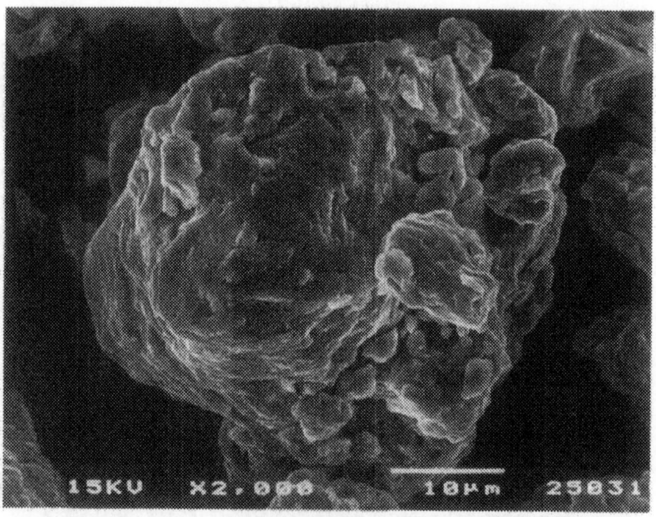


(b)

Figure 4.31. Morphologies of powder particles mechanically milled at 250 rpm.

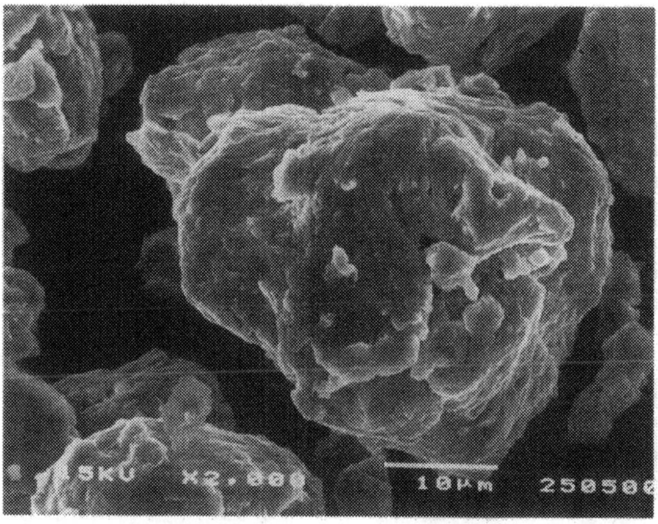


(c)

Figure 4.31. Morphologies of powder particles mechanically milled at 250 rpm (continued).

From analysis of the mechanical alloying process using a planetary ball mill, it has been noted that ball milling consists of two processes. One of which is relative rolling and friction between the balls and the inner surface of the vial, while the other is the collision between the balls, and between the balls and the inner wall of the vial. The ratio of rolling and friction to collision depends mostly on the speed of rotation. When low speed is used, rolling and friction dominate. Consequently, flattening of the particles may be the result. If high speed is employed, collision will be the main process during mechanical alloying. The powder particles can easily be cold welded together and be mechanically alloyed quickly. For soft materials, the powder particles may even be cold welded to the balls and to the inner wall of the vial. Large size particles may also be produced under such conditions. Dual speed mechanical alloying process could be employed for some materials to avoid excessive cold welding to obtain small size particles [201].

Powder size distribution analysis in Figure 4.32 shows that very fine powder particles can be obtained at 250 rpm if process control agent is used. The particle size continues to increase with prolonged mechanical alloying duration if no process control agent is utilized. The results therefore suggest that process control agent has to be used for soft materials if fine powder particles are to be obtained.

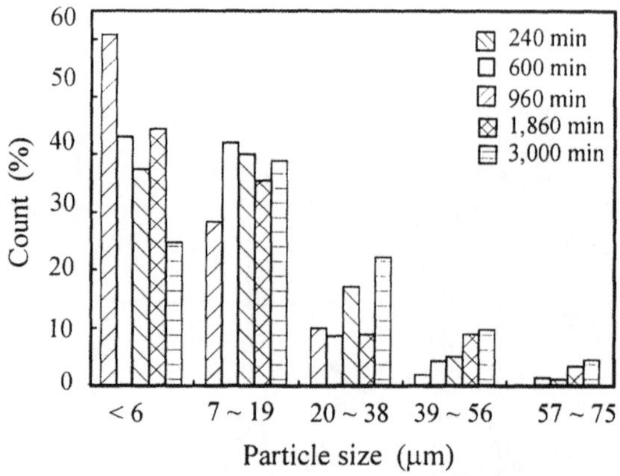


Figure 4.32. Distribution of particle size after different mechanical alloying durations.

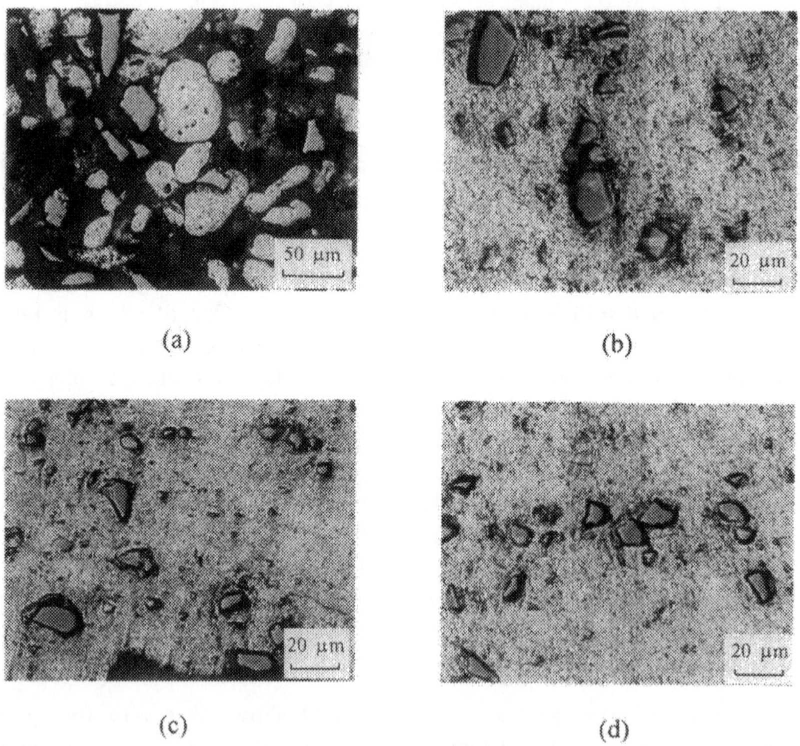


Figure 4.33. Microstructure of powder particles mechanically alloyed at 100 rpm for different durations: (a) 60 minutes, (b) 840 minutes, (c) 1,200 minutes and (d) 2,160 minutes.

Microstructural evolution of the mechanically alloyed powder particles at 100 rpm is shown in Figure 4.33. It can be observed from this figure that even after 1,200 minutes of mechanical alloying, only small size SiC particles are partially embedded into the matrix as individual SiC particle with large particles size can still be observed at this stage. The particle size of SiC is only reduced to about 15 μm after 2,160 minutes of mechanical alloying although the overall particle size of the composite is increased.

The microstructures of powder particles mechanically alloyed at 250 rpm are shown in Figures 4.34 (a) to (d). Besides the change in particle size, embedding of SiC into the matrix is found to be faster than that using low mechanical alloying speed since high speed milling contained more collision events while more rolling and friction were involved when low milling speed was used. Compared with Figure 4.33, it can be noted that the size of the SiC particles is much smaller. This is due to the high collision energy involved in fracturing the powder particles. Because SiC

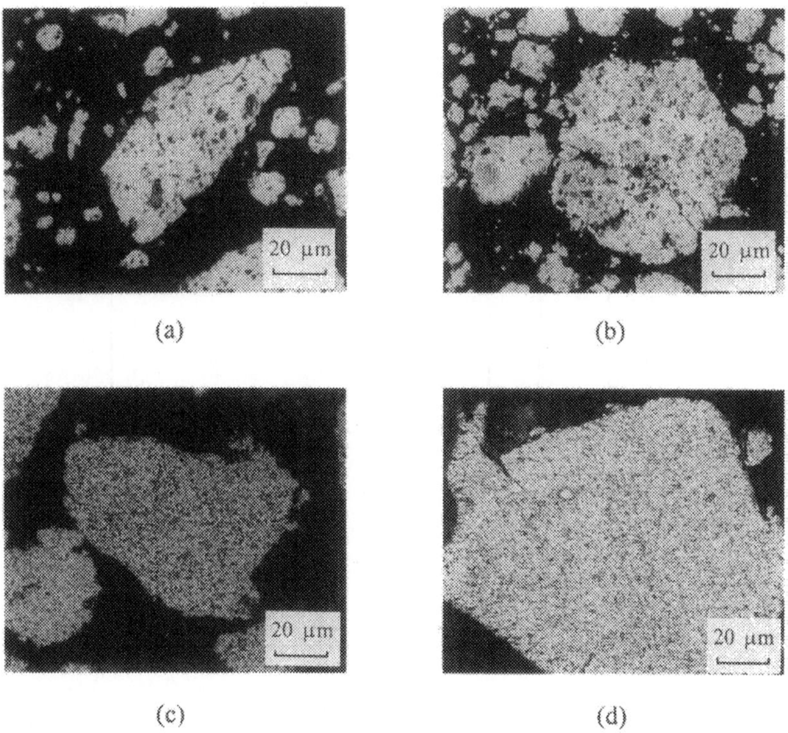


Figure 4.34. Microstructure of powder particles mechanically alloyed at 250 rpm (with process control agent) after different mechanical alloying durations: (a) 240 minutes, (b) 600 minutes, (c) 1,860 minutes and (d) 3,000 minutes.

particles with small size are easily embedded in the matrix, fracturing of SiC therefore promotes the rate of embedding. After 240 minutes of mechanical alloying, although SiC particles were not fully embedded into the matrix, the largest particles observed had been reduced to below 8 μm. They were further reduced to below 6 μm and were fully embedded into the matrix after 600 minutes of mechanical alloying. After this duration, the SiC particles were too small to be clearly observed under the optical microscope (Figure 4.34 (d)). One of the advantages of using mechanical alloying technique to produce metal matrix composite powders observed from the present work is the homogeneity of particle distribution which normally cannot be achieved by using blending processes. No clusters of SiC particles were here observed. Figures 4.35 (a) to (d) reveal the changes in x-ray diffraction pattern of the as-received Al, Cu and SiC powder mixture and powders mechanically alloyed at 250 rpm with process control agent after different mechanical alloying durations. The x-ray diffraction pattern shows a decrease in height and an increase in width at half peak width indicating some changes in crystalline size.

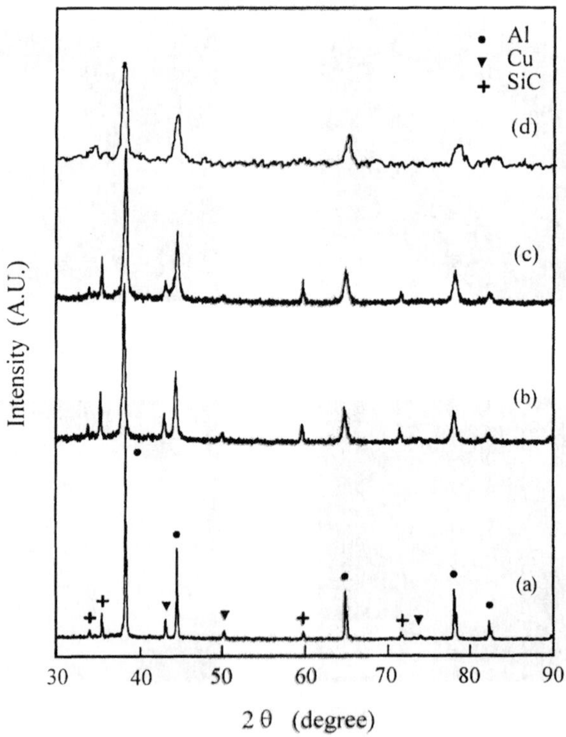


Figure 4.35. X-ray diffraction patterns of mechanically alloyed powder mixtures: (a) as-received powder, (b) 180 minutes, (b) 360 minutes and 3,000 minutes.

It can be seen that the x-ray diffraction pattern for Cu soon disappeared after a short duration of mechanical milling. The dissolution of Cu into Al is evident by the shift of 2θ angle of the x-ray diffraction peaks for Al from low value to high value during the mechanical alloying process. Another evidence is the change in x-ray diffraction peaks of SiC. The difference in width at half the maximum height indicates the change in the crystalline size during mechanical alloying. Because Al has fcc structure, crystalline size was hardly reduced although a slight change in width of the Al diffraction peaks was discerned.

No alloying effect could be achieved even after 2,100 minutes of mechanical alloying when a rotational speed of 100 rpm was used, implying that when rolling and friction are the predominating effects, alloying using ball mill cannot be performed.

### *4.4.1.2 Al-Ti-B system*

According to Al-Ti-B phase diagram, Al-$Al_3Ti$-$TiB_2$ composite powder particle will be formed if reactions between Ti and B and between Ti and Al take place during mechanical alloying of Al-rich Al-Ti-B system. The chemical reaction is expected to be:

$$3Al + 2Ti + 2B \rightarrow Al_3Ti + TiB_2$$

However, structural evolution during mechanical alloying is far more complicated than depicted in the reaction. Figure 4.36 shows the evolution of structure of the mechanically alloyed Al-18at.%Ti-36at.%B powder mixture. Only very slight shift of Al and Ti peaks is detected due to the dissolution in Al, and B atoms in Ti and Ti atoms in Al powder particles. Increase in peak width at half of maximum height indicates the reduction in crystalline size. After 600 minutes of mechanical alloying, x-ray diffraction peaks of Ti disappear while Al peaks are still clearly evident. No evidence of formation of $TiB_2$ and Ti aluminide structures may be discerned from the x-ray diffraction spectra. The broadening of Al diffraction peak indicates a graduate formation of nanostructural Al crystalline and high strain field induced by Ti and B atom dissolution. The disappearance of Ti diffraction peaks manifests the formation of Ti nanostructure after 1,800 minutes of mechanical alloying (Figure 4.36 (c)). Intensity of the Al diffraction peaks decreases continually while the width of the peaks increases with prolonged mechanical alloying. After 2,400 minutes of mechanical alloying, only (111), (200), (220), (311) and (222) Al peaks can be observed (Figure 4.36 (d)). As dissolution of Ti atoms in the Al is very limited, Ti may homogeneously be distributed as fine particles in the Al matrix during mechanical alloying.

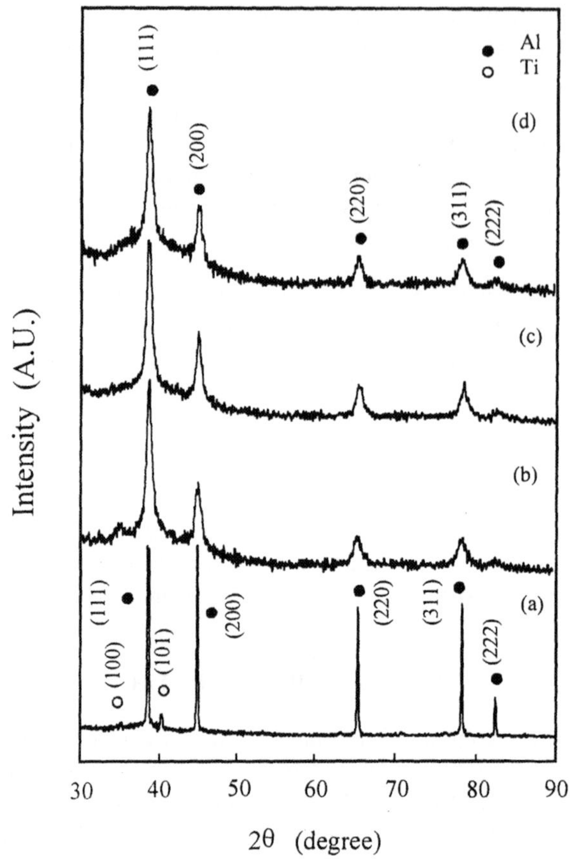


Figure 4.36. X-ray spectra of mechanically alloyed Al-Ti-B mixture.

Although the formation of $TiB_2$ phase from mechanical alloying of Ti and B elements involves a strong exothermic reaction, this strong exothermic reaction is moderated due to the presence of a large amount of Al. Dilution of Ti and B with Al prohibits the possibility of direct contact between pure Ti and B. Although x-ray diffraction spectra do not show any evidence of the formation of either $TiB_2$ or $Al_3Ti$ phase, scanning electron microscopy reveals some very small hexagonal shape particles in the powder mixture mechanically alloyed for 2,400 or 3,000 minutes, as shown in Figure 4.37. Since $TiB_2$ has a typical hexagonal shape, it is expected that some $TiB_2$ particulates may be formed during mechanical alloying. However, since the content of $TiB_2$ is very low and the crystalline size is extremely small, x-ray diffraction single is too low to be detected. The mechanism for the formation of $TiB_2$ within the Al matrix during mechanical alloying is schematically

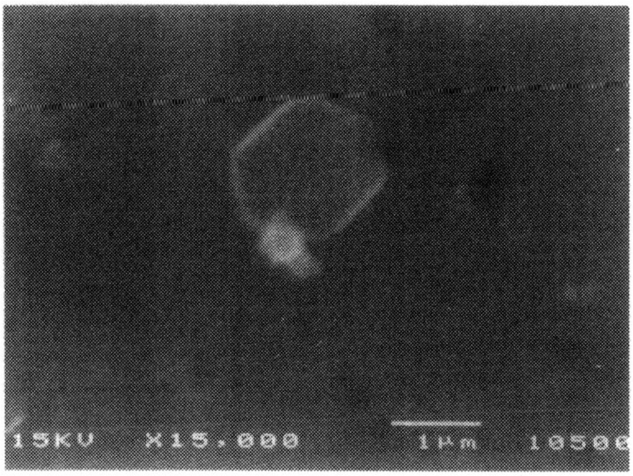


Figure 4.37. Microstructure of powder mechanically alloyed for 3,000 minutes.

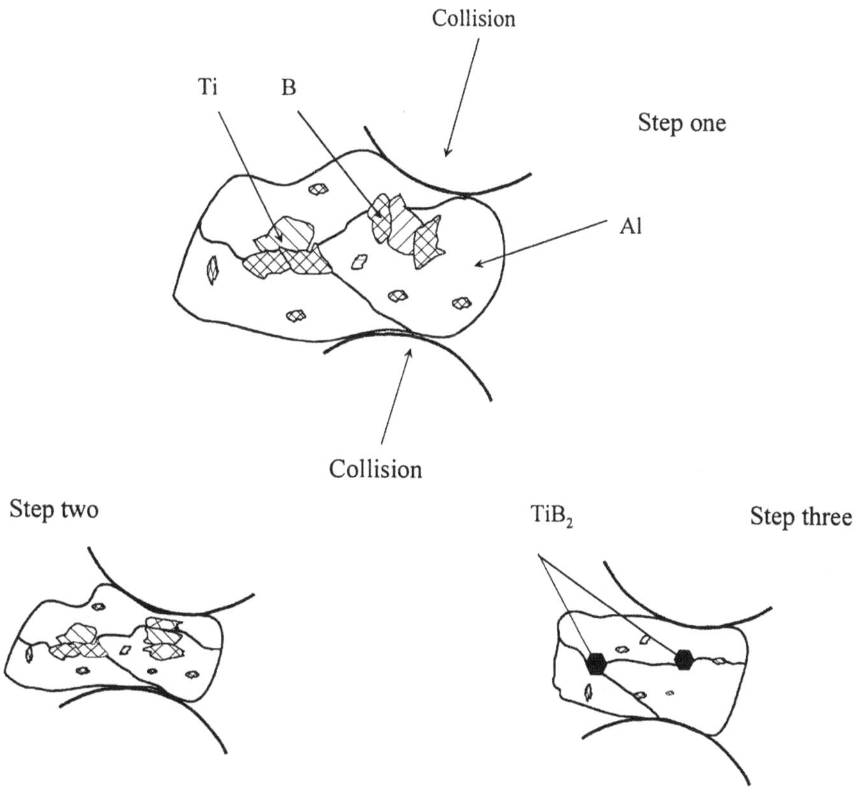


Figure 4.38. Formation of titanium diborite during mechanical alloying.

given in Figure 4.38. During alloying, Ti and B powder particles are mechanically distributed within the Al powder particles. At the early stage of the process, a large amount of Ti and B may be cold welded together. Although the big Ti and B particles may sooner or later be fractured within the Al particles, they may still be in contact with each other to form $TiB_2$ under collision.

If the mechanically alloyed Al-Ti-B mixture is thermally activated, namely, annealed, further chemical reaction may take place. Figures 4.39 (a) and (b) show the x-ray diffraction spectra of the powders annealed at temperatures of 723 and 873 K respectively. The x-ray spectrum of the powder annealed at 723 K as shown in Figures 4.39 (a) and (b) shows the presence of $Al_3Ti$. At this annealing temperature $TiB_2$ phase cannot be clearly discerned. However, if the mechanically alloyed powder is annealed at 873 K as shown in Figures 4.39 (a) and (b), $TiB_2$ phase may be detected in the 1,200and 2,400 minutes mechanically alloyed powder mixtures. At the same time, the relative intensity of $Al_3Ti$ decreases.

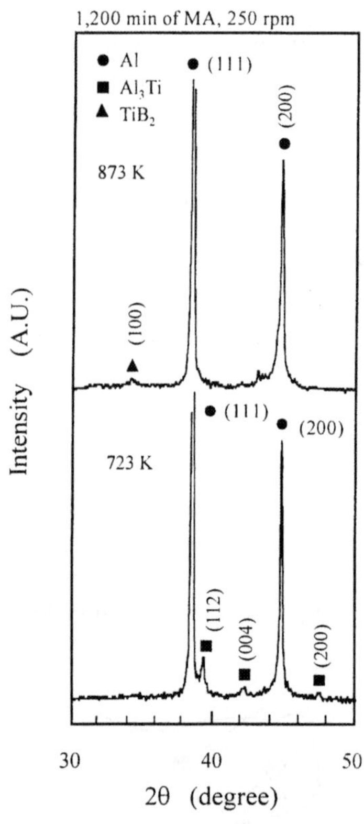


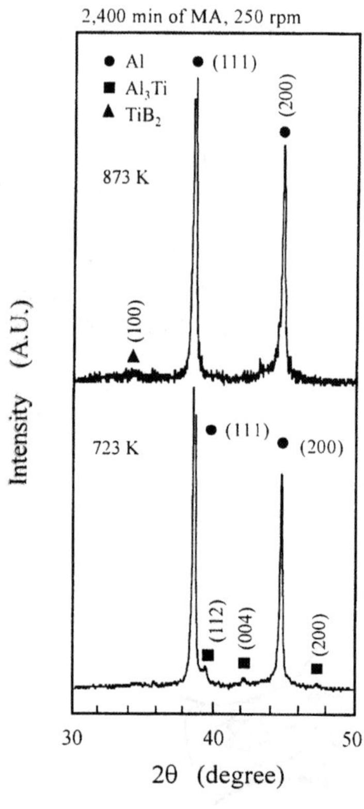


Figure 4.39. X-ray spectra of mechanically alloyed Al-Ti-B powder mixture milled (a) for 1,200 minutes and (b) 2,400 minutes.

### 4.4.2 Synthesis of titanium-based composites

Titanium borides are particularly attractive for high temperature applications owing to their high modulus, excellent refractory properties and chemical inertness [154]. From Figure 4.40, it can be seen that $TiB_2$ is more refractory than TiB while the latter is a stable compound in equilibrium only with Ti-rich alloy [172]. Hence, $TiB_2$ is an ideal reinforcement with titanium aluminides.

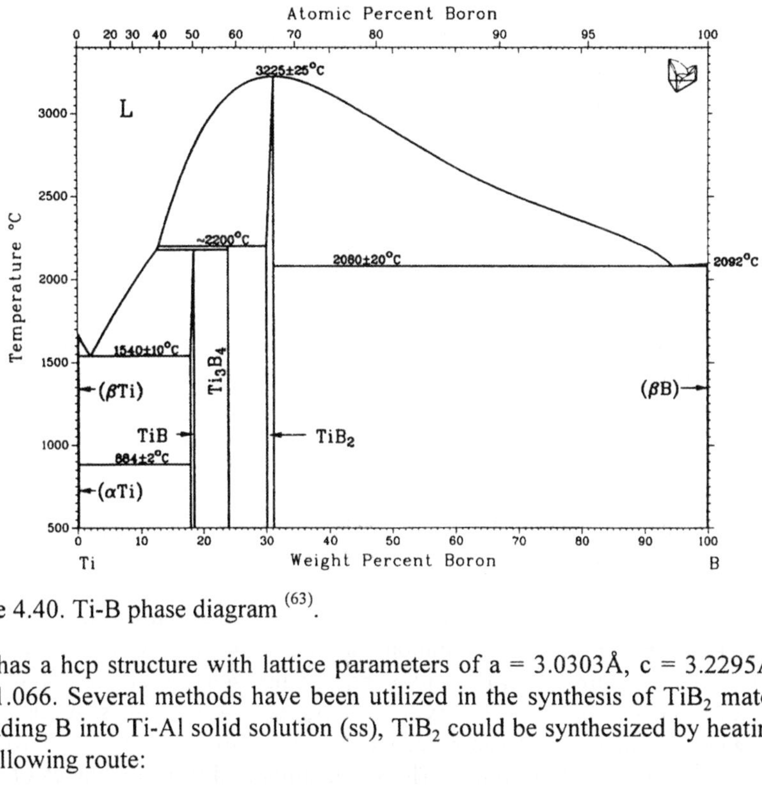


Figure 4.40. Ti-B phase diagram [63].

$TiB_2$ has a hcp structure with lattice parameters of a = 3.0303Å, c = 3.2295Å and c/a= 1.066. Several methods have been utilized in the synthesis of $TiB_2$ materials. By adding B into Ti-Al solid solution (ss), $TiB_2$ could be synthesized by heating via the following route:

$$Ti(Al)ss + B \rightarrow (Ti, Al)B_2$$

$$Ti + (Ti, Al)B_2 \rightarrow TiB_2 + (TiAl)ss$$

Although much attention has been devoted to a novel in-situ formation of composites in Ti-Al alloys, the reinforcements employed are generally in the form of fibres [173]. The major problem in fibre reinforced Ti alloys is the difference in coefficient of thermal expansion between the reinforcements and the matrices. Extensive cracks in the matrices due to thermal mismatch can be induced during thermal cycling [173]. Another problem is the chemical instability of the reinforcements if they are not carefully selected. Table 1 shows the adaptability of different reinforcements in Ti, of which the most stable one in pure Ti is TiB. As far as ternary Ti-Al-B system (Al-rich) is concerned, formation of $TiB_2$ is thermodynamically more stable. The phases evolved from the monovariant reaction have been found to change from $\beta$+TiB to $\beta+TiB_2$ to $\alpha+TiB_2$ and eventually to $\gamma+TiB_2$ as the Al content increases [(172)].

More recently, mechanical alloying technique has been used as a mechanochemical process by which fine grained ceramics can be synthesized [(174)]. This technique provides a unique means to synthesize compounds and alloys at room temperature [(175)]. Different types of materials have been successfully produced via the mechanical alloying technique. Formation of borides by combustion synthesis during mechanical alloying of Ti, Zr-B system has been reported [(154)]. The effects of size of milling ball and weight ratio of ball to powder on the process of formation of composites and combustion behaviour have been investigated. It was found that exothermic temperature spikes took place during mechanical alloying indicating the formation of new compounds.

*4.4.2.1 Ti-Al-B system*

With information on the formation of $TiB_2$ from Ti-B system, the formation of $\gamma$ phase composite reinforced with in-situ borite through mechanical alloying of Ti, Al and B elements can be expected. The intended reaction is of the form:

$$2Ti + Al + 2B \rightarrow TiAl + TiB_2$$

with a ratio of TiAl/15wt.%$TiB_2$. It is assumed that there is no reaction between the Al and the B elements.

Figure 4.41 shows the x-ray diffraction patterns of the mechanically alloyed Ti-Al-B mixture as a function of milling duration with that of the as-mixed elemental Ti, Al and B shown in Figure 4.41 (a) as a reference. Figures 4.41 (b) and (c) represent the initial stages of the mechanical alloying process up to 120 minutes. As there is no appearance of new peaks, it can be deduced that no structural changes have taken place during this period of milling. The slight shifts in diffraction angles of Ti and Al are attributed to the formation of solid solution between Ti, Al and B. Because the diffusivity of Al into Ti is large than that of Ti into Al, more Al diffuses into Ti. Among the three elements employed, the atomic radius of B is very much smaller than those of Ti and Al (0.98, 1.47 and 1.43 Å for B, Ti and Al respectively). Hence, B is the fastest diffuser into Ti and Al. Although formation of $TiB_2$ or TiB within a short time of mechanical alloying of the Ti-B system was expected, there was no evidence of the formation of borites. The reason may be attributed to the dilution of Al in Ti. Similar phenomenon has been observed by Malchere and Gaffet [(158)] when Si-C and Al-Si-C systems were mechanically alloyed. Apart from the absence of formation of borites, an increase in intensity of the (101) peak of Ti relative to the (200) peak of Al has also been observed. The weakening of the Al peak may be due to the dissolution of Al into Ti.

Figure 4.41 (d) represents the x-ray diffraction pattern after 300 minutes of milling. Considerable changes have taken place as evidenced by the broadening and overlapping of the crystalline peaks. It is estimated that the milled mixture has already reached a nano-size of about 27 nm.

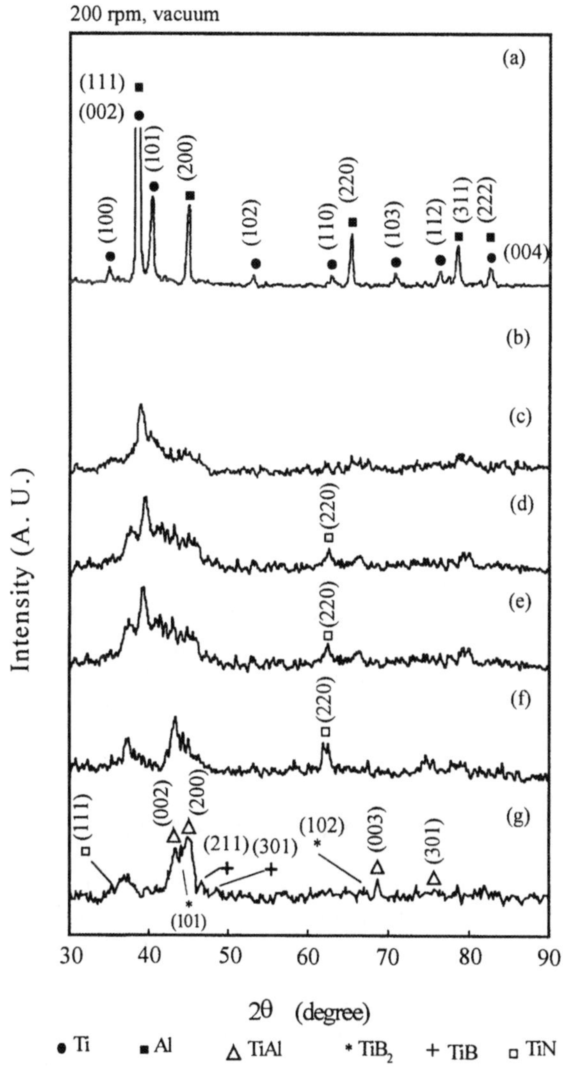


Figure 4.41. X-ray diffraction patterns of mechanically alloyed Ti, Al and B mixture.

Since the reaction between Ti and B elements is highly exothermic, it is speculated that the large energy released during the reaction can accelerate the mechanical alloying process. This leads to a faster nanocrystallization at an early stage of mechanical alloying. Another consequence may be the interstitial diffusion of B elements into the Ti(Al) solid solution. Similar to the case of milling in atmospheric conditions, segregation of B atoms to the grain boundaries of the lamellae structure results in accumulation of strain and hence high density of dislocations. This hinders the mobility of the defects and thus the rate of work hardening is higher.

X-ray diffraction patterns at milling duration longer than 300 minutes appear to be similar to those of the Ti-Al system. With the presence of B elements, however, it is expected that there would be formation of compounds of Ti and B during milling. Given the high diffusivity of Al into Ti and the relatively small amount of B, it is assumed that there would be a solid solution of Al in Ti with traces of B situated interstitially at the interfaces of the lamellaed layers of Ti and Al. Thus, the first step in the formation of metal matrix composite may be:

$$\text{Ti} + \text{Al} + \text{B} \rightarrow (\text{Ti,Al})(\text{B})\text{ss}$$

The reaction between Ti and B is expected to take place first before the reaction between Ti and Al. As far as the formation of $TiB_2$ is concerned, however, there is excess of Ti in the Ti-Al-B powder mixture. Consequently, any reaction between Ti and B elements would result in TiB being formed first rather than $TiB_2$. From the equilibrium phase diagram of Ti-B system, TiB is the first phase to exist in the Ti rich regions. Some of the peaks in Figure 4.41 (f) and (g) match those of TiB. More conclusive evidence is nevertheless needed given the difficulties in identifying the relevant peaks from the broad and overlapping profile of the x-ray diffraction pattern. There are two possible hypotheses to account for the formation of $TiB_2$. In the first hypothesis, it is assumed that all the B firstly react with Ti since the reaction between Ti and B is much faster than that between Ti and Al, and there is no reaction between B and Al:

$$\text{Ti} + \text{Al} + \text{B} \rightarrow \text{TiB} + \text{TiB}_2 + (\text{Ti,Al})\text{Bss}$$

With the formation of TiB in the initial stages, the remaining Ti is assumed to react with Al to form Ti(Al)ss (solid solution). It is believed that beside the formation of TiB and $TiB_2$, B still exists in the solid solution of Ti(Al)ss.

If the reaction is based on the amount of Ti remaining, any reaction between Ti and Al would form $TiAl_3$ due to the rich concentration of Al as compared to Ti. TiB and $TiAl_3$ are intermediate metastable products of the mechanical alloying process. As equilibrium compositional homogeneity requires the formation of $TiB_2$ and TiAl as the end products, the second hypothesis is

$$\text{Ti} + \text{Al} + \text{B} \rightarrow \text{TiAl}_3 + \text{TiB} + \text{Ti(Al)ss}$$

As milling continues, the reaction will proceed as follows:

$$TiAl_3 + TiB + Ti(Al)ss \rightarrow TiB_2 + TiAl$$

The second hypothesis does not appear so evident. Based on the work of Kim [(123)], it was found that the $TiAl_3$ phase was formed by precipitation from supersaturated Al(Ti) during annealing. The Ti in Al shares other merits of transition metal additions. The maximum solubility of Ti in Al obtained from the mechanical alloying process was limited to 2.46wt.% [(123)]. As this solubility is far from that of $TiAl_3$, it does not appear possible to synthesize $TiAl_3$ from a hyper-peritetic Ti-Al system.

After a mechanical alloying time of 1,800 minutes, the x-ray diffraction pattern shown in Figure 4.41 (f) depicts a significant change even though TiAl and $TiB_2$ had not clearly been detected. The inability to detect these compounds may be due to two possibilities. Firstly, TiB/$TiB_2$ may not have been formed at this stage. The second possibility may be that the formation of $TiB_2$ has taken place but in a nanocrystal form. Figure 4.41 (g) presents the x-ray diffraction patterns after 3,000 minutes of mechanical alloying. At least three different phases were found to be present at this stage of milling, namely TiAl, TiB and $TiB_2$. The main (101) peak of $TiB_2$ corresponding to a diffraction angle of 44.4° is overlapped by peaks corresponding to TiAl in regions of lower diffraction angle. Moreover, the amount of $TiB_2$ is small in relation to TiAl. Because of its fine grain size and small quantity, it was not detectable with x-ray diffraction measurement. It was evident, however, that there was a formation of fct phase of TiAl. Hence, based on the above data, the reaction sequences in Ti-Al-B system most likely follow the form:

$$Ti + Al + B \rightarrow (Ti, Al)Bss \rightarrow TiB + TiB_2 + TiAl(B)ss$$

TiAl/15%$TiB_2$ may be fabricated from Ti, Al and B. The most likely exothermic reaction route of this process may be

$$Ti+B+Al \rightarrow (Ti, Al)Bss \rightarrow TiB + TiB_2 + TiAl(B)ss \rightarrow TiB_2+TiAl$$

after annealing at 1,123 K. Because of the presence of Al in the Ti-Al-B system, the concentration of Ti or B was diluted leading to the exothermic reaction between Ti and B being delayed. However, grain refinement of Ti and Al in this system went down to nanometer more rapidly than that in Ti-Al system as a result of the contribution from B.

## 4.5 Nanostructured materials

### *4.5.1 Formation of nanostructure via mechanical alloying*

Studies on different materials have demonstrated the possibilities of forming of nanocrystalline via simple mechanical alloying technique [(202-207)]. A series of transmission electron microscopy study has shown that crystals in the deformed powder particles are heavily strained in a rather inhomogeneous manner. Shear bands which are typical of deformation mechanism occurring at high strain rates in contrast to slip and twinning mechanism at low and moderate strain rates, have been observed. These shear bands, separated by areas of similar lateral dimensions in the micron range, have low defect densities consisting of individual grains with a diameter of 20 nm and are slightly rotated with respect to each other at a rotation angle of less than 20° [(208)]. Fecht [(208)] studied AlRu system and found that with the increase in collision process, the shear bands grew over larger areas and eventually the entire powder particle disintegrated into subgrains with a final grain size of 5-7nm. The elemental processes leading to grain size refinement include the following three basic stages [(208)]:

(a) The deformation is localized in shear bands consisting of an array of dislocations with high density. This is accompanied by atomic level strains which can increase up to 3% for the compound phases.

(b) At a certain strain level, these dislocations annihilate and recombine as small angle grain boundaries separating the individual grains. The subgrains formed via this route are in the nanometer size range. During further collision the column having small grains extends throughout the powder particles.

(c) The orientations of the grains with respect to their neighbouring grains become completely random. The latter event can be understood in the following way. The yield stress required to deform a polycrystalline material by dislocation movement is related to the average grain size *d* by

$$\sigma=\sigma_0+kd^{-1/2}$$

where $\sigma_0$ and $k$ are constants (Hall-Petch relationship). An extrapolation to nanocrystalline dimensions shows that very high stresses are required to maintain plastic deformation. For a grain size of 10 nm the minimum yield stress is of the order of 5 GPa corresponding to 15% of the theoretical shear stress of a hexagonal metal which sets a limit to the grain size reduction achieved by plastic deformation during ball milling. Therefore, the reduction in grain size to a few nanometer is limited by the stresses applied during ball milling as long as no dramatic elastic softening of the crystal lattice occurs.

It has been realized that energy storage by mechanical deformation is only possible by an alternative mechanism. Grain boundary sliding has been observed in many cases at high temperature leading to superplastic behaviour. Alternatively, grain boundary sliding can also be achieved at very small grain size and low temperature by diffusional flow of atoms along the intercrystalline interfaces which allows ductile ceramics to be synthesized [(209)]. This provides a mechanism for self-organization and rotation of the grains which increase the energy of the grain boundaries proportional to their misorientation angle and excess volume.

The crystalline size is milling intensity independent. When milling intensity is increased from 5 to 9, the crystalline size of FeAl increases from 8 to 13 nm while internal strain decreases [(143)]. It appears that milling temperature does not significantly contribute to the change in crystalline size when the milling temperature is low. Shen *et al.* [(214)] milled elemental powders of Cu and Ni, and compounds of $Cu_{50}Ni_{50}$ and $Ni_{50}Co_{50}$ at 303 K cooled by fan and at 193 K cooled by liquid nitrogen respectively. It was found that only Cu exhibited a significant change in grain size from 17 nm at 193 K to 26 nm at 303 K. The crystalline sizes of the rest of the powders were almost identical at the two milling temperatures. This indicates that a dynamic recovery occurrs for Cu milled at 303 K. Another possibility is that grain refinement of Cu at low temperature may be more effective than that at relatively high temperature. The ultimate crystalline size achievable during mechanical milling is determined by the minimum crystalline size that can sustain a dislocation pile-up within a grain and by the rate of recovery [(215)]. The magnitude of the stacking fault energy of these alloys does not appear to influence the minimum grain size attainable. Shen *et al.* [(214)] showed that the crystalline size of Cu-Ni and Ni-Co alloys could be inversely correlated to their mechanical hardness and melting temperature. This suggests that a balance between defect creation and recovery is important in the determination of the crystalline size.

### *4.5.2 Formation of nanostructure via mechanical alloying and thermal treatment*

Coarse grained structures may be refined into very fine crystalline structure by a special treatment called hydrogenation, disproportionation, desorption and recombination process (HDDR) [(216-218)]. In the original HDDR process, nanocrystalline materials can be synthesized through two steps:

(a) decomposition of the intermetallic compound $A_{1-x}B_x$, where A and B are a hydride former and a non hydride former respectively, into a hydride $AH_x$ and the other B-rich $A_{1-x-y}B_x$ compound or pure element B via hydrogenation;

(b) realloying between element A which is produced by decomposition of $AH_x$, and B-rich $A_{1-x-y}B_x$ compound or pure element B.

These two steps can be completed via mechanical alloying and a suitable thermal treatment [(219)]. One example is the successful process to synthesize Ti-Al system because Ti-Al system consists of hydride former of Ti and non-hydride former of Al.

The first step of hydrogenation can be completed by mechanical alloying of TiAl in a hydrogen atmosphere of 2 MPa. Figure 4.42 shows the x-ray spectra of the mechanically alloyed TiAl powder at different milling durations. The original material consists mainly of TiAl and of small amount of $Ti_3Al$ structures [(219)]. After

360 minutes of milling, $Ti_3Al$ peak disappears while TiAl peaks are broadened due mainly to reduction in grain size. TiAl (002) diffraction spectrum overlaps with (200) while (202) with (220). Titanium hydride seems to appear at this stage as indicted by the two arrows. The content of titanium hydride of $TiH_2$ increases with the increase in milling duration. $TiH_2$ diffraction peaks are evident after 2,880 minutes of milling. TiAl compound is formed by the decomposition of $TiH_2$. The precipitation of TiAl follows the reaction of decomposition of Ti.

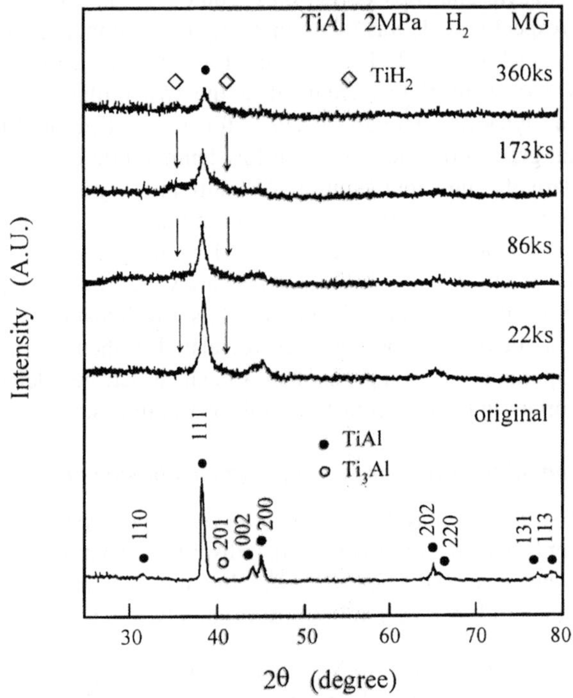


Figure 4.42. Structural evolution by hydrogenation process [219].

### 4.5.3 *Formation of nanostructure from amorphous phase*

Another method to fabricate more stable nanocrystalline phase is to anneal the amorphous structure at around the glass transition temperature [220] or to mechanically crystallize the amorphous phase [221].

Full polymeric crystallization to a nanocrystalline structure has been observed in the amorphous phase of $Cu_{50}Ti_{40}Ni_{10}$ after annealing at 686 K. Other materials are $Zr_{48}Ni_{27}Al_{25}$, $Cu_{50}Ti_{40}Al_{10}$, $Al_{92}Sm_8$, $Ni_{36}Zr_{34}$, $Fe_{73.5}Cu_1Nb_3Si_{13.5}B_9$, $Fe_{90}Zr_7B_3$ and $Pd_{60}Ti_{20}Si_{20}$. In this process, the powder is firstly milled to form amorphous structure followed by thermal analysis to obtain the glass transition temperature which is to be used for annealing. Sometime it is difficult to determine the glass transition temperature because of the overlap of the transition temperature and the crystallization temperature. A careful study using some thermal analysis instrument, for example differential scanning calorimeter, should therefore be carried out. The most important parameter in this process is to control the crystalline size and to avoid grain growth. The crystalline size during annealing depends on the nucleation frequency which may be expressed as [220]

$$f_v = N_v \frac{D_n}{a^2} \exp\left(\frac{Q}{kT}\right) \qquad [1]$$

where $Q$ is the activation energy for the formation of a nucleus of critical size, $D_n$, diffusivity, $a$, the size of the particle and $N_v$, the number of atom per unit volume. Activation energy can be calculated using

$$Q = \frac{16\pi}{3} \frac{\gamma_{a\to c}}{\Delta G_{a\to c}} \qquad [2]$$

where $\gamma_{a\to c}$ is the amorphous-crystal interfacial energy and $\Delta G_{a\to c}$, the free energy difference between amorphous and crystal phase.

Crystallization temperature changes with the duration of milling. Eckert *et al.* [222] found that the crystallization temperature of mechanically alloyed $Ni_{68}Zr_{32}$ was increased to about 853 K when milling duration was between 4 and 12 hours. The temperature decreased significantly after 12 hours of mechanical alloying. The crystallization temperature after 72 hours corresponded to the crystallization temperature of $Ni_{50}Zr_{50}$ in the amorphous phase, indicating the sustained precipitation of $Ni_3Zr$ that lowered the Ni content in the remaining amorphous phase.

It is important that the homogeneous nucleation frequency must be high and grain growth can be controlled.

To produce nanocrystalline materials below 6-7 nm is difficult using the annealing process. However, if the amorphous alloys are mechanically crystallized using mechanical milling, crystalline size of 2 to 3 nm can be achieved [221]. This is because mechanical crystallization destabilizes the amorphous structure. Impurities is introduced during mechanical milling to modify the short range composition of the material. Since destabilization during milling is localized on a very small scale, the crystallization process is slow. Slow crystallization results in extremely fine crystalline size which is smaller than the nanostructure formed directly from mechanical milling.

## 4.6 Formation of amorphous materials

Since the discovery of formation of amorphous materials in the solid state by Koch *et al.* [223], amorphization using mechanical alloying or milling has generated wide research interest [224-229]. In comparison to rapid solidification process, solid state amorphization yields a wider glass-forming composition range in many binary alloy systems while that for rapid solidification is restricted to deep eutectic regions due to kinetic restraints [230]. Solid state amorphization has the advantage that it may lead to techniques for the production of bulk amorphous materials. To synthesis amorphous materials, the following three basic criteria must be satisfied [110, 231]:

(a) The system is an asymmetric diffusion couple. Atoms of the metal $M_A$ diffuse through metal $M_B$ significantly faster than the atoms of $M_B$.

(b) The system exhibits a negative heat of mixing since a solid state amorphization reaction is thermodynamically driven.

(c) The system possesses sufficient diffusivity at temperatures well below the crystallization temperature.

Some of the features of amorphization from mechanical alloying or milling are similar to diffusion-induced grain boundary migration. During mechanical alloying, the powder particles are extensively deformed. A large amount of dislocations generates new interface boundaries between two metals. There are at least three mechanics for solid state amorphization, namely, diffusion-induced decay of fragment boundaries node disclinations; mis-spending of fragment boundaries; and intersection of moving dislocation with grain boundaries. The above conditions are essential but not sufficient for the formation of amorphous phases. An example is the mechanical alloying of Al-Ni system [232]. Although very high negative formation enthalpy for the ready amorphization, mechanical alloying leads only to the formation of NiAl intermetallic compound.

### *4.6.1 Amorphization*

Nucleation of the amorphization phase usually occurs at the grain boundaries. Hence, a system with thin alternate layers of different elements that are less than 100 μm in thickness is essential. Mechanical alloying provides this basic condition since repeated cold welding and fracturing dramatically reduce the thickness of alternate layers of different elements in the powder particles. Three types of amorphization reactions for mechanical alloying have been observed [110]:

(a) a continuous decrease in effective crystalline size and a shift in positions of the peaks of x-ray spectrum, resulting eventually in an amorphous alloy;

(b) a decrease in intensity of the Bragg reflections of the elements and separately, an increase in the intensity of the broad x-ray spectrum peak of the amorphous alloys; and

(c) the formation of crystalline intermetallic intermediate products with further milling resulting in a transformation of these products into a homogeneous amorphous alloy.

Alloys having a large negative heat of mixing show easy amorphization. It is important to note that amorphization is strongly affected by milling conditions. Schultz [(233)] investigated the influence of milling intensity on $Ni_xZr_{100-x}$ under two different milling intensities. Both intensities produced amorphous powders with some residual crystalline portions for X=80 and perfectly amorphous powder for X=65. For X=65, the lower intensity milling also formed amorphous powder while the higher intensity milling led to a partially crystallized material which cannot be intermetallic compound since Ni-Zr intermetallic compound can be amorphized by mechanical milling. Gaffet *et al.* [(234)] have also indicated the importance of milling intensity and constructed milling maps for different conditions.

The basic principle of formation of amorphous is that the free enthalpy of the equilibrium crystalline state $G_X$ is always lower than that of the amorphous state $G_a$ for metallic systems below the melting temperature [(235, 236)]. The amorphous materials are metastable phases. An energy barrier therefore exists to prevent amorphous materials from spontaneous crystallization. As shown in Figure 4.43, since $G_a$ is higher than $G_X$, it is necessary to create an initial crystalline state $G_0$ which is higher than both $G_a$ and $G_X$. Starting from $G_0$, the free enthalpy of the system can be lowered either by the formation of metastable amorphous phase or by the formation of crystalline intermetallic phases [(235)]. From Figure 4.43, it can be noted that equilibrium phases are energetically favoured. The actual phase present is, however, dependent on both the free energy and the kinetic barrier. Therefore, amorphization is possible if $G_a$ is lower than $G_0$ and the formation reaction for amorphous phases is much faster than that for crystalline phases, namely,

$$\tau_{0\to a} << \tau_{0\to x}$$

and

$$\tau_{0\to a} << \tau_{a\to x}$$

where $\tau$ is the time for reaction.

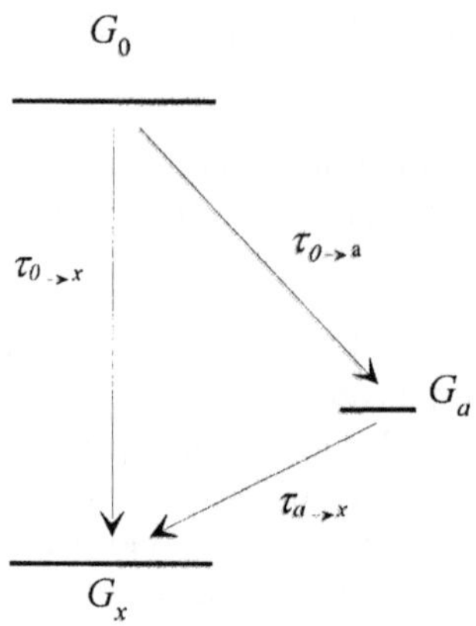


Figure 4.43. Basic principle of amorphization [235].

### *4.6.2 Formation enthalpy*

The possibility of formation of amorphous materials may be theoretically calculated by considering the formation enthalpy using Miedema's model [237, 238].

The enthalpy of different $\Delta H_i$ of the transition from a pure crystalline element to the amorphous, which is assumed to be independent of temperature, may be calculated using Miedema and Niessen equation [238]:

$$\Delta H_i = C_a T_i^m \qquad [4.3]$$

where $T_i^m$ is the melting temperature of the pure element *i* and $C_a$, constant to be equal to 3.5 $Jmol^{-1}K^{-1}$.

In this semi-empirical model, the enthalpy of mixing $\Delta H_c$ of an alloy of a transition material $M_A$ in a transition material $M_B$ is calculated by [110, 237, 238]:

$$\Delta H_c = x_A f_B^A \Delta H_{sol}^{A\ in\ B} \qquad [4.4]$$

where $\Delta H_{sol}^{A\ in\ B}$ is the solution enthalpy of $M_A$ in $M_B$ per mol $M_A$, $x_A$, the atomic fraction of $M_A$, $f_B^A$, the degree by which a $M_A$ atom is surrounded by $M_B$ atoms.

For the formation enthalpy of the alloys, a weighted average of the formation enthalpies of $M_A$ in $M_B$ and $M_B$ in $M_A$ is considered at 0 K by:

$$\Delta H_c = x_A x_B \left( f_B^A \Delta H_{sol}^{A\ in\ B} + f_A^B \Delta H_{sol}^{B\ in\ A} \right) \qquad [4.5]$$

where $\Delta H_{sol}^{B\ in\ A}$ is the solution enthalpy of $M_B$ in $M_A$ per mol $M_B$, $x_B$, the atomic fraction of $M_B$, $f_A^B$, the degree by which a $M_B$ atom is surrounded by $M_A$ atoms. $f_B^A$ or $f_A^B$ is defined by:

$$f_B^A = c_B^s\left[1 + k\left(c_A^s c_B^s\right)^2\right] \qquad [4.6]$$

where $c_A^s$ and $c_B^s$ are the surface concentrations of $M_A$ and $M_B$ respectively, $k$, constant (240) of value $k = 1$ for liquid, $k = 5$ for amorphous phase and $k = 8$ for ordered phases.

Surface concentrations of $c_A^s$ and $c_B^s$ may be calculated by (240):

$$c_B^s = \frac{x_B V_B^{2/3}}{x_A V_A^{2/3} + x_B V_B^{2/3}} \qquad [4.7]$$

and

$$c_A^s = 1 - c_B^s \qquad [4.8]$$

where $V_A$ and $V_B$ are the molar volumes of pure $M_A$ and $M_B$ respectively.

In solid solution, all chemical, elastic and structural contributions should be considered. Formation enthalpy $\Delta H_{ss}$ of a solid solution can be written as:

$$\Delta H_{ss} = \Delta H_c + \Delta H_e + \Delta H_s \qquad [4.9]$$

where subscripts $c$, $e$ and $s$ represent chemical, elastic and structural contributions, respectively. The elastic term in the enthalpy equation basically comes from the atomic size mismatch and can be written as (110):

$$\Delta H_e = x_A x_B\left(f_B^A E_e^{A\ in\ B} + f_A^B E_e^{B\ in\ A}\right) \qquad [4.10]$$

where $E_e$ is the elastic energy when mismatch takes place. It can be calculated using Eshelby's theory :

$$E_e^{A\ in\ B} = \frac{2\mu_B\left(V_A^* - V_B^*\right)^2}{3V_B^* + 4\mu_B K_A V_A^*} \qquad [4.11]$$

where $\mu_B$ is the shear modulus of the solvent, $V_A^*$ and $V_B^*$ are the molar volumes of solute and solvent respectively, and $K_A$ is the compressibility of the solute. The energy $E_e$ may also be found in a tabulated form from reference (239). Beke *et al.* [241, 242] have developed a general condition for solid state amorphization by mechanical alloying. Amorphization is possible if the maximum elastic energy stored during a chemical order-disorder transition is larger than the enthalpy of the topological change of amorphization. For the ternary solid solution, formation enthalpy of mixing can be calculated with references (243) and (244).

In the case of amorphous phase, the elastic and structural contributions to the enthalpy are absent. The enthalpy can therefore be written as [245]:

$$\Delta H_a = \Delta H_c + x_A \Delta H_A + x_B \Delta H_B \qquad [4.12]$$

Amorphization can also take place with mechanical alloying of crystalline alloys. During the milling of crystalline alloys, the enthalpy of the compounds is increased due to the introduction of defects and the increase in internal energy. When the free enthalpy of compound exceeds the free enthalpy of the amorphous phase, the compounds can transform to amorphous phases [240]. Figure 4.44 shows the range of formation of amorphous phase from about 13 to 87% of metal B.

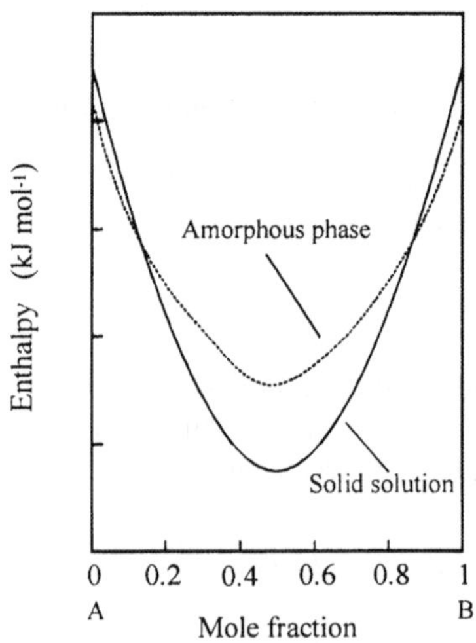


Figure 4.44. Free enthalpies of solid solution and amorphous phase.

## 4.7 Mechanochemical milling

Mechanochemistry is a branch of chemistry that deals with chemical reactions and physico-chemical transformation of substances induced by mechanical activation [246]. Mechanochemical activation has been reported as early as the 1890's when Carey-Lea discovered the formation of free metal and $Cl_2$ due to an apparently non-thermal chemistry associated with the grinding of metal chlorides [247]. During mechanical milling, impact gives rise to a stress field in the mechanically milled substance. This stress field may be relaxed by different mechanisms. The relaxation can directly or indirectly influence the progress of the solid state chemical reaction [248].

### *4.7.1 Processing*

Mechanochemical milling is basically a milling process. Experimental results have shown that mechanical milling can be used as a vehicle for reaction in the solid state [100, 249, 250]. One of such examples is the displacement reaction caused by mechanochemical milling of a simple metal oxide with a strong metallic reducing agent. This type of mechanochemical activation can be formulated as [100]:

$$A_xO + yB \rightarrow xA + B_yO \quad [4.13]$$

When the above mechanochemical reaction takes place, an abrupt increase in temperature occurs. After the increase, temperature decays to a steady value about the same as before the abrupt increase. This change is schematically depicted in Figure 4.45. X-ray diffraction spectra show a change in powder structures just after the temperature rise. It may be deduced that the combustion event indicates the occurrence of a chemical reaction. The combustion is a direct consequence of the increased reactivity of the systems. In highly exothermic systems, heat is generated by the reaction enthalpy faster than that dissipated through the milling tools. A series of mechanochemical milling has been investigated by Schaffer and McCormick [100]. Reduction reactions have been found for CuO+Al, CuO+Ca, CuO+Mg, CuO+Ti, CuO+Mn, CuO+Fe, CuO+Ni [100], ZnO+Ti, $Ag_2O$+Al, ZnO+$Al_2O_3$ [251] and $Fe_3O_4$+Al [252]. Results of their studies showed that two parameters are important in mechanochemical milling, namely, the precombustion milling period, $t_{ig}$, and the adiabatic temperature, $T_{ad}$. The theoretical adiabatic temperature associated with a reaction is often taken as measurement of the driving force. For systems which combust, the precombustion milling period generally decreases with increasing $T_{ad}$. Since diffusion, and thus reaction rates, depends on temperature, the overall rate of reaction is affected by the reaction enthalpy [100].

Interestingly, when some of the above systems are aged for centain periods after 80% of total milling duration at which the combustions take place, the systems are milled again. Ageing could significantly accelerate the onset of combustion, thereby substantially reducing the overall milling duration. The interrupted combustion effect is an indication that the enhanced chemical reactivity of the solids induced by mechanical milling is not due to dynamic maintenance of high reaction volumes. Impact increases the reactivity of the solids. However, if primary milling or ageing duration is too short, interupted combustion will not take place. In such a case, the total milling durations before and after ageing is equivalent to that of continuous milling [253].

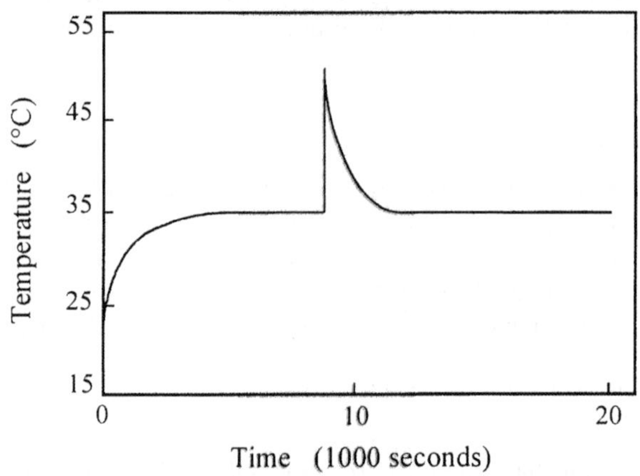


Figure 4.45 Change in temperature during mechanochemical milling [180].

Reaction rate in the solid state depends on temperature, surface area of the powders, defect density and diffusion path. In general, the reaction area of the powder particles is a function of the particle size. With an increase in milling duration, the particle size is reduced to a very fine size and thus greatly increases the reaction area. The high defects density further increases the reaction rates by providing short circuit diffusion paths [180]. The occurrence of combustion induced by milling is associated with the attainment of a critical reaction rate. The generation of heat associated with the reaction enthalpy is faster than that dessipated from the reaction volume [253]. The reaction enthalpy $\Delta H_r$ can be written as [100]:

$$\Delta H_r = \sum C_p^s (T_m - T_i) + \sum \Delta H_m + \sum C_p^l (T_{ad} - T_m) \qquad [4.14]$$

where $C_p^s$ and $C_p^l$ are respectively the mean heat capacities of the solid and liquid products, $\Delta H_m$, the heat of fusion of the products, $T_m$, the melting temperature of the products, $T_i$, the initial temperature and $T_{ad}$, the adiabatic temperature.

Multiple combustion induced by mechanical milling has been observed from Zr+S system [254]. A small amplitude of the temperature jump indicates that the reaction is restricted to a small fraction of the sample, most likely in the region where collision takes place. This region would not propagate immediately to the entire powder. The small jumps happen at rather unpredictable times. The series of small explosions is followed by a much larger temperature increase after about the same cumulative milling time.

### *4.7.2 Formation of nanostructure and nanocomposites*

Nanostructure and nanocomposites may be formed through Eq.[4.13]. Some reactions are highly exothermic. The collision of the milling ball with powders may initiate self propagating combustive reaction. To obtain a nanocomposite, the combustion has to be suppressed by diluting the reaction mixture with some inert materials sometime called additives [256, 257]. Small amount of inert additive like $SiO_2$ or the product phase delays the ignition of combustion while large quantities may entirely suppress the combustion. The presence of additives decreases the adiabatic temperature and consequently decreases the incubation time [258]. The reduction milling of metal oxides by more reactive metal usually takes place through a complex series of overlapping reactions, forming several intermediate and metastable phases. For example, when $Fe_3O_4$ or CuO are milled with Al, Zn, Cu, Ti or Fe [253, 257, 259], reduction takes place by

$$3Fe_3O_4 + 8Al \rightarrow 9Fe + 4Al_2O_3 \quad [4.15]$$
$$Fe_3O_4 + 4Zn \rightarrow 3Fe + 4ZnO \quad [4.16]$$
$$Fe_3O_4 + 2Ti \rightarrow 3Fe + 2TiO_2 \quad [4.17]$$
$$4CuO + 3Fe \rightarrow Fe_3O_4 + 4Cu \quad [4.18]$$

Since synthesis of $Fe\text{-}Al_2O_3$ composite involves a strong exothermic reaction, mechanochemcal milling normally induces a self propagating combustive reaction. To suppress combustion synthesis of $Fe\text{-}Al_2O_3$ nanocomposites, addition of some inert materials is required in the processing. It must be noted that properties of the nanocomposites may however be altered by such additions. If a less exothermic reaction is chosen, the inert materials may not be needed. In the case of formation of Fe-ZnO, since no combustion takes place during milling of the stoichiometry composition with 10% extra Zn, no diluting material is needed to be added into the milling system.

Schaffer *et al.* [255] have indicated the highly exothermic nature of a reaction as the most important condition for the initiation of combustion. $t_{ig}$ is normally used to indicate the time for initiation of ignition. The occurrence of combustion is associated with the attainment of a critical reaction rate above which the reaction accelerates to combustion as a consequence of the self-heating of the reactants [253]. If the partially milled powder mixture is heated, combustion may be thermally induced. Differential thermal analysis reveals this event as the ignition temperature

$T_{ig}$ [253]. Very interestingly, both $T_{ig}$ and $T_{crit}$ temperatures, where the latter is the critical temperature at which the temperature difference between the specimen and the reference first increases due to heat generated by the reaction, decrease with increased primary milling duration.

Because the rate of reaction of the solid-state displacement is diffusion controlled under thermal condition, the reaction kinetics depend on temperature and on a number of structural parameters. This observation indicates that mechanical milling may change the kinetics of the mechanically milled powder mixture and may increase the reaction rate [253].

## 4.8 Conclusions

Results from many researchers have demonstrated that mechanical alloying or milling can be used to synthesize not only oxide dispersion strengthened alloys but also other types of new materials such as amorphous alloys, non-equilibrium materials, ceramics, nanostructural materials and composites.

Nanostructural materials may be fabricated via the milling process or a combination of milling and annealing. Mechanical crystallization destabilizes the amorphous structure and hence may be employed to synthesize extremely fine crystalline structure. Mechanochemical milling can also provide a means for chemical reactions and physico-chemical transformation of substances. Particles with nano-size can be obtained by the process of mechanochemical reduction.

Since mechanical milling is carried out in the solid state, materials with very different melting temperatures can be easily alloyed. The process has therefore provided a new technique for scientists and engineers to develop a new generation of different engineering materials.

## 4.9 References

1. J.S. Benjamin, *Metall. Trans.*, Vol.1 (1970), 2943.
2. J.J. deBarbadillo and G.D. Smith, *Mater. Sci. Forum*, Vols.88-90 (1992), 167.
3. A. Wagendristel and E. Tschegg, *Aluminium*, Vol.48 (1972), 357.
4. J.S. Benjamin and M.J. Bomford, *Metall. Trans.*, Vol.8A (1977), 1301.
5. H.C. Neubing, J. Gradl and G. Jangg, *Aluminium*, Vol.67 (1991), 1112.
6. R.F Singer, W.C. Oliver and W.D. Nix, *Metall. Trans.*, Vol.11A (1980), 1895.
7. G. Jangg, J. Vasgyura, K. Schroder, M. Slesar and M. Besterci, *Horizons of Powder Metall.*, Ed: W.A. Kaysser and W.J. Huppmann, Verlagschmid GmbH, Freiburg, F.R. Germany, Part II (1986), 989.
8. G. Jangg, M. Slesar, M. Besterci, J. Durisin and K. Schroder, *Powder Metall. Inter.*, Vol.21 (1989), 25.
9. J. Durisin, M. Orolinova and K. Durusinova, *Adv. in Powder Metall. & Particulate Mater.*, Compiled: J.M. Capus and R.M. German, Metal Powder Industries Federation, Princeton, NJ, Vol.9 (1992), 195.
10. M.V. Gurjar, G.S. Murty and G.S. Upadhyaya, *J. Mater. Sci.*, Vol.28 (1993), 5654.
11. S. Enzo, G. Cocco and P.P. Macri, *Key Eng. Mater.*, Vol.81-83 (1993), 49.

12. Y.W. Kim and L.R. Bidwell, *Scripta Metall.*, Vol.16 (1982), 799.
13. A. Renard, A.S. Cheng, R. de la Veaux and C. Laird, *Mater. Sci. Eng.*, Vol.60 (1983), 113.
14. J.A. Hawk, W. Ruch and H.G.F. Wilsdoorf, *Dispersion Strengthened Aluminium Alloys*, Proc. the 6th Symp. , 25-29 Jan., 1988, Ed. Y.W. Kim and W.M. Griffith, Publ. TMS Minerals, Metals and Mater. Soc., 603.
15. P.S. Gilman and W.D. Nix, *Metall. Trans.*, Vol.12A (1981), 813.
16. R.C. Benn and P.K. Mirchandani, *New Materials by Mechanical alloying Techniques*, DGM Confer., Calw-Hirsau (FRG), Oct. 1988, Ed. E. Arzt and L. Schultz, Publ. Informationsgesellschaft Verlag, (1989), 19.
17. K. Uenishi and F.K. Kobayashi, *Mater. Trans, JIM*, Vol.36 (1995), 810.
18. K. Higashi, *Aspects of High Temperature Deformation and Fracture in Crystalline Materials*, Publ. Japan. Inst. Metals, Sendai, Japan, (1993), 447.
19. K. Tomita, M. Sugamata, J. Kaneko and R. Horiuchi, *65th Confer. of Japan. Inst. Light Metals*, Nov.1993, 2.
20. F. Li, K.N. Ishihara and P.H. Singu, *Metall. Trans.*, Vol.22A (1991), 2849.
21. M. Tatsuta, D.G. Kim, J. Kaneko and M. Sugamata, *Sci. Eng. of Ligh Metals*, Ed. K. Hirano, H. Oikawa and K. Ikeda, Publ. Japan. Inst. Light Metals, Tokyo, (1991), 685.
22. P.H. Shingu, *ibid*, 677.
23. J. Eckert, *Mater. Sci. Forum*, Vols.88-90 (1992), 679.
24. A.P. Tsai, K. Aoki, A. Inoue and T. Masumoto, *J. Mater. Res.*, Vol.8 (1993), 5.
25. T. Ishikawa, M. Kato, N. Yukawa and T. Jimma, *J. Japan Inst. Light Metals*, Vol.43 (1993), 725.
26. J. Eckert, L. Schultz and K. Urban, *Z. Metallkde*, Vol.81 (1990), 862.
27. J. Eckert, L. Schultz and K. Urban, *J. Less-Common Metals*, Vol.167 (1990), 143.
28. J. Eckert, L. Schultz and K. Urban, *Mater. Sci. Eng.*, Vol.A133 (1991), 393.
29. B. Huang, K.F. Kobayashi and P.H. Shingu, *J. Japan. Inst. Light Metal*, Vol.38 (1988), 165.
30. P.B. Desch, R.B. Schwarz and P. Nash, *Alloy Phase Stability and Design*, Ed: G.M. Stocks, D.P. Pope and A.F. Giamei, Publ. Mater. Res. Soc., Pittsburgh, PA, Vol.186 (1991), 439.
31. P.H. Shingu, B. Huang, S.R. Nishitani and S. Nasu, *Suppl. Trans. Japan. Inst. Metals*, Vol.29 (1988), 3.
32. H. Sugiyama, J. Kaneko and M. Sugamata, *Proc. 77th Confer. Japan. Light Metal Association*, Tokyo, Japan, 1989, 97.
33. V.I. Fadeeva and A.V. Leonov, *Mater. Sci. Forum*, Vol.88-90 (1992), 481.
34. P. Le Brun, L. Froyen, L. Delaey, *Adv. Mater. & Processes.*, Proc. of the 1st Europeqn Confer. on Adv. Mater. and Processes, EUROMAT'89, Vol.1, Ed. H.E. Exner and V. Schumacher, Publ. DGM Informationsgesellschaft Verlag, (1989), 231.
35. X. Niu, L. Froyen and L. Delaey, *J. Mater. Sci.*, Vol.29 (1994), 3724.
36. D.K. Mukhopadhyay, C. Suryanarayana and F.H. Froes, *Metall. Mater. Trans.*, Vol.26A (1995), 1939.
37. M.L. Ovecoglu and W.D. Nix, *New Materials by Mechanical alloying Techniques*, DGM Confer., Calw-Hirsau (FRG), Oct. 1988, Ed. E. Arzt and L. Schultz, Publ. Informationsgesellschaft Verlag, (1989), 287.
38. P. Le Brun, X.P. Niu, L. Froyen, B. Munar and L. Delaey, *Solid State Powder Processing.*, Proc. of the Confer., Indianapolis, IN, 1-5 Oct. 1989, Ed. A.H. Clauer and J.J. deBarbadillo, TMS, Warrendale, PA, (1990), 273.
39. P. Le Brun, L. Froyen and L. Delaey, *Mater. Sci. Eng.*, Vol.A157 (1992), 79.
40. X.P. Niu, P. Le Brun, L. Froyen, C. Peytour and L. Delaey, *Powder Metall. Inter.*, Vol.118 (1993), 120.
41. G.H. Narayanan, W.H. Graham, W.E. Quist, A.L. Wingert and T.M.F. Ronald, *Proc. of Structural Metals by Rapid Solidfication*, Ed. F.H. Froes, S.J. Savage, ASM Intern., Mater. Park, OH, 1987, 321.
42. S.K. Kang, *Proc. $39_{th}$ EMSA Annual Mt.*, Ed. G.W. Bailey, Publ. Claitor's Pub. Div., Baton Rouge, LA, (1981), 48.
43. P.S. Gilman, *Aluminium-Lithium Alloys* II, Ed. T.H. Sanders, Jr. and E.A. Starke, Jr., TMS, Warrendale, PA, 1984, 485.

44. A. Layyous, S. Nadiv and I.J. Lin, *Powder Metall. Inter.*, Vol.22 (1990), 21.
45. J.S. Benjamin and R.D. Schelleng, *Metall. Trans.*, Vol.12A (1981), 1827.
46. D.L. Zhang, T.B. Massalski and M.R. Paruchuri, *Metall. Trans.*, Vol.25A (1994), 73.
47. X.X. Yan, N. Bois and G. Cizeron, *J. Phys.* III, Vol.4 (1994), 1913.
48. P. Le Brun, L. Froyen and L. Delaey, *Structural Applications of Mechanical Alloying*, Proc. of an ASM Intern.. Confer., Myrtle Beach, South Carolina, 27-29 March 1990, Ed. F.H. Froes and J.J. deBarbadillo, Publ. ASM Intern.., Mater. Park, Ohio, 155.
49. H.T. Shin, Kaneko and M. Sugamata, *J. Japan Inst. Light Metals*, Vol.41 (1991), 607.
50. H.T. Shin, Kaneko and M. Sugamata, *J. Japan Inst. Light Metals*, Vol.55 (1991), 1159.
51. J.A. Hawk, J.K. Briggs and H.G.F. Wilsdorf, *Advances in Powder Metall.-1989*, Ed. T.G. Gasbarre and W.F. Jandeska, Jr., Metal Powder INdustries Federation, Princeton, NJ, Vol.3 (1989), 301.
52. M. Zdujic, K.F. Kobarashi and P.H. Shingu, *Z. Metallkde*, Vol.81 (1990), 380.
53. R. Sundaresan, R.S. Mishra, T. Raghu, A.G. Paradkar and M.C. Pandey, *Proc. of the 2nd Intern. Confer. on Structural Applications of Mechamical Alloying*, Vancouver, Canada, Ed.J.J. deBarladillo, F.H. Froes and R. Schwarz, Pul. ASM Intern., Materials Park, 20-22 Sep. (1993), 221.
54. R.B. Schwarz, P.B. Desch and S.R. Srinivasan, ibid, 227.
55. G. Janng, G. Korb and F. Kutner, *Aluminium*, Vol.51 (1975), 641.
56. G. Janng, F. Kutner and G. Korb, *Powder Metall. Inter.*, Vol.9 (1977), 24.
57. G. Janng, *New Materials by Mechanical alloying Techniques*, DGM Confer., Calw-Hirsau (FRG), Oct. 1988, Ed. E. Arzt and L. Schultz, Publ. Informationsgesellschaft Verlag, (1989), 39.
58. V. Arnhod and K. Hummert, *ibid*, 263.
59. M.A. Morris and D.G. Morris, *Mater. Sci. Forum*, Vols.88-90 (1992), 529.
60. G. Jangg, J. Zbiral, J. Wu, M. Slesar and M. Besterci, Aluminium, Vol.68 (1992), 238.
61. Z.Y. Ma, X.G. Ning, Y.X. Lu, J. Bi, L.S. Wen, S.J. Wu, G. Jangg and H. Daninger, *Scripta Metall. Mater.*, Vol.31 (1994), 131.
62. M.A. Morris and D.G. Morris, *Mater. Sci. Eng.*, Vol.A136 (1991), 59.
63. T.B Massalski, Binary Alloy Phase Diagrams, Publ. ASM, Metal Park, Ohio 44073, Vol.1 (1986).
64. J. Kumpfert, G. Staniek and W. Kleinekathofer, *Structural Applications of Mechanical Alloying*, Proc. of An ASM Inter. Confer., Myrtle Beach, South Carolina, 27-29 March 1990, Ed. F.H. Froes and J.J. deBarbadillo, Publ. ASM Inter., Mater. Park, OH, 163.
65. J.L. Murray, A. J. McAlister, R.J. Schaefer, L.A. Bendersky, F. S. Biancaniello and D.L. Moffat, *Metall. Trans.*, Vol.18A (1987), 385.
66. D. Shechtman, I. Blech, D. Gratias and J.W. Cahn, *Phys. Rev. Lett.*, Vol.53 (1984), 1951.
67. L.A. Bendersky, *Phys. Rev. Lett.*, Vol.55 (1985), 1461.
68. C. Suryanarayana and R. Sundaresant, *Mater. Sci. Eng.*, Vol.A131 (1991), 237.
69. G. Chen, K. Wang, J. Wang, H. Jiang and M. Quan, *Proc. of the 2nd Intern. Confer. on Structural Applications of Mechamical Alloying*, Vancouver, Canada, Ed.J.J. deBarladillo, F.H. Froes and R. Schwarz, Pul. ASM Intern., Materials Park, 20-22 Sep. (1993), 183.
70. K. Gururaj and G.R. Sambasiva, *Trans. Powder Metall. Assoc. India*, Vol.18 (1991), 11.
71. T. Takeuchi, Y. Yamada, T. Fukunaga and U. Mizutani, *J. Japan Soc. Powder Powder Metall.*, Vol.40 (1993), 951.
72. T. Takeuchi, Y. Yamada, T. Fukunaga and U. Mizutani, *J. Japan Soc. Powder Powder Metall.*, Vol.40 (1993), 307.
73. U. Mizutani, T. Takeuchi and T. Fukunaga, *Mater. Trans. JIM*, Vol.34 (1993), 102.
74. T. Takeuchi, T. Koyano, M. Utsumi, T. Fukunaga, K. Kaneko and U. Mizutani, *Mater. Sci. Eng.*, Vols.179/180 (1994), 224.
75. R.W. Cahn, *Metals, Mater. Processes*, Vol.1 (1989) 1.
76. C.T. Liu, *Ordered intermetallic alloys-brittle fracture and ductility improvement*, Sci. Adv. Mater., ASM Mater. Sci. Seminar, 26-29 Sep. 1988, Chicago, Ilinois, ASM Intern., Mater. Park, Ohio, 423.

77. J.H. Westbrook, *Mechanical Properties of Intermetallic Compounds*, Publ. Wiley, New York, (1960).
78. J.S. Benjamin, *Novel Powder Processing Advances in Powder Metall.*, Vol.7 (1992), Proc. the 1992 Powder Metall., San Francisco, CA, USA, 21-26 1992, pub: Metal Powder Inductries, 155-166.
79. G.B. Schaffer and P.G. McCormick, *Mater. Forum*, Vol.16 (1992), 91-97.
80. C. Suryanarayana, G.H. Chen, A. Frefer and F.H. Freos, *Mater. Sci. Eng.*, Vol.A158 (1992), 93.
81. D.K. Mukhopadhyay, C. Suryanarayana and F.H. Froes, *Proc. of the 2nd Intern. Confer. on Structural Applications of Mechamical Alloying*, Vancouver, Canada, Ed.J.J. deBarladillo, F.H. Froes and R. Schwarz, Pul. ASM Intern., Materials Park, 20-22 Sep. (1993), 131.
82. D.K. Mukhopadhyay, C. Suryanarayana and F.H. Froes, *Titanium'92*, Sci. & Techn., Ed. F.H. Froes and I. Caplan, Publ. The Minerals, Metals & Mater. Soc., (1993), 829.
83. R.B. Schwarz, P.B. Desch and S.R. Srinivasan, *ibid*, 227.
84 C.C. Kock, *Annu. Rev. Mater. Sci.*, Vol.19 (1989), 121.
85. S. Kobayashi and H. Kimura, *Mater. Sci. Forum*, Vols.88-90 (1992), 97.
86. W. Guo, S. Martelli, F. Padella, M. Magini, N. Burgio, E. Paradiso and U. Franzoni, ibid, 139.
87. J.H. Ahn, H.S. Chung, R. Watanabe and Y.H. Park, *ibid*, 347.
88. T. Suzuki, T. Ino and M. Nagumo, *ibid*, 639.
89. M. Miki, T. Yamasaki and Y. Ogino, *Mater. Trans., JIM*, Vol.34 (1993), 952.
90. W. Guo, S. Martelli, N. Burgio, M. Magini, F. Padella, E. Paradiso and I. Soletta, *J. Master. Sci.*, Vol.26 (1990), 6190.
91. G. Cocco, I. Soletta, L. Battezzati, M. Baricco and S. Enzo, *Phil. Mag.* B, Vol.61 (1990), 473.
92. J.H. Ahn, K.R. Lee and H.K. Cho, *Mater. Sci. Forum*, Vols.179-181 (1995), 153.
93. Mukhopadhyay, C. Suryanarayana and F.H. Froes, *Titanium'92, Sci. & Tech.*, Ed. F.H. Froes and I. Caplan, Publ. The Minerals, *Metals Mater. Soc.*, (1993), 829.
94. C. Suryanarayana, R. Sundaresan and F.H. Froes, *Mater. Sci. Eng.*, Vol.A150 (1992) 117.
95. R.B. Schwarz, P.B. Desch and S. Srinivasan, Statics and Dynamics of Alloy Phase Trans., Ed. P.E.A. Turchi and A. Gonis, Publ. Plenum Press, New York (1994), 81.
96. D. Shechtman, M. J. Blackburn and H.A. Lipsitt, *Metall. Trans.*, Vol.5 (1974), 1373.
97. S. Naka, M. Thomas and T. Khan, *Mater. Sci. Tech.*, Vol.8 (1992), 291.
98. J. Pouliquen, S. Offret and J.Fouquet, C.R. Acad, Sci, Ser.C,274(1972) 1760-1763.
99. R.B. Schwarz and W.L. Johnson, *Phys Rev Lett*, Vol.51, 1983, 415.
100. G.B.Schaffer and P.G. McCormick, *Mater. Forum*, Vol.16 (1992), 91.
101. Ihsan Barin, 'Thermochemical data of pure substances", VCH, Verlagsgesellschaft mbH, 1989.
102. P.S. Goodwin and C.W. Ward-Close, *Proc. of the 2nd Intern. Confer. on Structural Applications of Mechamical Alloying*, Vancouver, Canada, Ed.J.J. deBarladillo, F.H. Froes and R. Schwarz, Pul. ASM Intern., Materials Park, 20-22 Sep. (1993), 139.
103. C.Suryanarayana, Guo-Hao Chen and F.H. Froes, *Scripta Metall.*, Vol 26 (1989), 1727.
104. K.Y.Wang, G.L.Chen and J.G.Wang, *Scripta Metall.*, Vol 31 (1994), 87-92.
105. G.Chen and K.Wang, *Proc. of the 2nd Intern. Confer. on Structural Applications of Mechamical Alloying*, Vancouver, Canada, Ed.J.J. deBarladillo, F.H. Froes and R. Schwarz, Pul. ASM Intern., Materials Park, 20-22 Sep. (1993), 149.
106. Y.B. Liu, J.K.M. Kwok, S.C. Lim, L. Lu and M.O. Lai, *J. Mater. Proc. Tech.*, Vol.37 (1993), 441.
107. L. Lu, J.K.M. Kwok, M.O. Lai, Y.B. Liu and S.C. Lim, *J. Mater. Proc. Tech.*, Vol.37 (1993), 453.
108. A. Bhaduri, V. Gopinathan, P. Ramakrishnan, G. Ede and A.P. Miodownik, *Scripta Metall. Mater.*, Vol.28 (1993), 907.
109. A. Malchere and E. Gaffet, *Proc. of the 2nd Intern. Confer. on Structural Applications of Mechamical Alloying*, Vancouver, Canada, Ed.J.J. deBarladillo, F.H. Froes and R. Schwarz, Pul. ASM Intern., Materials Park, 20-22 Sep. (1993), 297.
110. A.W. Weeber and H. Bakker, *Physica* B153 (1988), 93.
111. E. Hellstern and L. Schultz, *Mater. Sci. Eng.*, Vol. 93 (1987), 213
112. E. Gaffet, F. Faudot and M. Harmelin, *Mater. Sci. Eng.*, Vol.A149 (1991), 85.
113. H.J. Fecht, E. Hellstrern, Z. Fu and W.L. Johnson, *Metall. Trans.*, Vol.21A (1990), 337.

114. M.N. Rittner, J.A. Eastman and J.R. Weertman, *Scripta Metall. Mater.*, Vol.31 (1994), 841-846.
115. A. Calka and W.A. Kaczmerk, *Scripta Metall.*, Vol. 26 (1992), 249-253.
116. C. Suryanarayana and F.H. Froes, *Mater. Sci. Forum*, Vols 88-90 (1992), 445-452.
117. P.G. McCormick, V.W. Wharton, M.M. Reyhani and G.B. Schaffer, *Microcomposites and Nanophase materials*, (eds D.C. Van Aken, G.S. Was and A.K. Gosh), TMS, Warvendale, Pennsylvania, 1991, 65.
118. G.B. Schaffer and P.G. McCormick, *Metall Trans.*, Vol.23A (1992), 1285.
119. G.B. Schaffer and P.G. McCormick, *Metall Trans.*, Vol.22A (1991), 3019.
120 .W. Guo, S. Martelli, N. Burgio, M. Magini, F. Padella, E. Paradiso and I. Soletta, *J. Mater. Sci.*, Vol.26 (1991), 6190.
121. E. Bonetti, G. Cocco, S. Enzo and G. Valdre, *Mater. Sci. Tech.*, Vol.6 (1990), 1258.
122. N. Burgio, W. Guo, M. Magini and F. Padella, *Structureal Applications of Mechanical Alloying*, Proc. an ASM Intern. Confer., Myrtle Beach, South Carolina, 27-29 March 1990, Ed: F.H.Froes and J.J. deBarbadillo, ASM Intern. Mater. Park, Ohio (1990), 175.
123. H.S. Kim, G. Kim and D.W. Kum, *Design Fund. of High Tem. Comp. Interm. and Metal-Crem. Systems*, Ed: R.Y. Liu, Y.A. Change, R.G. Reddy and C.T. Liu, the Mineral, Metals & Materials Society, (1996), 223.
124. G.H. Kim, H.S. Kim, Dong Wha Kum, *Scripta Metall.*, Vol.34 (1996), 421.
125. J. Tarnacki and Y.W. Kim, *Scripta Metall.*, Vol.22 (1988), 329.
126. K.S. Kumar and J.R. Pickens, *Scripta Metall.*, Vol.22 (1988), 1015.
127. W.O. Powers, J.A. Wert and C.D. Turner, *Philos. Mag.* A, Vol.60 (1989), 227.
128. H. Mabuchi, K. Hirukawa and Y. Nakayama, *Scripta Metall.*, Vol.23 (1989), 1761.
129. S. Zhang, J.P. Nic and D.E. Mikkola, *Scripta Metall.*, Vol.24 (1990), 57.
130. A.D. McQuillan, *Proc. Roy. Soc.* (London) A204, (1950), 309.
131. Y.E. Ivanov, T.F. Grigorieva, G.V Golubkova, V.V. Boldyrev, A.B. Fasman, S.D. Mikhailenko and O.T. Kalinina, *Mater Lett.*, Vol.7 (1988), 51.
132. S.J. Hwang, P. Nash, M. Dollar and S. Dymek, *High Temperature Ordered Intermetallic Alloys IV*, Ed. L.A. Johnson, D.P. Pope and J.O. Stiegler, Mater. Res. Soc., Pittsburgh, PA, Vol.213 (1991), 661.
133. S.G. Pyo, H. Yasuda and T. Yamane, *Mater Sci. Eng.*, Vols.179/180 (1994), 676.
134. A.K. Bhattacharya and E. Arzt, *Scripta Metall. Mater.*, Vol.l27 (1992), 635.
135. O. Abe, *InCoMe'93*, Proc. of The 1st Inter. Confer on Mechanochem., Kosice, Slovak Republic, 23-26 March, 1993, Ed. K. Tkacova, Publ. Cambridge Interscience Publ., UK, Vol.2, 27.
136. Y. Horie, R.A. Graham and I.K. Simonsen, *Mater. Let.*, Vol. 3 (1985), 354.
137. Y. Horie, R.A. Graham and I.K. Simonsen, *Metallurgical Applications of Shock-Wave and High-Strain-Rate Phenomena*, Ed: L.E. Murr, K.P. Staudhammer and M.A. Meyers, Marcel Dekker, Inc., New York, (1986), 1023.
138. C.C. Koch, J.S.C. Jang and P.Y. Lee, *New Materials by Mechanical Alloying*, DGM Confer. Calw-Hirsau, FRG, October (1988), Ed: E. Arzt and L. Schultz, DGM Informationsgesellschaft Verlag, 101.
139. G.B. Schaffer, *Scripta Metall.*, Vol. 27 (1992), 1.
140. Z.G. Liu, J.T. Guo, L.L. He and Z.Q. Hu, *NanoStructured Mater.*, Vol.4 (1994), 787.
141. F. Cardellini, G. Mazzone, A. Montone and M. Vitori Antisari, *Acta Metall. Mater.*, Vol.42 (1994), 2445.
142. M. Atzmon, *Phys. Re. Lett.*, Vol.64 (1990), 487.
143. C. Kuhrt, H. Schropf, L. Schultz and E. Arztz, *Proc. of the 2nd Intern. Confer. on Structural Applications of Mechamical Alloying*, Vancouver, Canada, Ed.J.J. deBarladillo, F.H. Froes and R. Schwarz, Pul. ASM Intern., Materials Park, 20-22 Sep. (1993), 269.
144. F. Cardellini, V. Contini and G. Mazzone, *Scripta Metall. Mater.*, Vol.32 (1994), 641.
145. J. Malagelada, S. Suriñach, M.D. Baró, S. Gialanella and R.W. Cahn, *Mater. Sci. Forum*, Vols.88-90 (1992), 497.

146. M.D. Baró, S. Suriñach and J. Malagelada, *Proc. of the 2nd Intern. Confer. on Structural Applications of Mechamical Alloying*, Vancouver, Canada, Ed.J.J. deBarladillo, F.H. Froes and R. Schwarz, Pul. ASM Intern., Materials Park, 20-22 Sep. (1993), 343.
147. S. Suriñach and J. Malagelada and M.D. Baró, *Mater. Sci. Eng.*, Vol.A168 (1993), 161.
148. G.T. Hida and I.L. Lin, *Combustion and Plasma Synthesis of High Temperature Materials*, Ed. J.B. Holt and Z.A. Munir, American Ceramic Soc., (1988), 246.
149. L. F. Mondolfo, Aluminium Alloys: *Structure and Properties*, (Butter Worths, London-Boston, 1976).
150. Y. Chen and J. Williams, *J. of Alloys and Compounds*, Vol.217 (1995) 181.
151. X. P. Niu, L. Froyen, L. Delaey and C. Peytour, *Scripta Metall. Mater.*, Vol.30 (1994) 13.
152. A.P. Radlinski and A. Calka, *Mater. Sci. Eng.*, Vol.A134, 1376.
153. D.D. Radev and D. Klisurski, *J. Alloys & Comp.*, Vol.206 (1994), 39.
154. Y.H. Park, H.Hashimoto and T. Abe, *Mater. Sci. Eng.*, Vol.A181/A182 (1994), 1291.
155. A. Teresiak, N. Mattern, H. Kubsch and B.F. Kieback, *NanoStructured Mater.*, Vol.4, (1994), 775.
156. J.Y. Huang, L.L. Ye, Y.K. Wu and H.Q. Ye, *Metall. Mater. Trans.*, Vol.26A (1995), 1.
157. Z.G. Liu, J.T. Guo, L.L. Ye, G.S. Li and Z.Q. Hu. *Appl Phys Lett*, Vol.65, 21 (1994), 2666.
158. A. Malchere and E. Gaffet, *Proc. of the 2nd Intern. Confer. on Structural Applications of Mechamical Alloying*, Vancouver, Canada, Ed.J.J. deBarladillo, F.H. Froes and R. Schwarz, Pul. ASM Intern., Materials Park, 20-22 Sep. (1993), 297.
159. A. Calka, J.I. Nikolov and B.W. Ninham, *ibid*, 189.
160. A. Calka, *Key Eng. Mater.*, Vols.81-83 (1993), 17.
161. A. Calka and J.S. Williams, *Mater. Sci. Forum*, Vols.88-90 (1992), 787.
162. Y. Ogino, T. Yamasaki, N. Atzumi and K. Yoshioka, *Mater. Trans., JIM*, Vol.34, 12 (1993), 1212.
163. Y. Chen and J.S. Williams, *Mater. Sci. Forum*, Vols.179-181 (1995), 301.
164. G. Broden, T.N. Rhodin, C. Brucker, R. Benbow and Z. Hurych, *Surface Sci.*, Vol.59 (1976), 593.
165. J.S.Benjamin and T.E. Volin, *Metall Trans*, Vol 5 (1974), 1930.
166. Ihsan Barin, *Thermochemical data of pure substances*, VCH, Verlagsagesellschaft, mbH, 1989.
167. Gero Walkowiak, Thomas Sell and Helmut Mehrer, *Z. Metalkd*, Vol.85 (1994) 5, 332.
168. William D.Callister, Jr, *Materials Science and Engineering*, 1991, John Wiley & Sons Inc.
169. L. Lu, M.O. Lai and S. Zhang, *J. Mater. Proc. Tech.*, Vol.48 (1995), 683.
170. T.Klassen, M.Ohering and R.Bormann, *J. Mater. Res*, Vol.9 (1994), 47.
171. Y.H. Park, H. Hashimoto, M. Nakamura, T. Abe and R. Watanabe, *Proc. of 1993 Powder Metall. World Congr.*, Ed. Y. Bando and K. Kosuge, Publ. Japan Soc. of Powder & Metall., (1993), 189.
172. J.J. Valencia, J.P. Lofvender, C. McCullough and C.G. Levi, *Mater. Sci. Eng.*, Vol.A144 (1991), 25.
173. F.H. Froes and C. Suryanarayana, *Physical Metall. and Proc. of Intermetallic Compounds*, Ed: N.S. Stoloff and V.K. Sika, Chapman & Hall (1994), 297.
174. A. Malchere and E. Gaffet, *Proc. of the 2nd Intern. Confer. on Structural Applications of Mechamical Alloying*, Vancouver, Canada, Ed.J.J. deBarladillo, F.H. Froes and R. Schwarz, Pul. ASM Intern., Materials Park, 20-22 Sep. (1993), 297.
175. R.B. Schwarz, *Scripta Mater.*, Vol. 34 (1996), 1.
176. A. Munir Zuhair, Ceram. Bulletin, Vol.67 (1988), 342.
177. R.M. Davis, B. McDermott and C.C. Koch, *Metall Trans.*, Vol 76 (1988), 281.
178. R.B.Schwartz and C.C. Koch, *Appl Phys Lett*, Vol.49 (1986), 146.
179. H.C. Yi and J.J. Moore, *J. Mater. Sci.*, Vol.25 (1990), 1159.
180. G.B. Schaffer and P.G. McCormick, *Metall. Trans.*, Vol.21A (1990), 2789.
181. A.K. Kuruvilla, K.S. Prasad, V.V. Bhanuprasad and Y.R. Mahajan, *Scripta Metall.*, Vol.24 (1990), 873.
182. M. Otsuka, T. Ishiara, M. Sugamata and J. Kaneko, *ibid*, 221.
183. E. Gaffet, P. Marco, M. Fedoroff and J.C. Rouchaud, *Mater. Sci. Forum*, Vols.88-90 (1992), 383.

184. E. Gaffet, F. Faudot and M. Harmelin, *Mater. Sci. Forum*, Vols.88-90 (1992), 375.
185. J.S. Benjamin, *Proc. of the Novel Powder Metall. World Congr.*, San Francisco, CA, USA, 21-26 Jane 1992, Pbl. Metal Powder Industries Federation, Princeton, NJ., Advances in Powder Metallurgy, Vol.7, 155.
186. W.E. Kuhn and A.N. Patel, *Modern Developments in Powder Metallurgy, Principles and Process*, Vol.12 (1980), Proc. of the 1980 Intern. Powder Metall. Confer., 22-27 June 1980, Ed. H.H. Hausner, H.W. Ants and G.D. Smith, Metal Powder Industries Federation, American Powder Metall. Inst., Princeton, NJ, 195.
187. *Metals Handbook*, Ninth edition, ASM Intern, Mater. Park, OH (1984), Vol.7, 56.
188. M.L. Ovecoglu and W.D. Nix, *High Strength of Powder Metallurgy Aluminium Alloys II*, Ed. G.J. Hildeman and M.J. Koczak, TMS, Warrendale, PA, (1986), 225.
189 X.P. Xiu, *Ph.D. thesis* at KULeuven, Belgium, (1991), 34.
190. A. Arias, *Chemical Reactions of Metal Powders with Organic and Inorganic Liquids during Ball Milling*, NASA TN D-8015, (1975).
191. L. Lu, M.O. Lai and S. Zhang, *Key Eng. Mater.*, Vol.104-107 (1995), 111
192. M.S.El-Eskandarany, K. Sumiyama, K. Aoki and K. Suzuki, *Mater. Sci. Forum*, Vol.88-90 (1992), 801.
193. F.R. de Boer, R. Broom, W.C.M. Mattens, A.R. Miedema and A.K. Niessen, *Cohesion in Metals*, North-Holland Physics Publ., Amsterdam, (1988), 110.
194. I.A. Ibrahim, F.A. Mohamed and E.J. Lavernia, *J. Mater. Sci.*, Vol.26 (1991), 1137.
195. V.B. Velidandla et al, *J. Powder Metall.*, Vol.27 (1991), 227-235.
196. Y.B. Liu, J.K.M. Kwok, S.C. Lim, L. Lu and M.O. Lai, *J. Mater. Proc. Tech.*, Vol.37 (1993), 441-451.
197. L. Lu, J.K.M. Kwok, M.O. Lai and S.C. Lim, Lai, *J. Mater. Proc. Tech.*, Vol.37 (1993), .453.
198. L. Lu and M.O. Lai, *Mater. Sci. Tech.*, (1996), Vol.13 (1997), 202.
199. F.H. Froes, *Structural Applications of Mechanical Alloying*, Ed: F.H. Froes and J.J. deBarbadillo, Proceedings of an ASM Intern. Conf., Myrtle Beach, South Carolina, 27-29 March 1990, publ. by: ASM Intern., Mater. Park, Ohio, 1.
200. S.J. Hong and P.W. Kao, *Mater. Sci. Eng.*, Vol. A148 (1991), 189.
201. L. Lu, M.O. Lai and S. Zhang, *Mater. Sci. Tech.* Vol.10 (1994), 319.
202. D.G. Morris and M.A. Morris, *Mater. Sci. Eng.*, Vol.A104 (1988), 201.
203. G. Cocco, S. Enzo, L. Schiffini and L. Battezzati, Mater. Sci Eng., Vol97 (1988), 43.
204. H.J. Fetch, E. Hellstern, Z. Fu and W.L. Johnson, Adv. in Powder Metall.-1989, Comp. T.G. Gasbarre and W.F. Jandeska, Jr., Publ. Metal Powder Industries Federation, Princeton, N.J., Vol.2 (1989), 111.
205. F.H. Froes and C. Suryanarayana, *JOM*, Vol.41 (1989), 12.
206. E. Hellstern, H.J. Fetcht, C. Garland and W.L. Johson, *Multicomponent Ultrafine Microstructures*, Ed. L.E. McCandlish, D.E. Polk, R.W. Siegel and B.H. Kear, Publ. Mater. Res. Soc., Pittsburgh, PA, Vol.132 (1989), 137.
207. A.R. Yavari, *Mater. Sci. Eng.*, Vol.A179/180 (1994), 20.
208. H.J. Fecht, *Nanophase Materials, Synthesis-Properties-Applications*, Ed: G.C. Hadjipanayis and R.W. Siegel, Pbl: Kluwer Academic Publishers, (1994), 125.
209. J. Karch, R. Birringer and H. Gleiter, *Nature*, Vol. 330 (1987), 556.
210. P.S. Gilman and W.D. Nix, *Metall. Trans. A*, Vol.12A (1981), 813.
211. N. Burgio, W. Guo, M. Magini and F. Padella, *Proc. an ASM Intern. Confer.*, Myrtle Beach, South Carolina, 27-29 March 1990, Ed: F.H.Froes and J.J. deBarbadillo, ASM Intern. Mater. Park, Ohio (1990), 175.
212. M. Otsuka, T. Ishiara, M. Sugamata and J. Kaneko, *ibid*, 221.
213. J.S. Benjamin, *Proc. of the Novel Powder Metall. World Congr.*, San Francisco, CA, USA, 21-26 Jne 1992, 155.
214. T.D. Shen and C.C. Koch, *Mater. Sci. Forum*, Vols.179-181 (1995), 17.
215. J. Echert, J.C. Holzer, C.E. Krill, III and W.L. Johnson, *J. Mater. Res.*, Vol.7 (1992), 1751.
216. T. Takeshita and R. Nakayama, *10th Intern. Workshop on Rare-Earth Magnets and Their Appl.*, Kyoto, (1989), 551.
217. I.R. Harris and P.J. McGuiness, *J. Less. Common Met.*, Vols.172-174 (1992), 243.

218. P.J. McGuiness, C. Short, A.F. Wilson and I.R. Harri, *J. Alloys Comp.*, Vol.184 (1992), 243.
219. K. Aoki, Y. Itoh and T. Masumoto, *Scripta Metall. Mater.*, Vol.31 (1994), 1271.
220. M. Baricco and L. Battezzati, *Mater. Sci. Forum*, Vols.179-181 (1955), 597.
221. M.L. Trudeau, *Nanophase Mater. -Synthesis-Properties-Applications*, Ed. G.C. Hadjipanayis and R.W. Siegel, Pul. Kluwer Academic Publishers, (1994), 153.
222. J. Eckert, L. Schultz and K. Urban, *J. Mater. Sci.*, Vol.26 (1991), 441.
223. C.C. Koch, O.B. Calvin, C.G. Mckamey and J.O. Scarbrough, *J. Appl. Phys. Lett.*, Vol.43 (1983), 1017.
224. R.B. Schwarz, R.R. Petrich and C.K. Saw, *J. Non-Cryst. Sol*, Vol.76 (1987), 281.
225. E. Hellstern and L. Schultz, *J. Appl. Phys.*, Vol.63 (1988), 1408.
226. R.B. Schwarz and W.L. Johnson, *Phys. Rev. Lett.*, Vol.41 (1983), 415.
227. E. Hellstern and L. Schultz, *Mater. Sci. Eng.*, Vol.93 (1987), 213.
228. J.A. Hunt, I. Soletta, S. Enzo, L. Meiya, R.L. Havill, L. Battezzati, G. Cocco and N. Cowlam, *Mater. Sci. Forum*, Vols.179-181 (1995), 255.
229. T. Nasu, K. Nagaoka, M. Sakurai and K. Suzuki, *Mater. Sci. Forum*, Vols.179-181 (1995), 97.
230. F. Petzoldt, *J. Less-Common Met.*, Vol.140 (1988), 85.
231. I.A. Ovid'ko, J. Phys. D: Appl. Phys., Vol.24 (1991), 2190.
232. G. Cocco, S. Enzo, L. Schiffini and L. Battezzati, *New Materials by Mechanical alloying Techniques*, DGM Confer., Calw-Hirsau (FRG), Oct. 1988, Ed. E. Arzt and L. Schultz, Publ. Informations-gesellschaft Verlag, (1989), 343.
233. L. Schultz, *Mater. Sci. Eng.*, Vol.97 (1988), 15.
234. E. Gaffet, *Mater. Sci. Eng.*, Vol.A135 (1991), 291.
235. W.L. Johnson, *Mater. Sci. Eng.*, Vol.97 (1988), 1.
236. L. Schultz, *New Materials by Mechanical alloying Techniques*, DGM Confer., Calw-Hirsau (FRG), Oct. 1988, Ed. E. Arzt and L. Schultz, Publ. Informations-gesellschaft Verlag, (1989), 53.
237. A.K. Niessen, F.R. de Boer, P.F. de Chatel, W.C.M.. Mattens and A.R. Miedema, *CALPHAD* 7, (1983), 51.
238. A.R. Miedema and A.K. Niessen, *Trans. Japan Inst. Metals*, Vol.29 (1988), 209.
239. A.K. Niessen and A.R. Miedema, *Ber. Bunsenges Phys. Chem.*, Vol.87 (1983), 717.
240. A.W. Weeber, *J. Phys. F.*, Vol.17 (1987), 809.
241. D.L. Beke, P.I. Loeff and H. Bakker, *Acta Metall.*, Vol.39 (1991), 1259.
242. D.L. Beke, H. Bakker and P.I. Loeff, *Acta Metall.*, Vol.39 (1991), 1267.
243. L.J. Gallego, J.A. Somoga and J.A. Alonso, *J. Phys.: Condens Matter*, Vol.2 (1990), 281.
244. B.S. Murty, S. Ranganathan and M. Mohan Rao, *Mater. Sci. Eng.*, Vol.A149 (1992), 231.
245. B.S. Murty, M.D. Naik, M. Mohan Rao and S. Ranganathan, *Mater. Forum*, Vol.16 (1992), 19.
246 . K. Tkacova, *InCoMe'93*, Proc. of The 1st Inter. Confer. on Mechanochemistry, Kosice, March 23-26, 1993, Ed. K. Tkacova, Publ. Cambridge Interscience Publ., UK , Vol.1 (1994), 9.
247. K.S. Suslick, *ibid*, 43.
248. V.V. Boldyrev, *ibid*, 18.
249. G.B. Schaffer and P.G. McCormick, *Appl. Phys. Lett.*, Vol.55 (1989), 45.
250. G.B. Schaffer and P.G. McCormick, *Scripta Metall.*, Vol.23 (1989), 835.
251. M.V. Zduji, O.B. Miloevi and Lj. S. Karanovi, *Mater. Lett.*, Vol.13 (1992), 125.
252. L. Takcas, *Mater. Lett.*, Vol.13 (1992), 119.
253. G.B. Schaffer and P.G. McCormick, *Metall. Trans.*, Vol.22A (1991), 3019.
254. L. Takcas, Appl. *Phys. Lett.*, Vol.69 (1996), 15.
255. G.B. Schaffer and P.G. McCormick, *J. Mater. Sci.*, Vol.9 (1990), 1014.
256. L. Takacs, *J. Appl. Phys.*, Vol.75 (1994), 5864.
257. L. Takacs, *Processing and Properties of Nanocrystalline Materials*, Ed. C. Suryanarayana, J. Singh and F.H. Froes, The Minerals, Metals and Mater. Soc., Warrendal, PA, 1996, 453.
258. L. Takacs, *Mater. Res. Soc. Symp. Proc.*, Vol.286 (1993), Publ. Mater. Res. Soc., 413.
259. L. Takacs, *Nanostructured Mater.*, Vol.2 (1993), 241.

# 5

# CHARACTERIZATION OF POWDERS

## 5.1 Particle analysis

Size, distribution of size and surface area of powder particles are important parameters in mechanical alloying and milling. They will influence the chemical reaction not only during milling but also post processing.

### *5.1.1 Particle size*

The size of a spherical particle is in general defined by its diameter. The mechanically milled powder particles, however, contain not just spherical particles but particles of other shapes as well. In the latter case, the size of the particles has to be defined differently. There exist several common methods to define such particle size [1]:

(a) Volume diameter, $d_V$:

This is the diameter of a sphere having the same volume, $V$, as the particle:

$$V = \frac{\pi}{6} d_v^3 \qquad [5.1]$$

(b) Surface diameter, $d_S$:

This is the diameter of a sphere having the same surface area, $S$, as the particle:

$$S = \pi d_S^2 \qquad [5.2]$$

(c) Surface volume diameter, $d_{sv}$:

This is defined as the diameter of a sphere having the same external surface to volume ratio as a sphere:

$$d_{sv} = \frac{d_v^3}{d_S^2} \qquad [5.3]$$

(d) Drag diameter, $d_d$,:

This is the diameter of a sphere having the same resistance to motion as the particle in a fluid of the same viscosity and at the same velocity:

$$F_D = C_D A \rho_f \frac{V^2}{2} \qquad [5.4]$$

where $\rho_f$ is the density of the fluid, $A = \frac{C_s}{C}$, $C_s$, the solid concentration in the sampled gas, $C$, the solids concentration in the gas stream and $C_D A = f(d_d)$.

(e) Sieve diameter:

This is the width of the minimum square aperture through which the particle will pass.

### *5.1.2 Measurement of particle size*

Several methods have generally been used for the measurement of particle size. Each method has its own advantages and disadvantages. Since each measurement technique is based on different principle, some discrepancy may arise. Especially for particles of irregular shapes, the assigned size is usually dependent upon the method of measurement, this is, the size of a particle is particle sizing technique dependent.

#### *5.1.2.1 Settling method*

If an irregularly shaped particle is allowed to settle in a liquid (which should be selected carefully), its terminal velocity may be compared with that of a standard spherical particle with the same density under the same flow condition. If their terminal velocities are the same, then the size of the irregularly shaped particle is considered equal to the diameter of the spherical particle [(1)]. This is one of the easiest methods to determine the effective size of a particle. In the laminar flow region, the particle moves with random orientation, but outside this region it orientates itself to give minimum resistance to motion so that the free-falling diameter for an irregular particle is greater in the intermediate region than that in the laminar flow region. A unique relationship between drag factor and Reynolds number can lead to a simple Stokes equation relating settling velocity and particle size applicable at low Reynolds number. The free-falling diameter in the laminar flow region becomes the Stokes diameter. Stokes' equation can be used for spherical particles up to a Reynolds number of 0.2 at which value it will give an under-estimation of the diameter of about 2%. At higher Reynolds numbers, corrections have to be applied. Corrections may also be needed for non-spherical particles, so that the derived diameter may be independent of the settling conditions and hence purely a function of the particle size.

The motion of a sphere can be written as a function of the drag force, $F_D$ [1]:

$$F_D = \frac{\pi}{6}\left(\rho_s - \rho_f\right) g D^3 \qquad [5.5]$$

where $\rho_s$, density of the powder particle, $D$, particle diameter, and $g$, the acceleration due to gravity.

The Reynolds number, $R_e$, is defined by:

$$R_e = \frac{\rho_f u D}{\eta} \qquad [5.6]$$

where $\eta$ is the viscosity of the fluid and $u$, the particle velocity

The drag coefficient can be given as:

$$C_D = \frac{F_D}{\pi \frac{D^2}{4} \rho_f \frac{u^2}{2}} \qquad [5.7]$$

where $C_D$ is the drag coefficient.

When the terminal velocity is reached, the drag on a spherical particle falling in a viscous fluid of infinite extent is due entirely to viscous forces. The relationship between the size of the particles and other parameters can be represented by Stokes' equation:

$$D = \left[ \frac{18\eta u_{St}}{\left(\rho_s - \rho_f\right)g} \right]^{1/2} \qquad [5.8]$$

where $u_{St}$ is the terminal velocity which can be represented by the settling height $H = u_{St}t$.

*5.1.2.2 Sieving*

Sieving is an effective means to measure particle size. For sieve measurement technique, certain quantity of particles and a variety of sieve apertures ranging from millimeter down to few microns are required. Sieve is often referred to by its mesh size determined by the number of wires per linear inch. Sieve size is the minimum square aperture through which the particle can pass. Fractionation by sieving is a function of two dimensions only, maximum breadth and maximum thickness. The apertures of a sieve may be regarded as a series of gauges which reject or pass the particles. Hence, errors may arise due to the size and shape of the particles, shaking method, duration of shaking and physical properties of the particles.

*5.1.2.3 Laser sizing*

In recent years, laser diffraction principle has become one of the most important methods of analysis for the determination of particle size distributions. The flexibility of the method makes it usable both in laboratory as well as in industrial scales. Laser diffraction is an optical measuring method where a laser beam emitted by laser diode is collimated and directed at sample particles suspended in the liquid flowing in a flow-thru cell. The beam diffracted or scattered by the sample particles is detected by different types of sensor. The method is suitable for use in rapid off-line and on line analysis and is usually applied to particle size range between submicron to millimeter.

### *5.1.3 Surface area analysis*

Understanding the surface area of the particles is of importance since it correlates with various of their kinetic and geometric characteristics. Surface area provides insight into the behaviour of the powder during chemical activity, catalysis, friction, adsorption and sintering [(2)]. Surface area measurement is an average measurement of the external condition of a large number of particles. For some powder particles, it may be more advantageous to measure the total surface area instead of the particle size.

*5.1.3.1 Measurement*

There are varieties of surface measuring apparatus even though the principle of the measurement remains approximately the same. To determine the surface area, the powder particles are pretreated by applying some combination of heat, vacuum and/or flowing gas to remove any adsorbed contaminants acquired from atmospheric exposure. The solid is then cooled under vacuum. An adsorbate gas is added incrementally in doses to achieve one or more data points. After each dose of adsorptive, the pressure is equilibrated and desorption isotherm data are obtained during the progressive removal of adsorbed gas. When the constant pressure, volume and temperature attain an equilibrium, the amount of gas is again calculated. The difference between the amount of gas present initially and finally represents the adsorbate lost from the gas phase to the adsorbed phase. The volume of gas adsorbed by the powder is then determined and expressed as a function of relative pressure. In order to characterize surface area, one or more adsorption data points are needed. Figure 5. 1 schematically shows an isotherm plot.

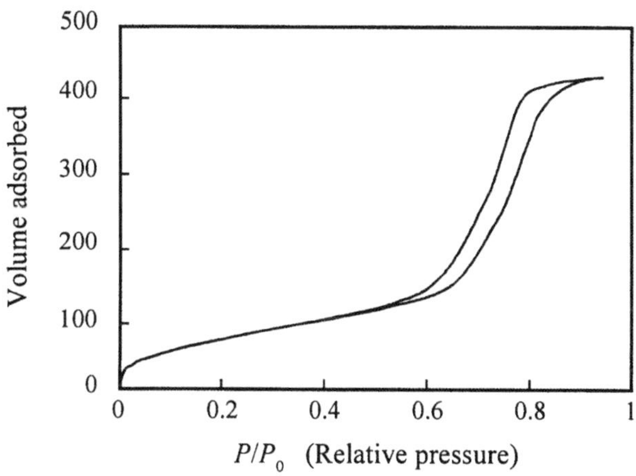


Figure 5.1. Adsorption-desorption isotherm plot.

Nitrogen is often used as an adsorptive while samples with low surface area may require the use of other gases such as krypton or argon.

There are usually four stages in the measurement. At stage 1, isolated sites on the sample surface begin to adsorb gas molecules at low pressure. At stage 2, as gas pressure increases, coverage of gas molecules increases to form a monolayer (one molecule thick). For this case, BET (Brunauer, Emmett and Teller) [3] equation is used to calculate the surface area. At stage 3, further increase in gas pressure causes the beginning of multilayer coverage. Smaller pores in the sample will be filled first. Further increase in pressure leads to the forth stage at which the gas pressure causes complete coverage of the sample and fill all the pores. Pore size, volume and distribution of the powder may then be determined using BJH (Barret, Joyner and Halenda) equation.

*5.1.3.2 Determination of surface area*

Gas adsorption approach is commonly used for the measurement of surface area. The amount of adsorption can be calculated using the BET equation which is based on the assumption that the forces producing condensation are responsible for the binding energy of multimolecular adsorption. Generalization of the ideal localized monolayer treatment is effected by assuming that each first layer of adsorbed molecule serves as a site for the adsorption of molecule into the second layer and so on (1). At equilibrium, the rate of adsorption equals the rate of evaporation. The surface area of the powder can be calculated from the adsorption behaviour, if each gas molecule occupies a precise area. The relationship between partial pressure, $P$, of adsorbate, the saturation pressure, $P_0$, of the adsorbate and the amount of gas $X$ adsorbed can be written as (2):

$$\frac{P}{X(P_0 - P)} = \frac{1}{X_m C}\left[1 + \frac{P}{P_0}(C - 1)\right] \qquad [5.9]$$

where $X_m$ is the monolayer capacity of the powder and $C$, a constant associated with the adsorption enthalpy. The ratio of $P/P_0$ is normally between 0.05 and 0.30.

The specific surface area, $S_t$, is obtained by summing the areas occupied by all of the adsorbed molecules needed to form the monolayer:

$$S_t = \frac{X_m N_A A_0}{wM} \qquad [5.10]$$

where $M$ is the molecular weight of the adsorbate, $A_0$, the average area occupied by an adsorbate molecule, $N_A$, Avogadro's number, $w$, the weight of powder measured and $X_m = (A+B)^{-1}$. $A$ and $B$ may be obtained from the following equation:

$$\frac{P}{X(P_0 - P)} = B + A\frac{P}{P_0} \qquad [5.11]$$

If $P/[X(P_0-P)]$ is plotted against $P/P_0$, $A$ and $B$ may be obtained from the intercept and the slope respectively.

Volumetric adsorption analysis may be subjected to many sources of errors. Although most of the systematic errors result in only small errors in the surface area measurement, some errors do significantly affect the accuracy and repeatability of both BET surface area and the pore size [4].

## 5.2 Thermal analysis

When a chemical reaction or a structural change occurs in the mechanically milled powder particles, the change may be monitored by thermal analysis using instruments such as differential thermal analysis (DTA) and differential scanning calorimeter (DSC). In principle, the DTA and DSC measure the temperature difference between a reference material and a sample. Therefore any heat released or absorbed from the measured sample can be recorded since there is no change in heat content in the reference material within the temperature range applied.

### *5.2.1 Application of thermal analysis*

Since ball milling is a process in which heavy plastic deformation, cold welding and fracturing take place simultaneously, extremely fine structure down to nano-size with highly disorder structures can be found. Thermal analysis is widely used in the study of reaction, stability and recrystallization of the mechanically milled powder particles. Very often, the analysis is employed to monitor the crystallization temperature of amorphous phases of mechanically milled powders [5-8]. Disordered structures have been known to occur in mechanically milled powder. The disordered structures are metastable which will experience an ordering transition during heating resulting in exothermic reaction. The thermodynamics and kinetics of the ordering transition from disordered structure may be measured with the help of thermal analysis [9-12]. Schwarz *et al.* [13] measured the mechanically milled $Al_3Ti$ powders using a DSC and observed an exothermic peak at about 400K during heating. However, if the annealed powder was measured, no peak could be detected. Since equilibrium $Al_3Ti$ phase has a $Ll_2$ ordered structure where as mechanically milled $Al_3Ti$ has disordered structure, heating the powder will drive the ordering transition with heat release. Further heating caused $Ll_2$ to transform $DO_{23}$ transition. Schwarz *et al.* [13] found that the heat of transformation for the reaction from $Ll_2$ to $DO_{23}$ structure decreased as milling duration increased. The same composition of material was also studied by many other researchers. When powder milled for short duration was measured, the DSC curves showed an endothermic peak at 933K. However, after a certain duration of milling, say, 960 minutes, a sharp exothermic peak appeared with the disappearance of the endothermic reaction. The endothermic peak indicated the melting of Al at low milling duration while the exothermic peaks indicated the ordering transition.

Malagelada *et al.* [10] studied the thermal behaviour of disordered $Ni_3Al$ milled from ordered $Ni_3Al$. They could not find any peak for the as-received powder. Two exothermic peaks were, however, observed from the milled powders. The enthalpy released was seen to increase up to 1,200 minutes of milling without saturation. The peak temperatures in the DSC measurement provide the necessary information for the isothermal or annealing treatment.

When formation of a new compound takes place from a mechanically alloyed powder during heating in a DTA or DSC cell, the heat released will give rise to exothermic peak. Figure 5.2 shows the DSC measurement of mechanically alloyed Mg-Si mixture in which an exothermic peak appears at about 460K. No exothermic reaction is observed for the as received Mg-Si mixture. To confirm the reaction of the alloyed powder, x-ray measurement can be carried out before and after the isothermal annealing at 463K for 60 minutes. X-ray diffraction spectra reveal weak $Mg_2Si$ peak at 24° for the powder mechanically alloyed for 600 minutes before annealing. This is an indication that the DSC peak at 460K is due to the formation of $Mg_2Si$. The activation energy of this transition can be calculated from to the shift of the peaks when heating rates are varied [(14-16)].

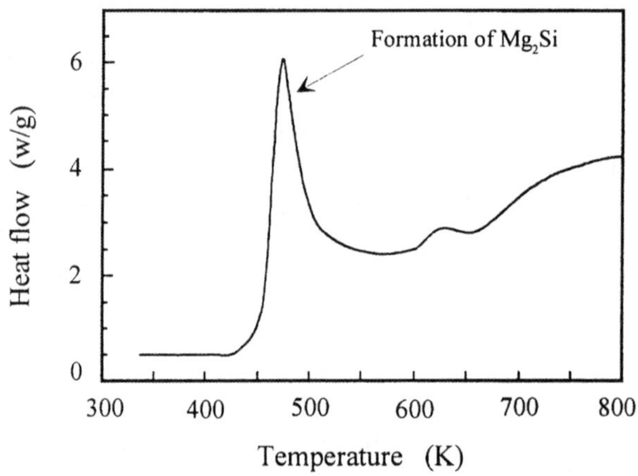


Figure 5.2. DSC measurement of mechanically alloyed Mg-Si mixture.

From the above discussions, it may be seen that thermal analysis is a powerful means to monitor as well as to study the change in physical properties of the mechanically alloyed and milled powder particles.

### *5.2.2 Activation energy*

Since 1955, innumerable methods have been published concerning the determination of data for reaction kinetics from thermodynamic measurements. The methods of Kissinger [17], Ozawa [18] and ASTM E698 standard have achieved particular recognition.

To determine the reaction mechanism, the values of the degree of transformation have been introduced into the reaction equations. There are several types of reaction models based on different considerations. Among them, Wilhelmy model is often used:

$$\frac{d\alpha}{dt} = k(1-\alpha)^n \qquad [5.12]$$

where $d\alpha/dt$ is the reaction rate in $s^{-1}$, $k$, rate constant in $s^{-1}$, $\alpha$, extent of reaction and $n$, the order of reaction. At the start of a chemical reaction ($\alpha$=0), the reaction rate is identical to the rate constant.

The equation required to calculate isothermal conversion as a function of the reaction time can be obtained from the integration of Eq.[5.12]. When $n$ = 1, the fraction of transition at a constant temperature is:

$$\alpha = 1 - \exp(-kt) \qquad [5.13]$$

For $n \neq 1$,

$$\alpha = 1\left(kt(n-1)+1\right)^{1/(1-n)} \qquad [5.14]$$

In most solid reactions, the rate constant $k$ can be described by the Arrhenius equation [19]:

$$k = k_0 \exp\left(-\frac{Q^*}{RT}\right) \qquad [5.15]$$

where $k_0$, in $s^{-1}$, is frequency factor or rate constant at infinite temperature, $Q^*$, activation energy in J $mol^{-1}$, $R$, gas constant and $T$, temperature of the reaction in K.

Activation energy, $Q^*$, may be calculated using measurements from differential scanning calorimeter or differential thermal analysis. Based on Kissinger [17] equation, the calculation procedures using a suitable thermal instrument may be carried out as follows:

(a) Run the thermal analysis using a DSC or DTA on the milled powder using a minimum of four different heating rates. Normally the heating rates can be selected from 1 to 40°/min. Measure the peak temperatures of the transition of the powders as shown in Figure 5.3.

(b) Plot the data on a $-\ln(\dot{T} / T_p^2)$ versus $1000/T_p$ plot where $\dot{T}$ is the heating rate and $T_p$, the peak temperature, as shown in Figure 5.4. The slope of the linear line that passes through the data is equal to $Q^*/R$, where $R$ is gas constant.

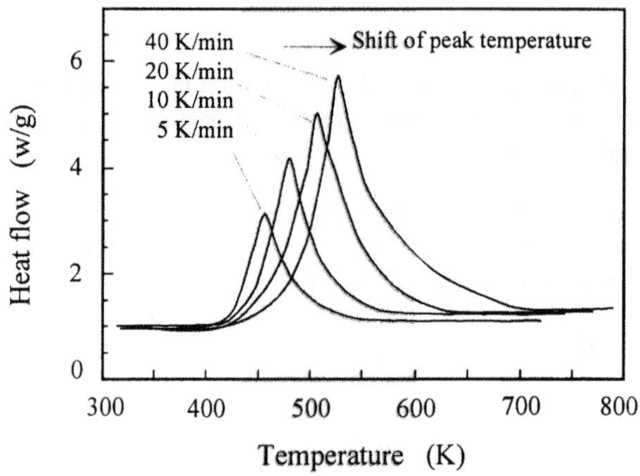


Figure 5.3. DSC measurement of mechanically milled powders.

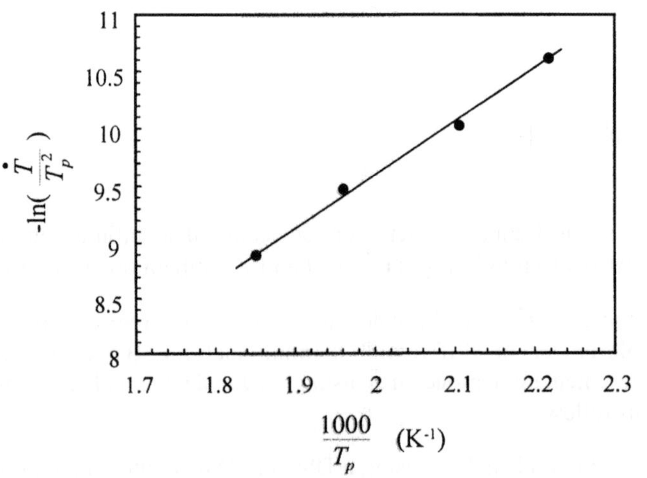


Figure 5.4. A plot for the estimation of activation energy of transition.

Accuracy of the kinetic calculations is limited by several factors such as:

(a) Systematic error of the instruments may often be observed.

(b) Two or more thermal reactions may overlap or even superimpose.

(c) The model has been developed for homogeneous reactions in solution and is of limited use in solid-state reactions.

The empirical activation energy $Q^*$ is an important experimental parameter. It gives the temperature dependence of transformation rate. Estimates of the magnitudes of activation energy are useful in understanding the mechanisms of transformation.

## 5.3 Determination of crystalline size using x-ray technique

### *5.3.1 Broadening of x-ray diffraction peaks*

X-ray diffraction method, a simple and easy way to identify crystalline size, is widely accepted in the study of crystalline size of mechanically alloyed powder particles. The determination of crystalline size is based on the concept of x-ray line broadening after mechanical alloying.

Peak broadening at half height intensity of the x-ray diffraction patterns is induced by a reduction in crystalline size, faulting and microstrains within the diffracting domains. In mechanical alloying or milling, the powder particles are subjected to plastic microforging. Dislocations and other defects are generated resulting in an increase in internal energy. At the same time dislocation arrays such as all angle boundaries which subdivide the original grains into small coherent domains or crystallites, are produced. The dislocations and their associated stress fields produce microstrains within the coherently diffracting domains. Cold work also produces faulting which may be single layer deformation stacking faults, n-layer deformation stacking faults or twin faults. They may be accompanied by long range elastic strains, changes in lattice parameters and layer spacing. In the calculation of crystalline size, it may be assumed that the long-range elastic stresses are zero insofar as the sensitivity of low order peaks is concerned [(20)]. The microstrains associated with the domains may be conceived as strain distributions about dislocations and dislocation networks which have an isotropic stress distribution. Figure 5.5 shows an example of a broadened x-ray spectrum.

Beside the influence of crystalline size and strain field on the broadening of an x-ray peak, the influence of x-ray diffractometer on the broadening has to be considered.

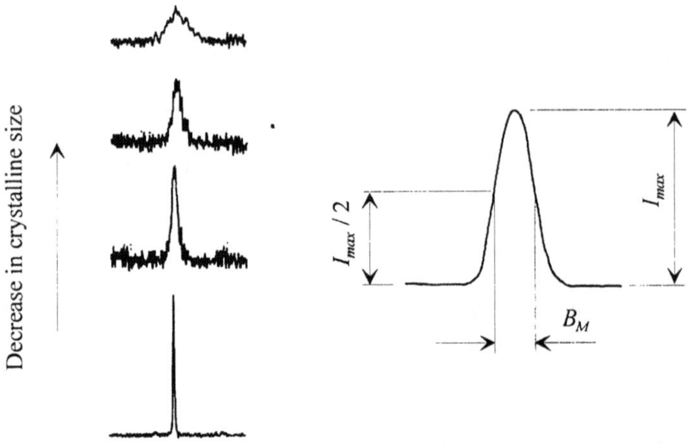


Figure 5.5. Broadened x-ray spectrum.

### *5.3.2 Scherrer equation*

Crystalline size can be calculated by the Scherrer equation [21]:

$$\beta = \frac{\varsigma\lambda}{d_c \cos\theta} \qquad [5.16]$$

where β [rad] is the broadening of the diffraction peak due purely to crystalline size measured at half its maximum intensity, $d_c$, diameter of crystalline size and ς, constant. The constant ς depends mainly on the crystalline shape and the indices. It has a value of about 0.9. The validity of Scherrer's equation has been subsequently borne out by different investigators [22-24]. In recent years, this equation is widely accepted for the evaluation of crystalline size down to nanometer.

The key in determining crystalline size of powder particles from peak broadening is to determine $d_c$ purely from the pure breadth β of the diffraction peak. However, it must be borne in mind that several factors can contribute to peak broadening, namely, decrease in crystalline size, internal strain and broadening due to the x-ray machine itself. After large plastic deformation of the powder particles, slip occurs in each grain and the grain changes its shape to become flattened, elongated or fractured. The change in shape of any one grain is determined not only by the forces applied to the power particles but also by the fact that each grain retains contact on its boundary surfaces with all its neighbours. Because of this interaction between grains, every grain in the particle is not free to deform the same way as in an isolated single crystal. As a result of the restraint by its neighbours, a plastically deformed grain in the particle aggregate usually has regions of its lattice left in an elastically bent or twisted condition or in a state of uniform tension or compression. The residual strain will remain in the powder particles after mechanical milling. The effect of uniform residual strain will lead to a parallel shift in the x-ray diffraction peaks, while the nonuniform residual strain will lead to the broadening of x-ray diffraction peaks.

In Scherrer's original investigation, the broadening parameter $B$ due purely to change in crystalline size and strain was obtained from the measured broadening of the diffraction peak, $B_M$ (Figure 5.5 (b)) and the measured breadth $B_I$ due to instrument at half maximum intensity of the peak. The broadening $B_I$ is measured under similar geometrical conditions using a material with a crystallite size well in excess of 100 nm. The broadening parameter $B$ can be obtained from [25]:

$$B = B_M - B_I \qquad [5.17]$$

### *5.3.3 Instrumental broadening*

Many methods have been proposed to make the correction on broadening due to instrument. Among these, Warren's is the simplest and a more accurate method. It states that,

$$B^2 = B_M^2 - B_I^2 \qquad [5.18]$$

Eq.[5.18] indicates that the square of the breadth of the measured peak at half maximum intensity is equal to the sum of the squares of the breadths of the peak due purely to diffraction broadening and the broadening due to the x-ray spectrometer.

In order to obtain $B_I$, a well-annealed specimen with grain size above 1 μm should be used for the x-ray diffraction measurement. Since there is no residual strain in a well-annealed specimen, the broadening measured can be assumed to be purely due to the instrument employed.

### *5.3.4 Residual strain broadening*

Non-uniform strain as a result of heavy deformation will lead to a broadening of x-ray diffraction peaks. The broadening due to non-uniform strain can be written as [26]:

$$B_s = -2\frac{\Delta d}{d}\tan\theta = C\tan\theta \quad [5.19]$$

where $B_S$ is the broadening of the x-ray peaks due to residual strain, $d$, the lattice spacing and $\Delta d$, the change in lattice spacing and $C = -2\Delta d/d$.

Since the broadening parameter, $B$ is due both to strain effect and size effect, $B$ can be written as:

$$B = B_S + \beta \quad [5.20]$$

Substituting Eqs.[5.16] and [5.19] into Eq.[5.20], it follows that [27]:

$$B = C\tan\theta + \frac{\varsigma\lambda}{d_c\cos\theta} \quad [5.21]$$

Substituting Eq.[5.18] into Eq.[5.21] and multiplying cosθ to both sides of the equation, Eq.[5.21] can be rewritten as:

$$\sqrt{B_M^2 - B_I^2}\cos\theta = C\sin\theta + \frac{\varsigma\lambda}{d_c} \quad [5.22]$$

### *5.3.5 Determination of crystalline size*

To accurately identify crystalline size, the following procedure is normally taken:

(a) Determination of broadening due to x-ray spectrometer

The powder should be well annealed to minimize strain field in the powder particles. After annealing, measurement of x-ray diffraction should be carried out on the well-annealed particles. The width broadening at half the maximum intensity of the peak is considered due to the influence of x-ray diffractometer itself.

(b) Measurement of x-ray diffraction of the mechanically alloyed powder

The second step is the measurement of x-ray diffraction on the mechanically alloyed powder particles. The width at half maximum intensity of the peak is then recorded for the calculation of crystalline size.

(c) Calculation of crystalline size

The values of breath measurement obtained from at least two to three peaks are taken to calculate the crystalline size using Eq.[5.22].

In Eq.[5.22], $B_M$ and $B_I$ can be measured from the breadth of the x-ray diffraction peaks while $C$ and $d_c$ are unknown. To solve the equation graphically, $\sqrt{B_M^2 - B_I^2}\cos\theta$ may be plotted as a function of $\sin\theta$, namely:

$$Y = C\sin\theta + A \qquad [5.23]$$

where $Y = \sqrt{B_M^2 - B_I^2}\cos\theta$ and $A = \dfrac{\varsigma\lambda}{d_c}$.

The crystalline size $d_c$ can then be calculated from the intercept $A$ of the straight line from Eq.[5.23] while the value of strain $C$ can be obtained from the slope of the straight line. Figure 5.6 shows the plot which is called Hall-Williamson plot. It is widely used in the analysis of crystalline size of mechanically milled powders.

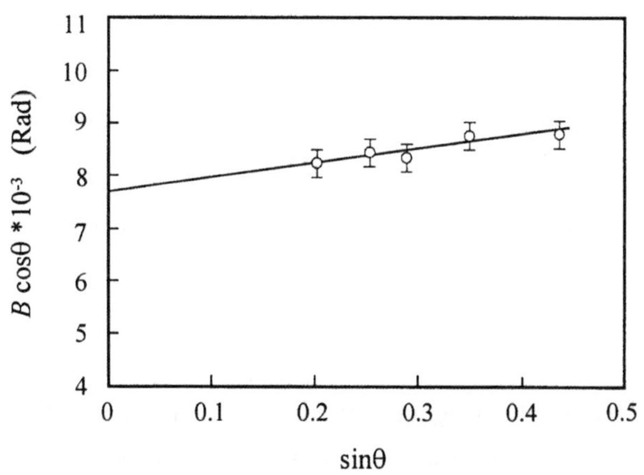


Figure 5.6. Hall-Williamson plot.

## 5.4 Quantitative analysis using x-ray diffractometry

Very often quantitative analysis of the mechanically milled powder may be required. One commonly used method is the use of x-ray diffraction technique. This technique is associated with the measurement of intensity of the diffraction pattern of a particular phase in a mixture of phases.

### *5.4.1 Fundamentals*

The intensity of x-ray diffraction peak of α phase may be given by [28]

$$I_{(hkl)\alpha} = \left[\frac{I_0\lambda^3}{64\pi r}\left(\frac{e^2}{m_e c^2}\right)^2\right]\left[\frac{M_{hkl}}{V_\alpha^2}\left|F_{(hkl)\alpha}\right|^2\left(\frac{1+\cos^2 2\theta\cos^2 2\theta_m}{\sin^2\theta\cos\theta}\right)\right]\left[\frac{X_\alpha}{\rho_\alpha(\mu/\rho)_s}\right]$$

[5.24]

where $I$ is the diffraction intensity of x-ray diffraction, $I_0$, intensity of incident-beam, $r$, distance from the specimen to the detector, λ, wave length of the x-ray radiation, $[e^2/(m_e c^2)]^2$, the square of the classical electron radius, μ, linear attenuation coefficient of the specimen, $M_{hkl}$, multiplicity for reflection $hkl$ of phase α, $V_\alpha$, value of the unit cell of phase α, $F_{(hkl)\alpha}$, the structure factor for reflection $hkl$ which includes anomalous scattering and temperature effects, θ, the angle of incident beam, $\theta_m$, the Bragg angle of the monochromator crystal, ρ, density and $X_\alpha$, the weight fraction of phase α. If the specimen contains $n$ phases, Eq.[5.24] may be written in a shorter form as

$$I_{(hkl)j} = \frac{C_j f_j}{\mu'} \qquad [5.25]$$

where $C_j$ is constant, $f_j$, fraction of $j$ phase and μ', linear attenuation coefficient of the $n$ phase mixture.

If the specimen contains only two phases, the relative proportion of the phases may be calculated by comparing the integrated intensities of the two phases:

$$\frac{I_A}{I_B} = \frac{R_A f_A}{R_B f_B} \qquad [5.26]$$

where $I_A$ and $I_B$ are the integrated intensities of phase A and B, $f_A$ and $f_B$ are the volume fraction of phase A and B, and $R_A$ and $R_B$ are constant.

This method has been used to quantitatively estimate the fraction of the phases in two phase materials [29].

### *5.4.2 Method of standard additions*

The method of standard additions has been widely used in quantitative analysis of powder particles. In this method, the diffraction peaks of two different phases in the powder mixture should not be overlapping. If the mixture has an α and a β phase, $Y_\alpha$

grams of pure α phase should be added to the powder mixture. The intensity ratio of α and β per gram of unknown after adding $Y_\alpha$ grams of α, can be written as:

$$\frac{I_{(hkl)\alpha}}{I_{(hkl)\beta}} = \frac{K_{(hkl)\alpha}\rho_\beta(X_\alpha + Y_\alpha)}{K_{(hkl)\beta}\rho_\alpha X_\beta} \qquad [5.27]$$

where $X_\alpha$ and $X_\beta$ are the initial weight fractions of phases α and β respectively, and $Y_\alpha$ is the number of grams of pure phase added per gram of the original sample.

By plotting $I_\alpha/I_\beta$ against $Y_\alpha$ as shown in Figure 5.7, $K = [K_{(hkl)\alpha}\rho_\beta]/[K_{(hkl)\beta}\rho_\alpha X_\beta]$ may be obtained from the slope of the line. From Eq.5.27, it is understood that the intersection of the line is $X_\alpha$.

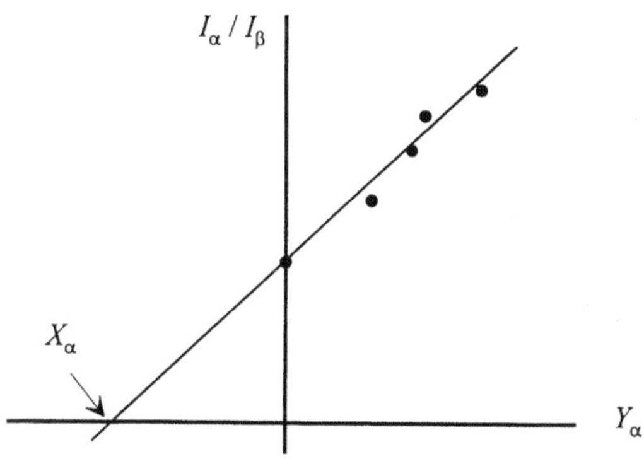


Figure 5.7. Quantitative measurement using the method of standard additions.

## 5.5 Conclusions

Size and surface area of the powder particles, which are important parameters, may directly influence the reaction kinetics of the mechanical milling process. These parameters may be measured using different techniques. Reaction kinetics and release of internal energy can be monitored using differential scanning calorimetry or differential thermal analysis.

Crystalline size may be simply detected using an x-ray diffractometer. Scherrer equation is widely used in the determination of crystalline size via the principle of x-ray peak broadening. Broadening can be used to determine crystalline size of less than 1μm. In the calculation of crystalline size, influence of instrumental broadening should be considered.

Quantitative measurement using x-ray diffraction may provide some useful information on the formation of new phases. The method is applicable to a wide range of phases. The method of standard additions can easily be applied to quantitative measurement of powders.

## 5.6 References


1. T. Allen, *Particle Size Measurement*, Publ. Chapman & Hall, (1990), 125.
2. R.M. German, *Powder Metallurgy Science*, Publ. Metal Powder Industries Federation, 1994, 65.
3. S. Brunauer, PH Emmett and E. Teller, *J. Amer. Chem. Soc.*, Vol.60 (1938), 309.
4. A. Richard, Wenman and E.S. Joseph, *Amer. Lab*, Nov. (1995).
5. W. Guo, A. Iasonna, M. Magini, S. Martelli and F. Padella, *J. Mater. Sci.*, Vol.29 (1994), 2436.
6. B.S. Murty, S. Ranganathan and M. Mohan Rao, *Mater. Sci. Eng.*, Vol.A149 (1992), 231.
7. Y.H. Kim, K. Hiraga, A. Inoue, T. Masumoto and H.H. Jo, *Mater. Trans., JIM*, Vol.35 (1994), 293.
8. Y.H. Park, H. Hashimoto and R. Watanabe, *Mater. Sci. Forum*, Vols.88-90 (1992), 59.
9. M.D. Barò, S. Surlñach and J. Malagelada, *Mechanical Alloying for Structural Applications*, Proc. of the 2nd intern. Confer., Vancouver, British Columbia, Canada, 20-22 September (1993), Ed. J.J. deBarbadillo, F.H. Froes and R. Scharz, ASM Intern., Mater. Park, OH, 343.
10. J. Malagelada, S. Surlñach and M.D. Barò, *Mater. Sci. Forum*, Vols.88-90 (1992), 497.
11. L. Lu, M.O. Lai and S. Zhang, *Mater. Research Bullt.*, Vol.29 (1994), 889.
12. F. Cardellini, V. Contini and G. Mazzone, *Scripta Metall. Mater.*, Vol.32 (1995), 641.
13. R.B. Schwarz, P.B. Desch and S. Srinivasan, *Statics and Dynamics of Alloy Phase Transformations*, Ed: P.E.A. Turchi and A. Gonis, Plenum Press, New York, (1994), 81.
14. P.Le Brun, L. Froyen and L. Delaey, *Structural Appl. of Mechanical Alloying*, Ed. F.H. Froes and J.J. deBarbadillo, ASM Intern., Mater. Park, OH, USA (1990), 155.
15. C. Suryanarayana and R. Sundaresan, *Mater. Sci. Eng.*, Vol.A131 (1991), 237.
16. G. Fan, M. Quan and Z. Hu, *Acta Metall. Sinica*, Vol.7 (1994), 124.
17. H.E. Kissinger, *Anal. Chem.*, Vol.29 (1957), 1702.
18. T. Ozawa, *J. Thermal Analysis*, Vol.2 (1970), 301.
19. S. Arrhenius, *Z. Physik. Chem.*, Vol.4 (1889), 226.
20. D.E. Mikkola and J.B. Cohen, *Local Atomic Arrangements Studies by x-Ray Diffraction*, Ed. J.B. Cohen and J.E. Hilbard, *Metall. Soc. Confer.*, Vol.36 (1965), 289.
21. P. Scherrer, *Göttinger Nachruchten*, Vol.2 (1918), 98.
22. M.V. Laue, *Z. Krist*, Vol.64 (1926), 115.
23. W.L. Bragg, *The Crystalline State*, Vol.I, A General Survey, Publ. G.Bell and Sons, London, Vol.1 (1919), 189.
24. A.L. Patterson, *Phys. Rev.*, Vol.49 (1936), 884.
25. H.P. Klug and L.E. Alexander, *X-Ray Diffraction Proceedures - For Polycrystalline and Amorphous Materials*, Publ. John Wiley & Sons, Inc, USA, 1992.
26 B.D. Cullity, *Elements of x-ray Diffraction*, Addison-Wesley Publishing Company, INC., (1973), 259.
27. G.K. Williamson and W.H. Hall, *Acta Metall.*, Vol.1 (1953) 22.
28. R. Jenkins and R.L. Snyder, *Intr. to X-Ray Powder Diffractometry*, Chemical Analysis, Vol.138 (1996), Ed. J.D. Winefordner, Publ. John Wiley & Sons, Inc.
29. C. Suryanarayana, R. Sundaresan and F.H. Froes, *Mater. Sci. Eng.*, Vol.A150 (1992), 117.

# 6

# DENSIFICATION

## 6.1 Cold compaction

### *6.1.1 Processing*

A simple densification method is to use cold compaction to obtain relative high density of the green compacts. Like ordinary powder metallurgy process, higher compaction densities may result in better green strength and smaller dimensional change during later thermal sintering process. Cold compaction can be carried out using a cold compaction die as illustrated in Figure 6.1.

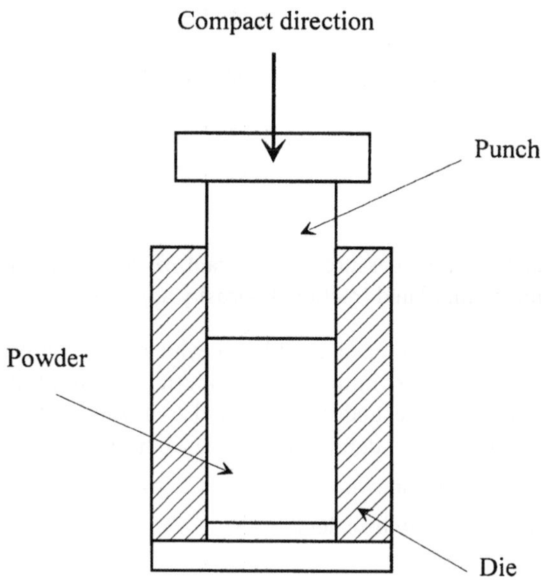


Figure 6.1. Cold compaction.

A cold compaction die consists of two parts: punch and die. Powder is loaded in the die and then a force is applied to the punch. For the fine powder particles after mechanical alloying, protection from contamination during loading may be necessary. Deformation of the powder takes place in the compaction die and the powder is repacked under the pressure applied. Frictional force between powder particles and the wall of the die increases due to the presence of the force in the radius direction. In general, the mechanically alloyed or milled power particles possess high hardness and yield strength in comparison with that in the unmilled state due to two factors: work hardening and dispersion strengthening. Consequently, high deformation pressure is required for the cold compaction. The key parameters controlling the density of the cold compacts are mainly the particle size which controls the packing density and the ability of the powders to plastically deform. Since densification is formability dependent, the increase in density for the soft and ductile powders is always higher. In general, compaction process may be divided into three stages:

(a) Particle rearrangement - where the particles move in the die without deformation;

(b) Deformation - where elastic and plastic deformation take place and powder particle are rearranged under the pressure applied; and

(c) Working hardening and fragmentation - which is not so dominating since the milled particles are already hardened.

### 6.1.2 *Compaction force*

The force required to compact the powder particles may be expressed by the well-known Konopicky's equation:

$$p = \frac{1}{C} \ln \frac{b_0}{b} \qquad [6.1]$$

where $p$ is the compaction pressure, $C$, constant, $b$, the logarithm of the fraction porosity and $b_0$, the extrapolated fractional porosity at $p = 0$.

Fischmeister and Arzt [1] developed a model to evaluate the compaction force based on microscopic considerations. In this model, an arrangement of monosize spherical particles is being compressed under an external hydrostatic pressure as shown in Figure 6.2 (a). There exists a deformation zone with an area of $a_s$ between two neighbouring particles. As shown in Figure 6.2 (b), the effective pressure $p_{eff}$ between the two particles may be calculated as:

$$p_{eff} = F_s / a_s \qquad [6.2]$$

where $F_s$ is the contacting force acting between the two particles.

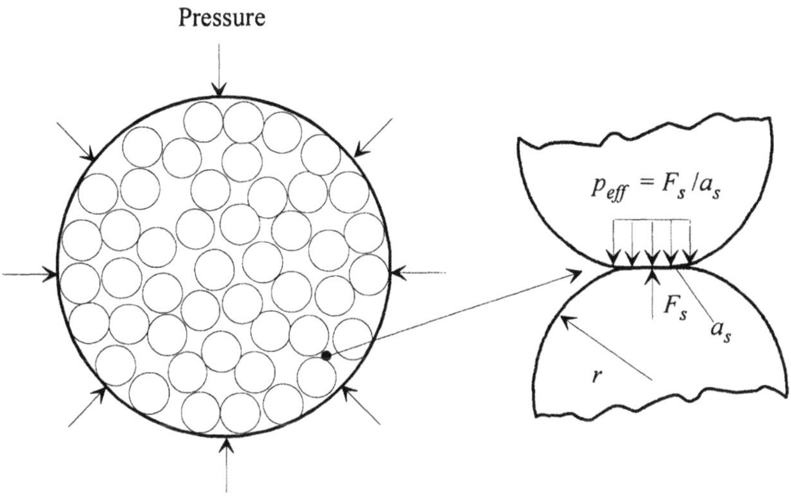


(a) (b)

Figure 6.2. Fischmeister and Arzt model [1].

If two spheres are compressed against each other, elastic deformation takes place. With the increase in contact pressure, the particles undergo plastic deformation. The contact force $F_s$ which causes the yielding of the two spherical particles may simply be given by

$$F_s = 3a_s\sigma_y \qquad [6.3]$$

where $a_s$ is the average contact area of the two spheres after deformation and $\sigma_y$, the yield stress.

Because of the changes in density and hardness of the powders, the pressure $p(\rho)$ is a faction of the above variables [2]:

$$p(\rho) = \frac{3}{4\pi}\sigma_y[1 + H(\rho)]\frac{a_s(\rho)Z(\rho)}{r^2}\rho \qquad [6.4]$$

where $H(\rho)$ is the work hardening function, $Z$, the average number of contacts, $r$, radius of the particle and $\rho$, the packing density of the powder compact.

The initial compact packing density $\rho_0$ is 0.63 for spherical particles. For axial pressing, due to the presence of frictional force, distribution of densification press is inhomogeneous. The axial pressure may be written as a function of the distance $z$ from upper punch [3]:

$$p(z) = p_u \exp\left[\frac{\left(-2\mu_f kz\right)}{r}\right] \qquad [6.5]$$

where $p_u$ is the pressure at $z = 0$, $k$, a constant and $\mu_f$, the coefficient of friction.

It can be seem from Eq.[6.5] that the bottom of the die has the lowest compact pressure and hence the lowest density. To avoid large difference in density of a cold compact, double-action pressing technology may be used. This technique requires the punches to move from two opposite directions as shown in Figure 6.3. The pressure distribution and thus the density of the cold compact are symmetrical with respect to the center line in the height direction.

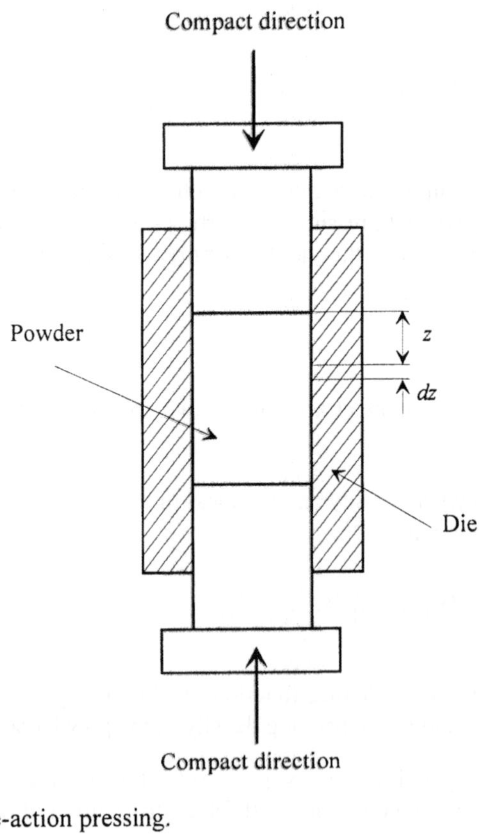


Figure 6.3. Double-action pressing.

## 6.2 Hot compaction

### *6.2.1 Processing*

Sine the mechanically alloyed or milled powders are in general too hard to be easily cold compacted, the compaction may be carried out in an elevated temperature by which high green density may be obtained. Compact at elevated temperature is called hot compaction and is schematically given in Figure 6.4.

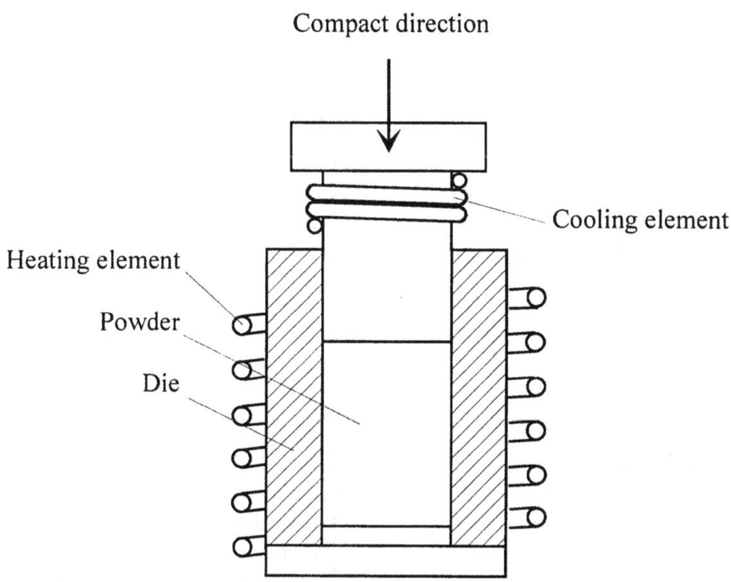


Figure 6.4. Hot compaction device.

The hot compaction device includes the compacting die, punch, heating element, and oxidation protection device. The punches are sometimes cooled by cooling water to avoid softening. In most cases, inert gases have to be used to prevent oxidation during the long period of hot compaction. In some cases, hot pressing in a vacuum system which enables efficient densification because of the reduction of gases in the possess, may also be employed [(4)]. Since the particle size of the mechanically alloyed powder is very fine, protection of oxidation is crucial.

Due to the high processing temperatures, the complexity of the dies is limited in comparison with that used in cold compaction. Pressure employed in the compaction is also low. Typical compaction pressure is up to 100 MPa for steel alloy dies and less than 50 MPa for graphite die [(3)].

Graphite is the most commonly used material for very high temperature heating. Other materials used are tungsten, TZM (dispersion-hardened molybdenum), stellites (cobalt alloys) and heat resistant alloys [3].

One critical parameter in hot compaction is the heating temperature which controls the final density and microstructure of the compacts [6, 7]. Too low a temperature may lead to low density while too high a temperature can lead to unexpected large grain sizes. For conventional materials, pores may be removed through sintering at high temperature for long time. The process is driven by the reduction in surface energy. Since mechanically alloyed powders contain nanostructure, the consolidation temperature needs to be kept low to maintain grain size at nano scale [8]. Full densification is possible at some intermediate elevated temperatures since nanophase powders are expected to show increased sinterability in comparison with their conventional coarse-grained counterparts [8-10]. Since the mechanically milled powder particles have very fine grain size which possesses high interfacial energy and may provide short diffusion path, densification rate can be increased. Hence, a reduction in sintering time becomes possible. The compact stress affects densification at high temperature through

$$\frac{\partial \varepsilon}{\partial t} = A\frac{\sigma^n}{RTd^q}\exp\left(-\frac{Q}{RT}\right)f(\rho) \qquad [6.6]$$

where $\sigma$ is the compact stress, $d$, the grain size, $\partial\varepsilon / \partial t$, the strain rate, $n$, the stress exponent, $q$, the grain size dependent exponent, $A$, coefficient associated with diffusivity factor, Burgers vector, shear modulus, etc., $Q$, activation energy, $\rho$, density, $R$, the gas constant and $T$, compact temperature. This equation indicates that high pressure applied may increase the rate of densification. The strong grain size dependence of creep deformation can contribute significantly to densification by plastic deformation at relatively low temperature [9, 10].

Results on densification of mechanically alloyed NiAl have shown that densification increases quickly at the first few ten seconds. Rate of densification then slows down progressively to a linear line, and finally reaches a value of almost zero, even though fully densed green compact may not have been achieved [5]. The final density of the hot pressed sample depends on the processing parameters, pressure and temperature. A higher density can be obtained by increasing both the temperature and the pressure, but the increase in temperature is more effective.

It has been observed that densification of the mechanically alloyed powders not only depends on the hot pressing temperature but also strongly on the milling duration. Yoshizawa *et al.* [4] hot-pressed mechanically alloyed $NbAl_3$. They found that the powder obtained after 2,160 minutes of mechanical alloying did not achieve satisfactory densification, whereas the powder milled for 4,320 minutes was able to attain 100% density. Fully densed compact may be obtainable for the powder milled for 2,160 minutes by increasing the heating temperature to as high as 1,623 K while the same result could be obtained for the compact alloyed for 4,320 minutes at a temperature of 1,320 K.

He *et al.* [8] consolidated mechanically milled Fe-26Al-2Cr nanostructural powder particles via different steps to avoid the possibility of significant grain growth. In their process, the first step is to cold compact the powder at a pressure of 1.2 GPa for about 16 hours. Since the particles have been work-hardened during milling, only 80% of the theoretical density could be obtained. In the second step, the cold compact is hot compacted at elevated temperature for several hours. The density of the hot-compressed compact increases with the temperature of hot compaction. When the temperature rises to 818 K which is below the recrystallization temperature of the material, almost fully densed compact of 99.5% can be fabricated. The pressure used in this process is as high as 1.25 GPa. Since the temperature is low, the final size of the grain is only about 32 nm. If pressureless sintering is used for the densification, only 90% of the theoretic density can be obtained at a sintering temperature of 1,273 K at which the grain size is far above nanometer. Since microhardness measurements indicate a significant strengthening effect due to nanophase grain size, the control of grain size is crucial in hot processes.

Schwarz *et al.* [11] hot pressed a two-phase Al-$Al_3Ti$ at 1,103 K in a graphite die. A similar temperature of 1,073 K and pressure of 100 MPa were used by Itsukaichi and co-workers. The die was heated by eddy current in an argon atmosphere. Upon reaching the compaction temperature, the pressure was increased to 12.5 MPa and maintained for 15 minutes. The two-phase Al-$Al_3Ti$ powder consolidated by hot pressing has a room-temperature compressive strength of 1.2 GPa, a value exceeding that of cast $Al_3Ti$ by a factor of 10. Very high hot pressing temperature of 1,773 K for $MoSi_2$, 1,963 K for $MoSi_2$-20vol.%YPSZ and 2,108 K for $MoSi_2$-20vol.%SiC have been used. The measured densities as a percentage of the theoretical densities of the hot pressed compacts obtained were respectively 96%, 93% and 93% [12].

### *6.2.2 Grain growth*

Since large amount of stored energy is in the form of grain boundaries, an important factor that limits the application of nanocrystalline powder is its stability. Consequently, hot processing at elevated temperature may cause serious grain growth. There are two basic approaches to stabilize grain growth: one involves further modifications to the processing and the other involves modifications to the chemistry of the alloy. The latter approach is based on Zener pinning where an inert second phase prevents grain boundary migration [13, 14]. Mechanical alloying provides the opportunities for the formation of oxide dispersoids as the second phase and hence can reduce the tendency of grain growth.

In general, the rate of grain growth may be written as:

$$\frac{dD}{dt} = \frac{K}{D} \qquad [6.7]$$

where $D$ is the mean grain size, $t$, time and $K$, a constant of proportionality which can be expressed by the temperature of heating:

$$K = K_0 \exp\left(\frac{-Q}{RT}\right) \qquad [6.8]$$

where $K_0$ is constant, $Q$, an empirical heat of activation, $T$, temperature and $R$, gas constant.

Based on the above equations, Kawanishi *et al.* (15) measured the kinetics of the mechanically alloyed $NbAl_3$ nanophase powder and found that the grain growth kinetics can be described by

$$\ln\left(\frac{dD}{dt}\right) = \ln\left(\frac{K}{3}\right) - 2\ln D. \qquad [6.9]$$

The activation energy for grain growth was 162±32 kJ/mol. Kumpmann *et al.* (16) found that the grain size of the ultrafine-grained Cu and Ag powder compacts may grow at very low temperature. However, if an oxide phase was present, abnormal grain growth was found to be increased due to grain boundary pinning by the oxide phase.

In single phase material, grain growth can occur by grain boundary migration controlled by short range diffusion across the interface. In mechanically alloyed alloys, grain coarsens more slowly during high temperature processing. This implies that short range diffusion cannot be rate controlling. Schaffer *et al.* (14) proposed that grain growth in mechanically alloyed materials is controlled by Ostwald ripening which typically occurs in precipitation hardened materials. During ripening, grain growth is controlled by long range diffusion of the alloying elements which partition between the phases. Therefore, grain growth is determined by the solubility of the diffusing elements of the matrix phase, the diffusion rates of these elements and the interfacial energy between the phases. Although mechanically alloyed materials provide fine dispersoids in the matrix, grain growth is still a major consideration in hot processing.

## 6.3 Cold isostatic processing

Because of the influence of frictional force between the particles and the dies in cold die compaction, homogeneous densification is almost impossible to achieve. Although more densed and homogeneous compacts can be produced using hot compaction, homogeneity of density and the ratio of height to diameter are still limited as a result of the frictional forces between the wall of the dies and the powder being compacted. By using cold isostatic press, it is possible to eliminate fractional force to produce compacts with homogenous density. As shown in Figure 6.5, a cold isostatic press generally consists of a pressure vessel with a high pressure pumping system. The mechanically milled powder is firstly placed into a soft container which may be rubber, silicone rubber, PVC or polyurethane. Sometimes the contains are often referred to as 'wet-bag' or 'dry bag'. After that, the bag should be vacuumed as shown in Figure 6. 5 (a) followed by isostatic pressing as shown in figure 6. 5 (b).

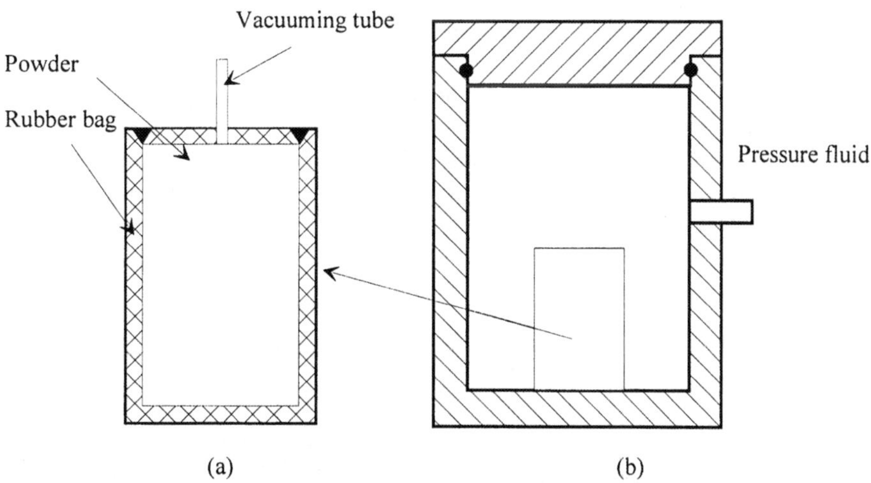


Figure 6.5. Cold isostatic pressing.

The disadvantage of cold isostatic processing is mainly the dimensional change after pressing since the container changes its shape when pressure is applied.

## 6.4 Hot isostatic processing

### *6.4.1 Processing*

Hot isostatic pressing was initially used for diffusion bonding of nuclear fuel elements. It has been used in the consolidation of power metallurgy. This technique can also be used in the consolidation process of mechanically alloyed materials. Hot isostatic pressing normally consists of several components, such as pressure vessel, heating element, compressor and temperature and pressure controllers.

In general, hot isostatic processing may be divided into two different categories based on different pressing methods. The first category is called capsule method. In this method, the powder is enclosed in a container (capsule) using a suitable canning system. Sometimes it is called encapsulated hot isostatic pressing which is the only processing route that allows pressure-added densification over the complete range from green compact density to theoretical density [(3)]. Encapsulation containers are normally made of low carbon steel, stainless steel, Ti, Ni and Cu. It must be noted that thermodynamic reduction of sheet metal grain boundaries and weld metal oxides may occur in high-pressure Ar which may subsequently lead to leakage [(17)]. Pyrex, Vycor and quartz glass containers are also very often employed [(18, 19)]. The container should be well vacuumed to at least $10^{-2} \sim 10^{-4}$ Torr. Sometimes the vacuum process must be carried out at an elevated temperature. Under such condition, the container should be hermetically sealed by welding to make the container gas tight. For glass containers which can be produced by glass blowing or casting a slurry of ground glass into the required form, they will become soft and continuously decrease in viscosity with increasing temperature. Although glass containers will become soft, they may be used at as high a temperature as 1,500 to 2,000°C where sheet metal container cannot be employed [(19)]. Sheet metal containers may be removed by machining or chemical milling while glass containers tend to spall from the surface after hot isostatic pressing. The common route for hot isostatic pressing is given in Figure 6.6.

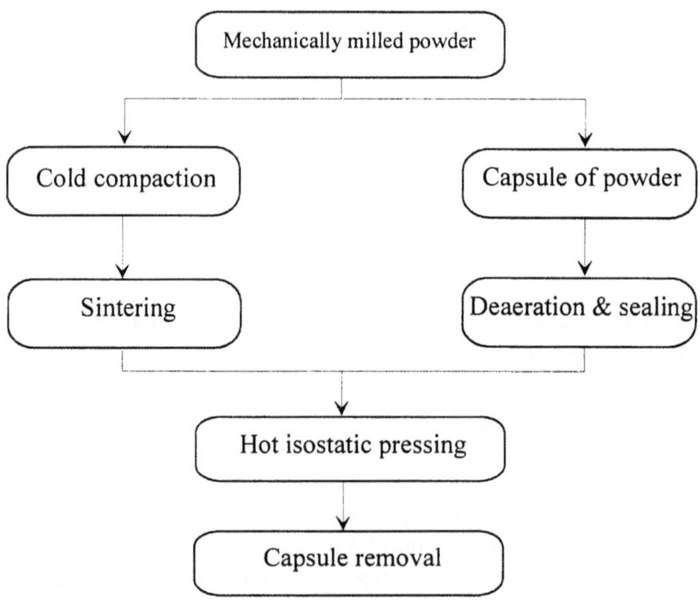


Figure 6.6. Hot isostatic pressing.

For the second category of hot isostatic processing, the powder is firstly cold compacted and sintered to achieve a state of closed porosity. The sintered compact is then hot isostatically pressed. Pores may be completely eliminated from the sintered compacts after the pressing process. The advantages are that sometimes no canning is required, easy handling and also a large quantity of compacts can be hot isostatically pressed at the same time.

The most often used operating temperature is between 773 and 1773 K although the graphite furnace can reach as high as 2730 K. The process has been widely employed to densify mechanically milled powders [(22-24)]. Since it is performed at high temperature, crystallization of amorphous and grain growth for nanocrystalline materials are the major concern during processing. Experimental results show that for mechanically alloyed Al-65at.%Ni hot isostatically pressed at 1523 K for 120 minutes, the crystalline size grew from nanosize to 30 μm [(23)]. However, if the same powder was sintered at 1373 K, the crystalline size was only 20 μm. Kimura *et al.* [(25)] studied the densification of mechanically alloyed $Co_{79.5}Nb_{15}Zr_{5.5}$ amorphous material and found that the onset of glass transition temperature was at about 800 K. The amorphous was pressed under a gas pressure of 196 MPa at 853 K for two different durations of 5 and 30 minutes. The compact pressed for 5 minutes still showed amorphous structure with 98% of theoretical density. The compact pressed for 30 minutes, however, showed partial crystallization. This investigation indicates that both pressing temperature and duration are important parameters.

Table 6.1. Temperature and pressure used in hot isostatic pressing.

| Materials | Temperature (K) | Pressure (MPa) | Ref. |
|---|---|---|---|
| Al alloys | 725-805 | 100 | 3 |
| Ti alloys<br>Ti aluminides | 1123-1223<br>1473-1573 | 100<br>150, 198 | 3<br>18, 28 |
| Ni alloys<br>Ni alumnide | 1373-1473<br>1523 | 100-150<br>100 | 3<br>23 |
| Fe | 1273 | 175 | 27 |
| $Co_{79.5}Nb_{15}Zr_{5.5}$ amorphous | 853 K | 196 MPa | 25 |

Mahmood *et al.* [26] investigated hot isostatic pressing of nanocrystalline $MoSi_2$ powders. In their process, the $MoSi_2$ powder was first cold isostatically pressed at a pressure of 345 MPa to obtain a preform. The cold pressed compact was encapsulated using either Pyrex which can be used to a maximum temperature of 1,273 K or Vycor for higher temperature. The capsules were vacuumed for at least four hours at the degassing temperature of 623 K. The crystalline size of the hot isostatically pressed compact was to be from 35 to 54 nm. No dramatic increase in the crystalline size was observed. The densities were found to range from about 82% at pressing temperature of 1073 K to 91% of the theoretical density at 1773 K. It was relatively easy to process the nanocrystalline $MoSi_2$ powders as compared to large grained $MoSi_2$ powders. Table 6.1 gives some examples of hot isostatic pressing temperatures and pressures used in processing different materials.

### 6.4.2 Fundamentals

In pressure sintering, the external pressure acting on the compact via a gas medium is transmitted to the particle contacts and superimposed on the capillary stress. This results in an increase in contact stresses between the particles. The important mechanisms in hot pressing densification in solid state systems are [3]:

(a) plastic yielding
(b) power law creep
(c) diffusional creep
(d) grain boundary sliding.

Plastic flow takes place when surface stress of the powder particle exceeds the yield stress of the powder material. Yielding is time independent until the particles become flattened. Creep such as Nabarro-Herring creep may follow. Power law creep, diffusional creep and grain boundary sliding are time dependent. At the final stage of densification, when only isolated pores are present, the surfaces of the pores are not simply compressed together to develop a planar crack. Bonding occurs because atoms diffuse in both directions across the interface.

The densification function of isostatic pressing can be written as [20, 21]:

$$F(\sigma_{ij},\rho) = \frac{3}{2}\sigma'_{ij}\sigma'_{ij} + (\sigma_m / f)^2 - (\rho^n\bar{\sigma})^2 \quad [6.10]$$

where $\sigma'_{ij}$ is the deviatoric stress, $\rho$, the density ratio, $n$, constant and $f$, a function of $\rho$ which can be expressed as:

$$f = 1/a(1 - \rho)^m \quad [6.11]$$

where *a* and *m* are constants dependent on the characteristics of the powder. In the principal stress space, the surface of an ellipsoid is given by $F = 0$. It has been shown experimentally that stress for a particular density ratio lies on the same elliptical curve [20,21,29].

One important tool for the selection of pressing condition is the hot isostatic maps which give the predominant mechanisms for densification under various conditions of pressure and temperature [30,31].

## 6.5 Other methods

Other methods used for densification include isothermal forging [32] which has been used to form oxide dispersion strengthened Ni base superalloy turbine blades, explosive compaction and hydrostatic extrusion [33] which are being employed to fabricate Ti aluminide rods, hot hydrostatic extrusion [34] which has been applied in oxide dispersion strengthened Al alloys and hot extrusion [35,36] which is often used to synthesize many different materials.

Another method used in the consolidation of nanocrystalline powder particles is dynamic compaction which is a viable method to retain the initial grain size of a material after compaction [37,38]. In dynamic compaction, pressure above the yield strength of the material is readily achieved to deform and densify the powder particles into bulk material. Consolidation is achieved by high dynamic pressures in a very short time of about 1 μs. The fast time scale offers the opportunity to consolidate ultrafine-sized powders which have no time to increase their grain size [39]. Densification process at a very short duration can also be carried out using electro-discharge compaction that can consolidate powders in a few hundred micro-seconds (about 300 μs) [40].

## 6.6 Conclusions

Several densification techniques can provide the means to consolidate the mechanically milled powders into bulk materials. Different methods may possess different advantages. Fully densed compacts with homogeneous structure may be easily obtained via hot isostatic pressing but cost involved may be high. Since large amount of energy stored in the powder particles is in the form of grain boundaries, the major difficulty in densification is how to avoid excessive grain growth which may totally damage the desired microstructure of the mechanically milled powder particles. Densification temperature should therefore be kept and controlled as low as possible.

## 6.7 References

1. H.F. Fischmeister and E. Arzt, *Powder Metall.*, Vol.26 (1983), 82.
2. E. Arzt, *Acta Metall.*, Vol.30 (1982), 1883.
3. F. Thümmler and R. Oberacker, *An Introduction to Powder Metall.*, Ed. I. Jenkins and J.V. Wood, Publ. The Inst. Mater., London (1993).

4. H. Yoshizawa and Y. Saito, *Mater. Sci. Forum*, Vols.88-90 (1992), 655.
5. H. Ardy, *Hot Pressing Study of Mechanically Alloyed Intermetallic NiAl*, Illinois of Inst. Tech., UMI, ABell & Howell Info. Com. (1994).
6. P. Nash, H. Kim, H. Choo, A. Ardy, S.J. Hwang and A.S. Nash, *Mater. Sci. Forum*, Vols.88-90 (1992), 603.
7. S.J. Hwang, P. Nash, M. Dollar and S. Dymek, *Mater. Sci. Forum*, Vols.88-90 (1992), 611.
8. L. He and E. Ma, *J. Mater. Res.*, Vol.11 (1996), 72.
9. H. Hahn, J. Logas and R.S. Averback, *J. Mater. Res.*, Vol.5 (1990), 609.
10. H.J. Höfler and R.S. Averback, *Nanophase and Nanocomposite Mater.*, Mater. Res. Soc. Symp. Proc., Ed. S. Komarneei, J.C. Parker and G.J. Thomas, Pittsburgh, PA, (1993), 9.
11. R.B. Schwarz, S. Srinivasan and P.B. Desch, *Mater. Sci. Forum*, Vol.88-90 (1992), 595.
12. S.R. Srinivasan and R.B. Schwarz, *Novel Powder Proc.*, Advs. in Powder Metall. & Particulate Mater. -1992, Vol.7, June 21-26, 1992, Compl. Joseph M. Capus and Randall M. German, Publ. MPIF, APMI, Princeton, New Jersey, 345.
13. J.W. Martin and R.D. Doherty, *Stability of Microstructure in Metallic Systems*, CUP, (1980).
14. G.B. Schaffer and A.J. Heron, *Mechanical Alloying for Structural Applications*, Proc. of the 2nd Intl, Confer., Vancouver, British Columbia, Canada, 20-22 September (1993), Ed. J.J. deBarbadillo, F.H. Froes and R. Scharz, ASM Intl., Mater. Park, OH, 197.
15. S. Kawanish, K. Isonishi and K. Okazaki, *Mater. Trans. JIM*, Vol.34 (1993), 44.
16. A. Kumpmann, B. Günther and H.-D. Kunze, *Mater. Sci. Eng.*, Vol.168A (1993), 165.
17. Metal Handbook, *Powder Metall.*, Vol.7, 9th Edition, 419.
18. Y.H. Park, H. Hashimoto, R. Watanabe, J.H. Ahn and H.S. Chung, *Mater. Sci. Forum*, Vol.88-90 (1992), 155.
19. H.V. Atkinson and B.A. Rickinson, *Hot Isostatic Processing*, Publ. Adam HilgerImprint, IOP Publ. Ltd., 1991.
20. S. Shima, T. Inoue, M. Oyane and K. Okimoto, *J. Japan Soc. Powder and Powder Metall.*, Vol.22 (1976), 257.
21. S. Shima and M. Oyane, *Inter. J. Mech. Sci.*, Vol.18 (1976), 285.
22. J.M. Sánchez, M.G. Barandika, J. Gil-Sevillano and F. Castro, *Scripta Metall. Mater.*, Vol.26 (1992), 957.
23. J. Wolska and J. Maas, *Mater. Sci. Forum*, Vols.179-181 (1995), 369.
24. L.P. Zhang, J. Shi and J.Z. Jin, *Mater. Letts.*, Vol.21 (1994), 303.
25. H. Kimura, K. Toda and T. Yuine, *Hot Isostatic Pressing-Theory and Applications*, Proc. of the Third Intern. Confer., Osaka, Japan, 10-14 June 1991, Ed. M. Koizumim, Publ. Elsevier Applied Sic., 223.
26. M.S. Haji-Mahmood and L.S. Chumbley, *Nanostructured Mater.*, Vol.7 (1996), 95.
27. B.R, Murphy and T.H. Courtney, *NanoStructured Mater.*, Vol.4 (1994), 365.
28. A. Kakitsuji and H. Miyamoto, *Hot Isostatic Pressing - Theory and Applications*, Proc. of The 3nd Intern. Confer., Osaka, Japan 10-14 June 1991, Ed. M. Koizumi, Publ. Elsevier Appl. Sci., 295.
29. S. Shima and K. Mimura, Intern. *J. Mech. Sci.*, Vol.28 (1986), 53.
30. E. Arzt, M.F. Ashby and K.E. Easterling, *Metall. Trans. A*, Vol.14 (1983), 211.
31. M.F. Ashby, Proc. *Hot Isostatic Pressing-Theory and Applications*, Ed. T. Garvare, Publ. Centek Publishers, Lulea, (1988), 29.
32. O. Tsuda, T. Matsushita, N. Kanamaru and K. Nishioka, *Stru. Appls. of Mechanical Alloying*, Proc. ASM Intern. Confer., Myrtle Beach, South Carolina, 27-29 Marh 1990, Ed. F.H. Froes and J.J. deBarbadillo, Publ. ASM Intern., Mater. Park, (1990), 49.
33. C.G. Li, W.H. Yang, A. Frefer and F.H. Froes, *Stru. Appls. of Mechanical Alloying*, Proc. of 2nd Intern. Confer. on Stru. Appls. of Mechanical Alloying, Vancouver, British Columbia, Canada, 20-22 Sep. 1993, Ed. J.J. deBarbadillo, F.H. Froes and R. Schwarz, Publ. ASM Intern., Mater. Park, (1993), 83.
34. G. Liang and Z. Li and E. Wang, *J. Mater. Sci. Techn.*, Vol.10 (1994), 285.
35. P. Nash, S.C. Ur and M. Dollar, *Stru. Appls. of Mechanical Alloying*, Proc. of 2nd Intern. Confer. on Stru. Appls. of Mechanical Alloying, Vancouver, British Columbia, Canada, 20-22

Sep. 1993, Ed. J.J. deBarbadillo, F.H. Froes and R. Schwarz, Publ. ASM Intern., Mater. Park, (1993), 291.

36. *Metall. Trans.* A, Vol.24A (1993), 1993.
37. M.L. Wilkins, A.S. Kusubov and C.F. Cline, *Metallurgical Application of Shockwave and High-Strain-Rate Phenomenon,* Ed. L.E. Murr, K.P. Staudhammer and M.A. Meyers, Publ. Marcel Dekker. Inc., New York (1986), 57.
38. D.A. Hoke, M.A. Meyers, L.W. Meyer and G.T. Gray, III, *Metall. Trans.*, Vol.23 A (1992), 77.
39. T.G. Nieh, P. Luo, W. Nellis, D. Lesuer and D. Benson, *Acta Mater.*, Vol.44 (1996), 3781.
40. D.K. Kim, J.H. Lee and K. Okazaki, Sci. and Eng. of Light Matels, Ed. K. Hirano, H. Oikawa and K. Ikeda, Japan Inst. Light Metals, Tokyo, (1991), 715.

# 7

# MECHANICAL PROPERTIES

## 7.1 Mechanical properties

A potential application of the mechanical alloying technique is to strengthen a material via extremely fine oxide dispersoids. Oxide dispersion strengthened alloys possess higher creep resistance and greater stability at elevated temperatures than conventional alloys [1-3]. Typical examples of such commercially available alloys are Inconel MA754 (Ni-20Cr-0.2Al-0.5Ti-0.6$Y_2O$), Inconel MA758 (Ni-20Cr-0.3Al-0.5Ti-0.6$Y_2O$), Inconel MA760 (Ni-20Cr-6Al-2Mo-3.5W-0.95$Y_2O$), Inconel MA6000 (Ni-15Cr-4.5Al-2.5Ti-4W-2Mo-2Ta-1.1$Y_2O$), Inconel MA956 (Fe-20Cr-4.5Al-0.5Ti-0.5$Y_2O$) [4] and IncoMAP AL-905XL (Al-4Mg-1.3Li-0.4O-1.1C) [5].

Materials synthesized via mechanical alloying have the potential to be used in high temperature applications. Incoloy MA957, a mechanically alloyed ferritic stainless steel, has been developed as a nuclear fuel cladding material for fast breeder reactors [6]. After processing into the bulk form, MA957 has an ultrafine microstructure containing submicrometer size grain of ferrite [7]. To avoid degradation of its mechanical properties due to grain growth, the stored energy before recrystallization may be controlled by 'preannealing' at a temperature high enough to permit recovery but not recrystallization. Since most of the stored energy of the alloy is in the form of grain boundaries, the stored energy can be so high that moving grain boundaries can readily overcome the drag resulting from the dispersoids. Hence, relief of the stored energy becomes important.

Early study has shown that Inconel MA6000 combines the high strength of the cast alloy at lower temperature of about 1,033 K and stability of the strength at high temperature of up to 1,373 K [8]. It is a stronger material at high temperatures than TD Ni because of substantial solid solution hardening from heavy elements. Thermal fatigue testing of several dispersion strengthened alloys in the longitudinal direction has shown that Ni-base materials possess excellent mechanical properties.

Fatigue behaviour of Inconel MA754 and MA6000 is approximately equal to that of directionally solidified Ni-base superalloys [(9)].

Another type of alloy is the aluminium base oxide dispersion strengthened alloys in which the major dispersoids are $Al_2O_3$, $Al_4C_3$ and $Y_2O_3$ [(7-11)]. Other types of desired precipitate phases may be produced by adding the appropriate combination of elemental pre-alloyed powders [(12)]. IncoMAP alloy Al-905XL with composition of Al-4.0Mg-1.3Li-1.1C-0.4O is an Al base alloy developed more toward structural applications [(13)]. The yield and tensile strengths of IncoMAP alloy Al-905XL are approximately 14% and 10% higher than those of AA7075-T73. Fracture toughness value of Al-905XL is also higher.

The mechanically alloyed materials generally show higher mechanical properties in comparison with their counterparts fabricated using other conventional methods. For example, in comparing mechanically alloyed Al-5Fe-4Mn and rapidly solidified Al-5Fe-7Mn alloy was, it was found [(14)] that their tensile properties at room temperature were almost identical to each other. However, when heating temperature is above 423 K, mechanically alloyed alloy shows a clear advantage in its remarkable temperature resistance. At 573 K, the mechanically alloyed material is 90 MPa higher than that of its rapid solidified counterpart. The ductility of the alloyed material is 3.2% at room temperature while rapidly solidified material 2.4%. At heating temperature of 473 K, the latter material decreases to only 1% while former, 1.5%. The increase in mechanical properties at elevated temperature is mainly due to the equiaxed dispersoids that are uniformly distributed in the mechanically alloyed aluminium alloy. These dispersoids, almost spherical shape were generated from several sources [(17)]: the aluminium oxide film that covers the surface of the original aluminium powder particles, oxide generated from oxidation during milling, and the dispersoids formed as a result of decomposition of the process control agents and the reaction with carbon.

The superior mechanical properties of the mechanically alloyed materials over the conventional ones can be seen from the greater equivalent amount of non-metallic dispersoids over the conventional composites [(10)]. Figure 7.1 shows the tensile strengths of both the mechanically milled Al and the unalloyed Al with non-metallic dispersoids (SAP) at equal amount of dispersoid. With the increase in volume of $Al_2O_3$ dispersoid, the strengths for both materials show linear functions with the volume of dispersoid. However, the mechanically milled Al is significantly stronger than SAP. The strength of the former Al at 5vol.% dispersoid is equivalent to the latter at 10vol.%. A similar behaviour is also observed when yield strengths of the two materials are compared. Besides superior strength, the mechanically milled Al also shows greater ductility at all strength levels as shown in Figure 7.2. The microstructure of the $Al_2O_3$ dispersoid in the mechanically milled Al is primarily equiaxed of about 30 nm in diameter [(15)]. The $Al_2O_3$ in SAP on the other

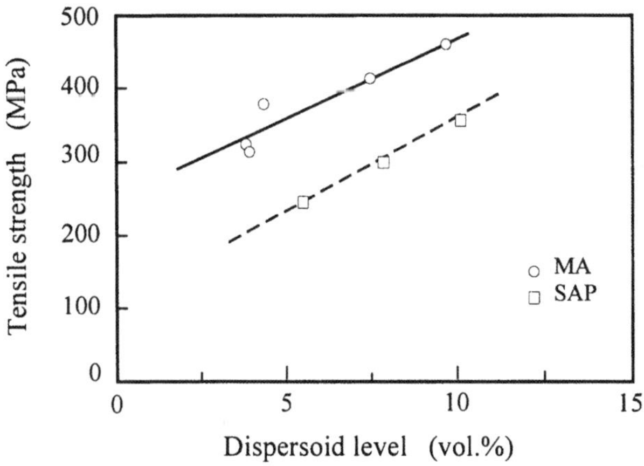


Figure 7.1. Tensile strength of mechanically alloyed Al and SAP material (10).

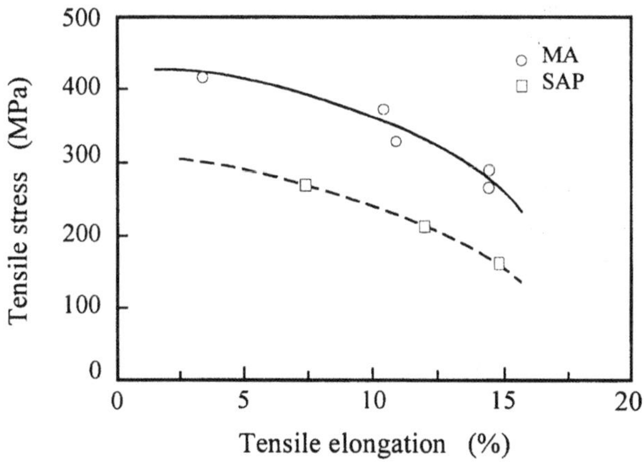


Figure 7.2. Relationship between yield strength and elongation of mechanically milled and SAP materials (10).

hand acts as an inefficient dispersion as it is in a form of flakes aligned parallel to the extrusion direction (16). These flaky particles are not dispersed evenly throughout the matrix and hence lowers the strength of the SAP material.

Figure 7.3 shows a comparison between DISPAL, a reaction milled Al with composition of Al-C-O, and other materials fabricated using conventional processes. The reaction milling which forms $Al_2O_3$ and $Al_4C_3$ as fine dispersoids generally leads to materials with excellent high temperature properties (18-20). Although the ultimate tensile stress of DISPAL at low temperature is lower than both Ti and $AlCuMg_2$, both the latter materials dramatically lose their strength when the application temperature is above about 448 K. The extremely fine microstructure of DISPAL with grain size between 0.6 and 1.5 μm and dispersoid size between 20 and 100 nm provides high density of dislocations [18].

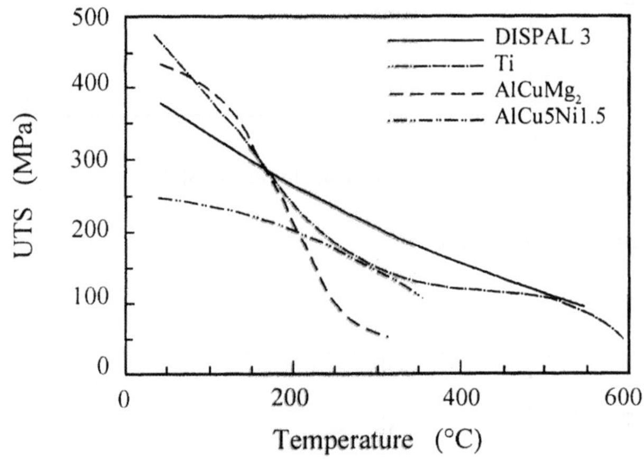


Figure 7.3. Tensile stresses of DISPAL and other conventionally processed materials [18].

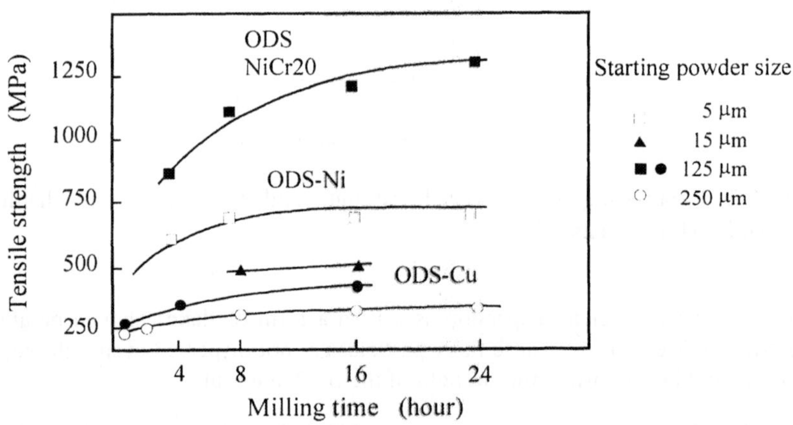


Figure 7.4. Influence of milling duration on strength of different materials [21].

Tensile measurement shows that the strength of mechanically alloyed or milled materials increases as milling duration is prolonged. For larger starting powder sizes, a longer milling time is necessary to reach constant strengths. Comparing two batches of Cu with different starting powder size as shown in Figure 7.4 [21], it can be seen that the increase in strength with milling duration for materials with smaller starting particle size is faster than those with larger big particle size. This is because the strength of the mechanically milled Cu is determined by it grain size and the dispersoid spacing while the partition of the dispersoids occurs by repeated fracturing and welding. When large particle size is used, cold welding may become dominant and hence may hinder the process of dispersion. The same trance can also be found for Ni and $NiCr_{20}$.

Mechanically alloyed aluminium alloys possess high strength caused by complex dislocation cell structures which are stable up to a high temperature of about 773 K as well as by the presence of dispersoids such as carbides and/or oxides [22, 23]. Due to the existence of dislocation cells and dispersoids, resistance to creep is also increased. Figure 7.5 shows the creep behaviours of Al-alloy IN-9052 with chemical composition of Al-3.86wt.%Mg-0.94wt.%C-0.65wt.%O [24], in terms of steady-state creep rate (minimum creep rate $\dot{\varepsilon}_s$) and modulus compensated stress $\sigma/E$, where $\sigma$ is the applied stress and $E$, the Young's modulus. The critical stress $\sigma_C$ depends on the creep temperature and decreases as the temperature is increased. When the creep temperature is above 773 K, the creep rate increases monotonically with a stress exponent of 20 to 30 as the stress is increased.

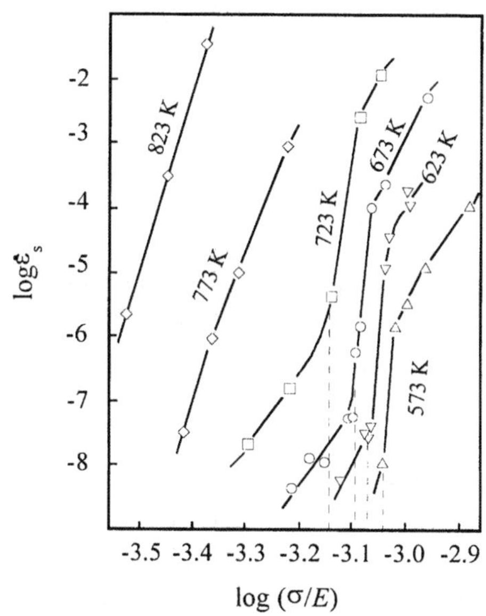


Figure 7.5. Creep behaviour of Al-alloy IN-9052

The resistance to creep is also dependent upon the type of dispersoid involved. A better creep property from a spinel oxide dispersion strengthened Al alloy than α-alumina strengthened one has been observed (25). The reason is that MgO and $Al_2O_3$ react to form spinel by lowering free energy of Al-Mg alloys. Spinel is stable over a wide high temperature range . The spinel has a lattice parameter almost twice that of Al, which might produce a good lattice match and low interfacial energy with Al in contrast to hexagonal $Al_2O_3$ where the interface is highly incoherent. Another reason is that the particle size of the spinel is smaller than that of α-$Al_2O_3$ even though the volume fractions of dispersoids are the same in two cases. Therefore, there are many more particles of $MgAl_2O_4$ thereby reducing the spacing between the particles and consequently, the creep rate. Fatigue crack propagation resistance in the lower $\Delta K$ region of $MgAl_2O_4$ dispersed Al-4Mg alloy was found to be better than that of $Al_2O_3$ dispersed material. The lower interfacial energy in $MgAl_2O_4$ alloy seems to prevent the occurrence of easy cross slip to restrict dislocation mobility, hence results resulting in higher $\Delta K_{th}$ value (26).

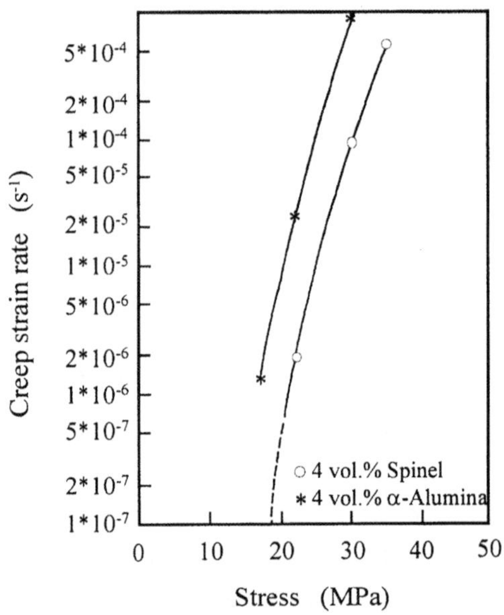


Figure 7.6. Steady-state creep rates at 683 K (25).

Figure 7.6 shows a comparison between the spinel and α-$Al_2O_3$ dispersion strengthened Al-3%Mg alloys (25). Based on the creep equation usually employed for oxide dispersion strengthened alloys in terms of a threshold stress (27):

$$\dot{\varepsilon} = A(\sigma - \sigma_t)^m \exp\left(\frac{-Q}{RT}\right), \qquad [7.1]$$

where $\dot{\varepsilon}$ is the creep strain rate, $A$ and $m$, constant, $R$, the gas constant, $T$, temperature, $\sigma$, applied stress, $\sigma_t$, threshold stress and $Q$, activation energy. The values of $\sigma_t$ and $m$ are respectively 14.2 MPa and 5.7 for spinel Al alloy, and 8.7 MPa and 7.0 for $Al_2O_3$ Al alloy. It can be seen that spinel strengened Al alloy has better creep resistance.

In dislocation climb models, dispersoid particles are considered as impenetrable obstacles which force gliding dislocations to climb a certain distance until they can continue to glide. In this case, the dislocation has to increase its line length in order to surmount the dispersoids and hence a retardation in creep rate is predicted [(28)]. To overcome the dispersoids, the dislocation line must pass an energy barrier which requires the stress to exceed a threshold stress for local climb [(28)]. This stress is proportional to the Orowan stress. Arzt [(28)] proposed a constitutive equation for equilibrium climb of dislocations:

$$\dot{\varepsilon} = \frac{\rho_d G b^4 B}{K_B T d^3}\left(a_p D_p + \pi D_v l d\right)\left(\frac{\sigma - \sigma_t^r}{\sigma_{\mathrm{Or}}}\right)^n \qquad [7.2]$$

where the stress exponent $n$ lies between 3 and 4, $\rho_d$, the density of mobile dislocations, $d$, particle width, $l$, space between particles, $a_p$, the cross section of a dislocation core, $D_p$, diffusivity, $D_v$, the volume diffusivity, $G$, shear modulus, $b$, Burgers vector and $\sigma_{\mathrm{Or}}$, the Orowan stress.

Since the maximum climb-related stress exponent predicted by the above model is only 6 which is far below the values of typical dispersion strengthened materials, only dislocation climb model cannot be used to explain the high stress sensitivity for creep [(28)]. Based on the evidence of an attractive dislocation-dispersoid interaction, a kinetic model for "detachment-controlled creep" was developed by Arzt [(28)]. The constitutive equation is given as:

$$\dot{\varepsilon} = \dot{\varepsilon}_0 \exp\left[\frac{Gb^2 r}{K_B T}(1-k)^{3/2}\left(1 - \frac{\sigma}{\sigma_{\mathrm{d}}}\right)\right] \qquad [7.3]$$

where $\sigma_d = (1-k^2)^{1/2}\sigma_{\mathrm{Or}}$, and $\dot{\varepsilon}_0 = 3D_v l \rho_d / b$.

Comparison between experimental creep data of Al alloys with theoretical predictions suggested that dislocation creep in dispersion strengthened alloys is controlled by thermally activated dislocation detachment from the dispersoids and not by a climb-related threshold stress [(29)].

## 7.2 Mechanisms of strengthening

### *7.2.1 Grain boundary strengthening*

The presence of large amount of grain boundary has a strong effect on the deformation behaviour of mechanically alloyed materials. The fine grain structure also affects the transformation temperature from brittle-to-ductile. This is of particular interest to high strength metallic materials and ceramic materials. Grain boundary serves as an effective barrier to the movement of dislocations. When the grain size is in the range of 20 nm, which is normally called nanocrystalline materials [30], the atoms in the grain boundary become a non-negligible fraction of 10% or more of the total number of atoms. Material behaviours are strongly affected by the diffusion of grain boundary atoms [31].

It is well-established that grain boundaries tend to strengthen materials at low temperature. The effects of grain boundary can be best described by Hall-Petch relationship [32, 33]:

$$\sigma_y = \sigma_0 + kd^{-1/2} \qquad [7.4]$$

where $\sigma_y$ is the yield strength, $\sigma_0$, resistance of the lattice to dislocation movement, $k$, locking parameter and $d$, grain size.

In the original Hall-Petch model [32], the grain boundaries are treated as barriers to the motion of dislocations. The model was derived for large-grained materials with high dislocation densities [34]. Based on Hall-Petch model, a variety of theories have been proposed to explain the hardening and softening phenomena of nanostructure. When grain size is below a critical value, the dislocation networks are bypassed by moving dislocations leading to softening of the nanostructure [35-37].

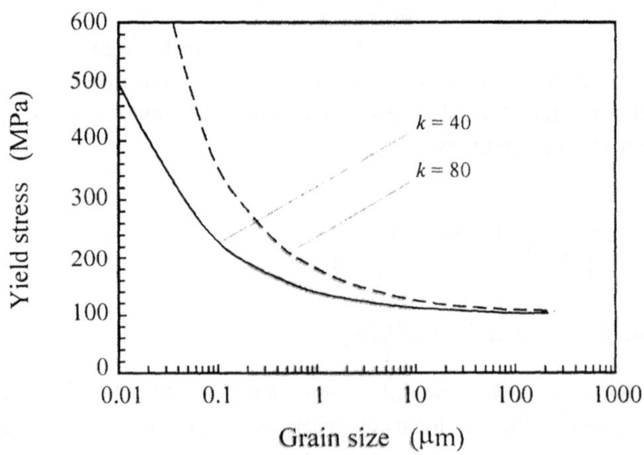


Figure 7.7. Effect of grain size on yield stress.

Eq.[7.4] indicates that a decrease in grain size can lead to an increase in yield strength. Strengthening effect is not significant when the grain size is relative large, but when the grain size is refined to below submicron level, the strength of the material can be dramatically increased. This phenomenon is depicted in Figure 7.7 where it can be seen that the yield stress is also strongly dependent upon the value of *k*.

### *7.2.2 Dispersoid strengthening*

Beside the influence of grain size, strength of a mechanically alloyed material is dependent on the distribution of fine dispersoids since yield strength is basically related to dispersoid and dislocation interaction by means of the Orowan bowing mechanism [38]. Dispersion strengthening takes place when the dispersoids as shown in Figure 7.8 cannot be cut by dislocations because they are incoherent. Fully coherent dispersoids may be looped or bypassed if the stress required to force the dislocation through the dispersoids is larger than the stress required to operate a bypass mechanism.

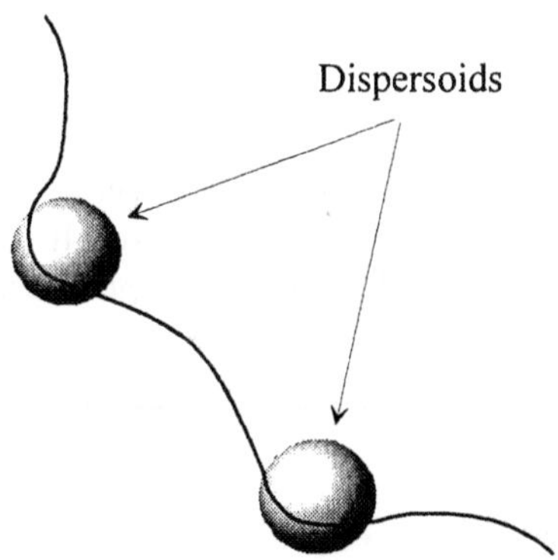


Figure 7.8. Dislocation and particle interaction.

The degree of strengthening resulting from dispersoids depends on the distribution of the dispersoids in the matrix. In addition to the shape of dispersoids, strengthening can be described by specifying the volume fraction, average diameter of the dispersoids and the mean interparticle spacing [39]. Orowan strengthening considers the metals to contain non-shareable particles where the dislocation can move forward without decreasing its radius of curvature. The flow stress $\tau_{Or}$ a function of Burgers vector, *b*, and space, *l*, between the dispersoids:

$$\tau_{Or} = \frac{Gb}{l} \quad [7.5]$$

where $G$ is the shear modulus. Eq.[7.5] defines the shear stress required to expand a dislocation loop between two fixed obstacles. This calculation does not take into account the effect of tension of the dislocation line and loop configuration [7]. The modified Orowan relation considers the influence of tension of the dislocation lines by [40-43]:

$$\tau_{Or} = \frac{0.81Gb}{2\pi(1-\nu)^{1/2}} \frac{\ln(2r/r_0)}{l-2r} \quad [7.6]$$

where $r$ is the radius of the particle, $\nu$, Poisson's ratio, $r_0$, the inner cut-off radius of the dislocation.

A critical parameter of dispersion strengthened materials is the interparticle spacing $l$. Interparticle pacing has been the subject of many interpretations. A simple expression for the linear mean free path is:

$$l = \frac{4(1-f)r}{3f} \quad [7.7]$$

where $f$ is the volume fraction of spherical dispersoids.

Figure 7.9 shows the Orowan strengthening effect at a constant volume fraction of dispersoids but with different dispersoid size. It can be seen that Orowan strengthening is only effective when the size of the dispersoid is very small.

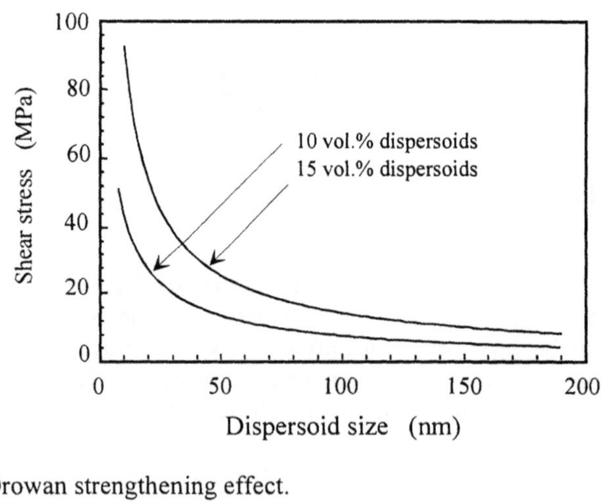


Figure 7.9. Orowan strengthening effect.

### *7.2.3 Influence on fracture toughness*

Hanh and Rosenfield (44, 45) proposed a model to predict the fracture toughness of aluminum alloys. In this model, the fracture toughness is dependent on: (a) the extend of the heavily strained region ahead of the crack tip, which is a function of the yield strength and modulus, (b) the size of the ligament which is related to $f$, the volume fraction of cracked particles, and (c) the work of rupturing the ligaments. When the extent of the heavily deformed region ahead of the crack tip is comparable with the width of the unbroken ligament separating the cracked particles, crack extension occurs.

From a metallurgical point of view, the following approximate failure criterion is instructive: crack extension proceeds when $\delta$, the extent of the heavily stretched region is comparable to the width of the unbroken ligament which is approximated by $l$, the spacing of cracked particles:

$$l \approx \delta \qquad [7.8]$$

According to the fundamentals of fracture mechanics, fracture toughness model can be given as:

$$K_{IC} = \sqrt{2\sigma_y E\delta} \approx \sqrt{2\sigma_y E l} \qquad [7.9]$$

In an ideal case of a regular cubic array of uniform spherical particles, $l$ can be obtained from:

$$f = \frac{\text{volume of particle}}{\text{volume of the cubic}} = \frac{\frac{1}{6}\pi d^3}{l^3} \qquad [7.10]$$

$$l = \sqrt[3]{\frac{\pi d^3}{6f}} = d\sqrt[3]{\frac{\pi}{6f}} \qquad [7.11]$$

where $d$ is the diameter of the second phase particle. Eq.[7.9] can be further written as (44, 45):

$$K_{IC} \approx \left[2\sigma_y E d\left(\frac{\pi}{6}\right)^{1/3}\right]^{1/2} f^{-1/6} \qquad [7.12]$$

Eq.[7.12] has been derived under the conditions of large plastic deformation, and has been used to analyze the fracture toughness of Al alloys. It can be seen that fracture toughness decreases with the decrease in particle size as a consequence of the reduction of spacing in particle.

### 7.3 Conclusions

Mechanical properties can be effectively improved by fine dispersoids which are produced by mechanical alloying and appropriate thermal processing. Smaller grain size makes mechanically alloyed materials stronger and more ductile. Oxide dispersion strengthened superalloys and other types of alloys exhibit a combination of excellent high temperature strength and oxidation resistance. The presence of dispersoids increases the resistance to creep. In general, the mechanically alloyed materials are better than the unprocessed materials.

### 7.4 References

1. J.S. Benjamin, *Metall. Trans.*, Vol.1 (1970), 2943.
2. V.C. Nardone, D.E. Matejczyk and J.K. Tiem, *Metall. Trans.*, Vol.14 A (1983), 1435.
3. T.S. Chou and H.K.D.H. Bhadeshia, *Mater. Sci. Tech.*, Vol.9 (1993), 890.
4. G.M. McColvin and M.J. Shaw, *Mater. Sci. Forum*, Vols.88-90 (1992), 235.
5. R.D. Schelleng, P/M in Aerospace and Defense Tech., Ed. F.H. Froes, Publ. Metal Powder Industries Federation, NJ, (1990), 67.
6. K. Asano, Y. Kohno, A. Kohyama, T. Suzuki and H. Kusanagi, *J. Nucl. Mater.*, Vol.155 (1988), 928.
7. H.G.F. Wilsdorf, *Dispersion Strengthened Aluminium Alloys*, Proc. of six-session Symposium, TMS Annual Mt, Phoenix, Arizona, Jan.25-29, 1988, Ed. Y.W. Kim and W.M. Griffith, Publ. TMS, (1988), 3.
8. J.S. Benjamin, *New Matr. by Mechanical Alloying*, DGM Confer., Calw-Hirsau, Germany, Oct. 1988, Ed. E. Arzt and L. Schultz, Publ. Deutsche Gesellschaft für Metallkunde eV., (1989), 3.
9. J. Daniel Whitterberger, *ibid*, 201.
10. J.H. Weber and R.D. Schelleng, *Dispersion Strengthened Aluminium Alloys*, Proc. of six-session Symposium, TMS Annual Mt, Phoenix, Arizona, Jan.25-29, 1988, Ed. Y.W. Kim and W.M. Griffith, Publ. TMS, (1988), 467.
11. V. Arnhold and K. Hummert, *ibid*, 483.
12. J.A. Hawk, P.K. Mirchandani, R.C. Benn and H.G.F. Wilsdorf, *ibid*, 551.
13. J.H. Weber and D.J. Chellman, *Structual Appls. of Mechanical ALloying*, Proc. of An ASM Intern. Confer., Myrtl Beach, South Carolina, 27-29 March 1990, Ed. F.H. Froes and J.J. deBarbadillo, Publ. ASM Intern. Mater. Park, OH, (1990), 147.
14. P. Le Brun, X.P. Niu, L. Froyen, B. Munar and L. Delaey, *Solid State Powder Processing*, Proc. of TMS Confer. Indianaplis, 2-5 Oct. 1989, Ed. A.H. Clauer and J.J. deBarbadillo, Publ. The Minerals, Metals & Mater. Soc., (1990), 273.
15. J.S. Benjamin and M.J. Bomford, Metall. Trans. A, Vol.8A (1977), 1301.
16. N. Hansen, Trans. TMS-AIME, Vol.245 (1969), 1305.
17. P.S. Gilman and W.D. Nix, *Metall. Trans. A*, Vol.12A (1981), 813.
18. V. Arnhold and K. Hummert, *New Matr. by Mechanical Alloying*, DGM Confer., Calw-Hirsau, Germany, Oct. 1988, Ed. E. Arzt and L. Schultz, Publ. Deutsche Gesellschaft für Metallkunde eV., (1989), 263.
19. V. Arnhold and J. Baumgarten, *Powder Metall. Intern.*, Vol.17 (1985), 168.
20. G. Jangg, J. Vasgyura, K. Schröder, M. Slesar and M. Besterci, *Horizons of Powder Metall.*, Ed. W.A. Kaysser and W.J. Huppmann, Publ. Verlagshmid GmbH, Freiburg, Germany, (1986), 989.
21. J. Zbiral, G. Jangg and G. Korb, *Mechanical Alloying for Structural Appls.*, Proc. of the 2nd Intern. Confer. on Structural Appls. of Mechanical alloying, 20-22 Sep. 1993, Vancouver, British Columbia, Canada, Ed. J.J. deBarbadillo, F.H. Froes and R. Schwarz, Publ. ASM Intern. Mater. Park, Ohio, (1993), 59.
22. J.S. Benjamin and R.D. Schelleng, *Metal. Trans. A*, Vol.12A (1981), 1827.
23. Y.W. Kim and L.R. Bidwell, *Scripta Metall.*, Vol.15 (1981), 483.

24. K. Matsuura and N. Matsuuda, Sintering'87, Proc. of The Intern. Inst. for The Sci. of Sintering (IISS), Tokyo, Japan, Nov. 4-6, 1987, 587.
25. T. Creasy, J.R. Weertman and M.E. Fine, *Dispersion Strengthened Aluminium Alloys*, Proc. of six-session Symposium, TMS Annual Mtg, Phoenix, Arizona, Jan.25-29, 1988, Ed. Y.W. Kim and W.M. Griffith, Publ. TMS, (1988), 539.
26. J.S. Cho and S.I. Kwun, *Scripta Metall. Mater.*, Vol.27 (1992), 789.
27. L.M. Brown, *Fatigue and Creep of Composite Mater*, Ed. H. Liliholt and R. T.alreda, Publ. Røskilde, Denmark, Risø Natl. Lab, (1982), 1.
28. E. Arzt, *New Matr. by Mechanical Alloying*, DGM Confer., Calw-Hirsau, Germany, Oct. 1988, Ed. E. Arzt and L. Schultz, Publ. Deutsche Gesellschaft für Metallkunde eV., (1989), 185.
29. J. Rosler and E. Arzt, *ibid*, 279.
30. H. Gleiter and P. Marquardt, *Zeit. Mtlkund*, Vol.75 (1984), 263.
31. B. Günther, H.-D Kunze and B. Scholz, *Proc. 1st European East-West Symp. on Mater. and Procs.*, Publ. Inderscience Enterprises Ltd, UK, (1990), 351.
32. N.J. Petch, *J. Iron Steel Inst.*, Vol.174 (1953), 25.
33. E.O. Hall, *Proc. Phys. Soc. B*, Vol.64 (1951), 747.
34. R. W. Siegel and G. E. Fougers, *Nanophase Mater., Synthesis-Properties-Applications*, Ed. G.C. Hadjipanayis and R.W. Siegel, Publ. Kluwer Academic Publishers, (1994), 233.
35. T.G. Nieh and J. Wadsworth, *Scripta Metall. Mater.*, Vol.25 (1991), 955.
36. A.M. El-Sherik, U. Erb, G. Palumbo and K.T. Aust, *Scripta Metall. Mater.*, Vol.27 (1992), 1185.
37. R.O. Scattergood and C.C. Koch, *Scripta Metall. Mater.*, Vol.27 (1992), 1195.
38. E. Orowan, *Discussion Symp. on Internal Stress in Metals and Alloys*, Monograph and Rept. Series No.5, Publ. Insti. of Metals, London (1948), 451.
39. G.E. Dieter, *Mech. Metall.*, Publ. McGraw-Hill Inc., (1988).
40. J.A. Martin, *Micromechanisms in Particle Hardened Alloys*, Publ. Cambridge Univ. Press, (1980).
41. U.F. Kock, Phil. Meg., Vol.13 (1966), 541.
42. A.J.E. Foreman and M.J. Makin, Phil. Meg., Vol.14 (1966), 911.
43. M.F. Ashby, *Oxide Dispersion Strenthening*, Proc. 2nd Bolton Landing Confer., Ed. G.S. Ansell, T.O. Cooper and F.V. Lenel, Publ. Gordon and Breach, (1968).
44. G.T. Hahn and A.R. Rosenfield, 3rd Intern. Confer. on Fracture, Munich (1973), III-211.
45. G.T. Hahn and A.R. Rosenfield, Metall. Trans., Vol.6A (1975), 653.

# 8

# MECHANISMS OF MECHANICAL ALLOYING

## 8.1 Mechanical activation

Unlike conventional processes in which materials are produced either by melting or sintering at high temperature, synthesis of different types of materials using the mechanical alloying or milling process is generally carried our at room temperature although temperature may rise during milling.

For thermal reaction, the atoms in the reaction system jump over the energy barrier by thermofluctuation while in mechanochemical processes [(1)], the atoms overcome the barrier by both the actions of mechanical forces and thermal activation. The mechanical milling process is, however, very different from the thermal process. Several models have been proposed to understand the mechanisms involved in the formation of materials. Macro-kinetic factors that are important in the process include particle size, temperature and deformation of the particles. The most common model is a mechanochemical system consisting of electrochemical reactions of stressed metallic electrode [(2)]. Under high stress fields, chemical potential and chemical activities of the system change with the change in potential applied to the interface and pressure. Since the powder is highly energized, the internal energy due to mechanical activation is extremely high. High strain value indicates high intensity dislocation flow causing formation of various structural imperfections [(3)]. These changes in thermodynamic parameters finally promote kinetic reaction in the mechanically milled powder particles [(4)]. Goldberg *et al.* [(3)] suggested that distortion of the deformed powder structure is so high that the original lattice is substantially changed. In this consideration, the concept of defect may not be able to provide a suitable solution to the modeling of the mechanical alloying process.

Besides thermodynamic models, some phenomenological kinetic models have also been proposed to understand the reaction process [(5, 6)]. The latter models may provide some information on the activation process as well as the relationship between difference parameters.

### *8.1.1 Diffusion*

Due to thermal energy, all the atoms in a solid vibrate about their rest positions. Occasionally, a particularly violent oscillation of an interstitial atom, or some chanced coincidence of the movements of the matrix and interstitial atoms, will cause an atom to result in a jump. Such atomic movements generate atomic fluxes and are known as diffusion. Although atoms spend most of their time at the lattice sites, a small fraction of the time is spent as atomic flexes. Fluxes exist in both homogeneous and inhomogeneous materials. The net flux at equilibrium in all directions is zero, and therefore, the atomic fluxes across a plane in the forward and reverse directions are the same. On the other hand, the net flux in a phase which is not in equilibrium, is not zero. Consequently, the system tends to return to its equilibrium state. Such diffusional fluxes determine the rates of solute transfer and hence the rates of transformations [(7)].

Diffusion flux is the amount of diffusion substance which passes through a plane of unit area perpendicular to the diffusion direction per unit time. At constant temperature and pressure, atoms migrate from regions of high chemical potential to those of low chemical potential. Chemical potential gradients induce fluxes; the fluxes vanish when the gradients become zero. When temperature, pressure and potential are not constant, fluxes can take place. Therefore, under constant operating parameters, fluxes are functions of the chemical potential gradients. In a binary system, the two fluxes $J_1$ and $J_2$ of two components 1 and 2 can be written as:

$$J_1 = -D_1(dC_1 / dx) \qquad [8.1]$$

$$J_2 = -D_2(dC_2 / dx) \qquad [8.2]$$

where $D_1$ and $D_2$ are partial chemical diffusion coefficients of component 1 and 2 respectively. The chemical interdiffusion coefficient $D$ can be expressed as:

$$D = N_1 D_1 + N_2 D_2 \qquad [8.3]$$

where $N_1$ and $N_2$ are the fractional concentrations of the two components 1 and 2 respectively.

For steady state diffusion, concentration of the two components 1 and 2 gradually changes at the interfaces as illustrated in Figure 8.1 (a). Diffusion in mechanical alloying process, however, differs from the steady state diffusion since the balance of atom concentration at the interface between two different components may be destroyed by subsequent fracturing of the powder particles. Consequently, new surfaces with very different composition meet each other to form new diffusion couples when different powder particles are cold welded together. This situation is depicted in Figure 8.1 (b). The concentration of the diffusing species is important in that it affects their diffusivity. Large difference in composition at the interface therefore promotes interdiffusion.

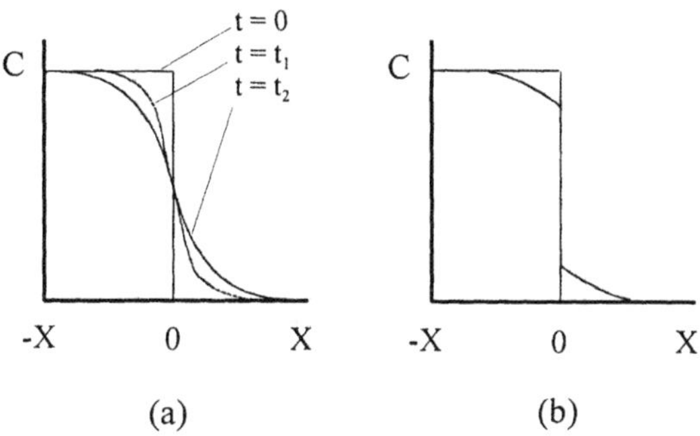


Figure 8.1. Change in atom concentration for (a) steady state diffusion; (b) diffusion after fracturing and rewelding of the powder particles.

### *8.1.2 Influence of activation energy*

In order to move an atom to an adjacent location, the atoms of the parent lattice must be forced apart into higher energy positions as shown in Figure 8.2. The increase in free energy is referred to as activation energy. Activation energy for diffusion is equal to the sum of the activation energy to form a vacancy and the activation energy to move the vacancy. The activation energy, $Q$, is therefore,

$$Q = Q_f + Q_m \quad [8.4]$$

where $Q_f$ is the activation energy for creating vacancies, and $Q_m$, the activation energy for moving vacancies. It appears that in most of the mechanical alloying processes, temperature is not a dominating factor. The temperature generated by collision is far from reaching the diffusion temperature. Therefore, factors other than temperature must significantly contribute to the diffusion process.

On microscopic level, atoms will rearrange corresponding to the limit of elasticity. There exists various mechanisms on such rearrangement. One such mechanism is the rupture of interatomic bonds and the rearrangement of atoms in the nucleus of dislocation [8]. Dislocation movement and formation of dislocations are obvious manifestation of heavy plastic deformation. The rearrangement of atoms during milling can be analyzed using classical chemical kinetics.

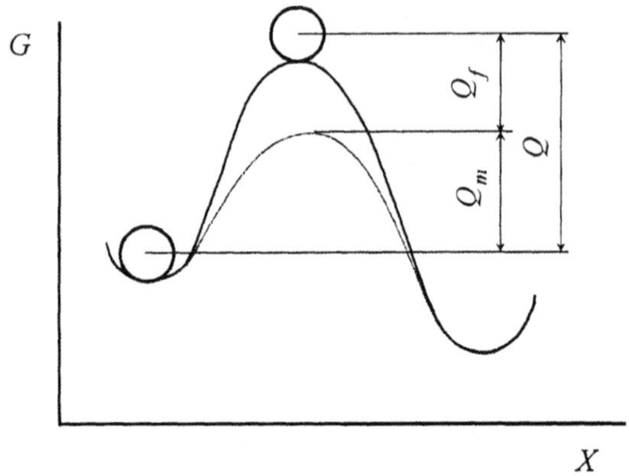


Figure 8.2. Activation energy for diffusion.

During mechanical alloying, mechanical energy can partially be stored by the creation of dislocations and grain boundaries in the mechanically alloyed material. There are two primary classes of point defects, vacancies and interstitials. If there is sufficient activation energy present, atom can move in crystal lattices from one atomic site to another. When dislocations move along a slip plane, the mechanical energy is transformed into kinetic energy of the atoms which excites the translational mobility of the atoms. All atoms receive additional energy and mobility [(1)]. As a result of the formation of large amount of defects (dislocations and vacancies are normally generated by thermal energy in thermally induced diffusion process) due to high energy collision of the powder particles, the total activation energy required by diffusion is lower because part of the activation energy required to form vacancies may or may not be required completely as graphically shown in Figure 8.2. In general, diffusivity [(13)] can be written as:

$$D = D_0 \exp\left(\frac{-Q}{RT}\right) \qquad [8.5]$$

where $D$ is the diffusion coefficient, $D_0$, material constant, $Q$, activation energy, $R$, universal gas constant and, $T$, temperature.

For diffusion in pure metal, $D$ varies inversely as the melting temperature of the material. $D$ is a linear function of the vacancy concentration and the vibrational frequency. An empirical correlation between $Q$ and $T_m$, the melting temperature, may be described by the following equation [18]:

$$Q = 33.7T_m \quad [8.6]$$

Substituting Eq.[8.4] into Eq.[8.5], Eq.[8.5] can be rewritten as:

$$D = D_0 \exp\left[\frac{-(Q_f + Q_m)}{RT}\right] \quad [8.7]$$

Eq.[8.7] establishes that at the same value of $D$, decrease in activation energy, such as that to create vacancies, is equivalent to increase in temperature. Therefore, it is possible to lower activation energy significantly by lowing $Q_f$. It is believed that the lowering in activation energy plays an important role in the mechanical alloying process. In thermal induced diffusion, lattice defects may be annealed out very rapidly to result in a decrease in diffusion coefficient. Defects probably contribute little to the increase in homogenization kinetics in such diffusion process. However, defect density during mechanical alloying increases with mechanical alloying duration and therefore it significantly contributes to homogenization kinetics.

Taking mechanical alloying of Al-Cu system as an example, the following parameters may be used [9, 10]: material constant $D_0$ = $1.5*10^{-5}$ $m^2/s$, activation energy for lattice diffusion $Q_l$ = 126 kJ/mole, and gas constant 8.31 J/K•mol. diffusivities $D$ are $1.0307*10^{-18} m^2/s$ at 500K and $8.8968*10^{-14}$ $m^2/s$ at 800 K. If defect diffusion is considered where activation energy is only 82 kJ/mole, the diffusivity will be $8.693*10^{-14}$ $m^2/s$ at 521 K which is equivalent to a diffusion process at 800 K. From this example, it may be better understood that the decrease in activation energy may lead to dramatic decrease in diffusion temperature.

Bhattacharya and Arzt [11] proposed a model for the calculation of diffusion. In this model, the particles are considered to be an alternate plate-like fashion. This simplification of alternate structure has experimentally been observed as layered composite structure during intermediate stage of milling. In the equation, diffusivity includes two parts as formulated in the following equation:

$$D = D_l \exp\left(-\frac{Q_l}{RT}\right) + \beta b^2 \rho D_c \exp\left(-\frac{Q_c}{RT}\right) \quad [8.8]$$

where $D_l$ and $D_c$ are respectively the material constants for the lattice and the core, $Q_l$ and $Q_c$ are respectively the activation energies of the lattice and core diffusion, $b$, Burgers vector, ρ, dislocation density and β, a core diffusivity factor. In principle, the model considers the effect of dislocation accumulation which takes place as a result of large amount of plastic deformation causing enhancement of the diffusion coefficient due to extra mobility along the dislocation cores. The rate of change in dislocation density with impact strain is [11, 12]:

$$\frac{d\rho}{d\varepsilon} = \frac{1}{100b}\sqrt{\rho} \qquad [8.9]$$

where ε is the deformation strain. Diffusivity can be calculated if dislocation density is known. The dislocations formed during collision effectively act as pipes which allow atoms to diffuse along them with diffusion coefficient $D_p$. The contribution of dislocations to the total diffusive flux through the powder particles depends on the density of dislocations. The apparent diffusivity $D_{app}$ is a function of the cross-sectional area of the dislocation pipes. The ratio of $D_{app}$ to $D_l$ can be written as [13]:

$$\frac{D_{app}}{D_l} = 1 + A_p \frac{D_p}{D_l} \qquad [8.10]$$

where $A_p$ is the cross-sectional area of pipe per unit area of particle.

Butyagin [8] has indicated that there are two main differences in mechanical treatment and chemical processes. The first difference is that the activation energy is influenced by the elastic stress while the second one is that the energy liberated after the transition of the barrier is the sum of the enthalpy of the process and activation energy of the barrier. Mechanical milling promotes the atoms to jump over the barrier by an exothermic process. In many cases, the height of the barrier is commensurable with the energy of interatomic bonds and energy released during the rearrangement of atoms. Some factors are important for the mechanochemical solid state reaction. These factors can be divided into two groups. The first group may be characterized by an "extensive", namely, specific surface area, particle size distribution and shape of particles, etc. Hilda [14] and Lu [15] have also found that the stored enthalpy increases with prolonged mechanical alloying duration. The decrease in transition temperature from disordered γ' to ordered γ' when mechanical alloying duration is increased [15] is the result of the increase in stored energy.

### *8.1.3 Thermodynamics of Defects*

Defects generated during collision between the milling balls and the powder particles can be in the forms of vacancies and dislocations. The removal of atoms from their sites not only increases the internal energy of the powder particles due to the broken bonds around the vacancy, but also increases the randomness of the system. Enthalpy has been found to increase with the increase in milling duration. The change in entropy can simply be written as [(13)]:

$$\Delta S_m = X_v \Delta S_v - R(X_a \ln X_a + X_v \ln X_v) \qquad [8.11]$$

where $X_v$ and $X_a$ are the mole fractions of vacancies and atoms respectively, $\Delta S_v$, the thermal entropy of vacancies due to changes in the vibrational frequencies of the atoms around a vacancy. The large contribution to the entropy is due to configuration entropy $\Delta S_{con}$ which can be given by

$$\Delta S_{con} = k \ln \Omega \qquad [8.12]$$

where $\Omega$ can be derived by considering the random distribution of $N_v$, the number of vacant lattice sites upon $N_l$, the total number of sites,

$$\Omega = \frac{N_l!}{N_v(N_l - N_v)!} \qquad [8.13]$$

If $X_v << 1$, Eq.[8.11] can be written as:

$$\Delta S_m = X_v \Delta S_v - R X_v \ln X_v \qquad [8.14]$$

Differentiation of the free energy term gives:

$$\frac{\partial G}{\partial X_v} = \Delta H_v - T \Delta S_v + RT \ln X_v \qquad [8.15]$$

The equilibrium concentration of the vacancies may be given by

$$X_v^e = \exp\frac{-\Delta G_v}{RT} = \exp\frac{S_v}{R}\exp\frac{-\Delta H}{RT} \qquad [8.16]$$

Since randomness is high while the enthalpy for the formation of vacancies is low, the number of vacancies may be high in the case of mechanical milling. The presence of a large number of vacancies may assist interdiffusion. Diffusivity of the powder can be related to the jump frequency Γ of the atoms by

$$D = \frac{1}{6}\alpha^2\Gamma \qquad [8.17]$$

where α is the jump distance. The number of successful jump per second may be expressed as

$$\Gamma = vzX_v \exp\frac{-\Delta G_m}{RT} \qquad [8.18]$$

where $v$ is mean frequency, $z$, the number of nearest adjacent sites and $\Delta G_m$, the activation energy barrier to migration. By substituting Eqs.[8.7] and [8.18] into Eq.[8.17], it can be shown that,

$$\begin{aligned} D &= \frac{1}{6}\alpha^2 zvX_v \exp\frac{-\Delta G_m}{RT} \\ &= \frac{1}{6}\alpha^2 zv \exp\frac{-(\Delta G_m + \Delta G_v)}{RT}\,. \end{aligned} \qquad [8.19]$$

Since $\Delta G = \Delta H - T\Delta S$, the above equation may be rewritten as

$$\begin{aligned} D &= \frac{1}{6}\alpha^2 zv \exp\frac{\Delta S_m + \Delta S_v}{R} \exp\left(-\frac{\Delta H_m + \Delta H_v}{RT}\right) \\ &= D_0 \exp\left(-\frac{\Delta H_m + \Delta H_v}{RT}\right) \end{aligned} \qquad [8.20]$$

where

$$D_0 = \frac{1}{6}\alpha^2 zv \exp\frac{\Delta S_m + \Delta S_v}{R}\,.$$

It can be seen that Eq.[8.20] is the same as Eq.[8.7]. It is clear that if $\Delta H_m$ and $\Delta H_v$ can be lowered, diffusion may be made easier at low temperature. This mechanically activated diffusion is one of the dominating factors in mechanical alloying and milling. If a large number of vacancies are formed during mechanical milling, it is expected that the temperature for the subsequent diffusion process will be lower.

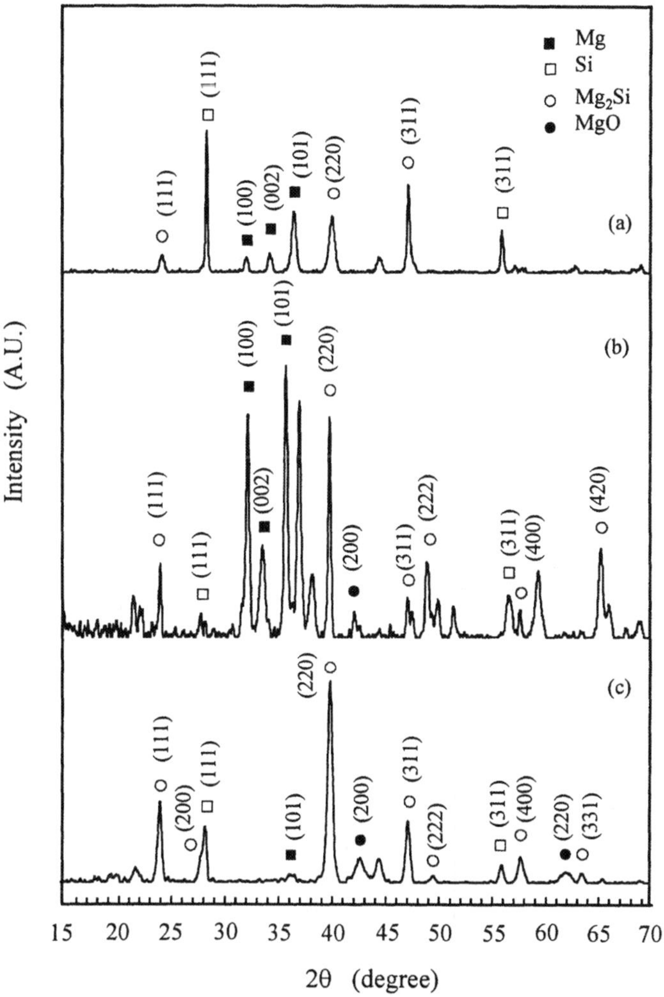


Figure 8.3. X-ray spectra of annealed Mg and Si mixture: (a) 600 minutes of mechanical milling, (b) blended Mg-Si mixture annealed at 823K and (c) 600 minutes of mechanical milling and annealed at 463K for 60 minutes.

An interesting example is the formation of $Mg_2Si$ at low temperature via mechanical milling of the powder mixture of Mg and Si. In the study, powder mixture of Mg and Si was mechanically alloyed for 600 minutes as shown in Figure 8.3 (a). The milled powders were subsequently annealed at 463K while the unmilled powder mixture was annealed in a compact form at 823K. After mechanical alloying for 600 minutes, strangely, the x-ray diffraction pattern of the compact annealed at 823 K as can be observed in Figure 8.3 (b) clearly shows that Mg and Si peaks are both present. Although there is an indication of formation of $Mg_2Si$ in Figure 8.3 (b), the amount is very small while almost 85% of the mechanically milled powder has already been transformed to $Mg_2Si$ intermetallic compound after the annealing of the milled mixture (Figure 8.3 (c)).

Study on mechanochemical reaction shows the decrease in critical temperature $T_{crit}$ (where $d\Delta T/dT > 0$) and $T_{ig}$, the temperature of onset of combustion [(16)]. Figure 8.4 shows the changes of the temperatures. The ignition temperature $T_{ig}$ drops from the original 850 K of the mixed powder to only about 460 K after 2 hours of mechanical milling. Since the rate of solid-state displacement reaction depends on temperature, interfacial area, internal energy, length of diffusion path and density of defects, the decrease in $T_{crit}$ and $T_{ig}$ indicate that mechanical milling can modify structural parameters so as to increase the reaction rate and lower the reaction temperature.

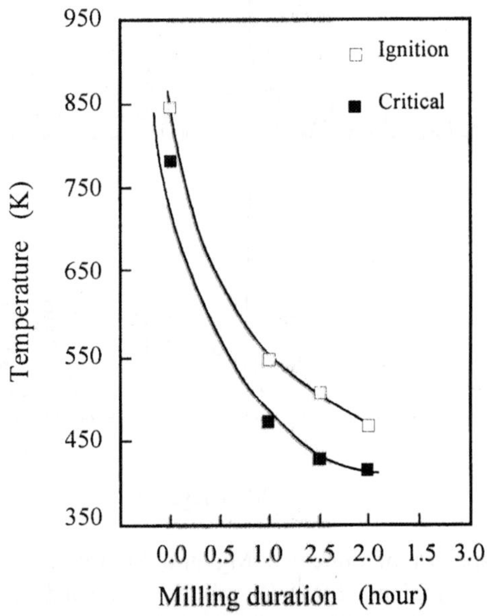


Figure 8.4. Change in critical and ignition temperatures [(16)].

### *8.1.4 Influence of collision pressure*

The free energy of a system changes with external pressure. Generally, the response of diffusivity to the application of relatively high pressure is small compared to small change in temperature. However, in the case of mechanical milling, the influence of pressure on diffusivity can be high due to the very high collision pressure involved. The relationship between diffusivity and externally applied pressure can be made by differentiating Eq.[8.19] with respect to pressure $P$ at constant temperature [18].

$$\left(\frac{\partial \ln D}{\partial P}\right)_T = -\Delta V_p - \Delta V_v^0 \qquad [8.21]$$

where $\Delta V_p$ is the change in the volume when one mol of diffusion atoms is transferred from the equilibrium lattice position to the saddle point at a pressure of 1 atmosphere, $\Delta V_v^0$ is the change in the volume produced by the creation of one mol of vacancies with an external pressure of 1 atmosphere.

The change in volume of a solid $\Delta V_v$ due to the creation of a vacancy at pressure greater than zero can be written as:

$$\Delta V_v = \Delta V_v^0 + P\left(\frac{\partial \Delta V_v}{\partial P}\right)_T. \qquad [8.22]$$

Therefore,

$$\Delta G_v = \Delta G_v^0 + P\Delta V_v \qquad [8.23]$$

and

$$\Delta G_m = \Delta G_m^0 + P\Delta V_p \qquad [8.24]$$

where $\Delta G_v^0$ and $\Delta G_m^0$ are respectively the free energies of vacancy formation and migration of atoms at 1 atmospheric pressure.

Substituting Eqs.[8.23] and [8.24] into Eq.[8.19], the relationship between diffusivity and pressure may be obtained by:

$$D_p = \frac{1}{6}\alpha^2 z\nu \exp\frac{-\left(\Delta G_m^0 + \Delta G_v^0 + P\Delta V_p + P\Delta V_v\right)}{RT} \qquad [8.25]$$

where $D_p$ is diffusivity under pressure. Eq.[8.25] may be written in a form similar to Eq.[8.20] as:

$$D_p = D_0 \exp\left(-\frac{\Delta H_m + \Delta H_v}{RT}\right) \exp\left(-P\frac{\Delta V_p + \Delta V_v}{RT}\right) \qquad [8.26]$$

It can be further simplified to

$$D_p = D \exp\left(-P\frac{\Delta V_p + \Delta V_v}{RT}\right). \qquad [8.27]$$

It can be seen from Eq.[8.27] that diffusivity changes with the change in pressure and volume.

## 8.2 Formation of new materials

### *8.2.1 Effect of temperature*

The increase in temperature during mechanical alloying is of particular concern. It has been argued that the large plastic deformation that takes place in the process induces local melting leading to the formation of new alloys (19). According to this hypothesis, new alloys are produced through a melting mechanism and/or diffusion at relatively high temperature. For some materials with low melting temperature, it is quite possible that local melting is induced by high collision energy. For most materials, however, melting cannot take place during mechanical alloying. To achieve cold-welding which is an essential condition for mechanical alloying, at least one material should normally be ductile. The latter acts as binder to bind powder particles together (20). However, it has been found that brittle materials such as solid solution of Si + Ge, intermetallic compounds like Mn + Bi and amorphous alloys of $NiZr_2 + Ni_{11}Zr_9$ can also be mechanically alloyed (21, 22).

A key feature of the mechanical alloying of brittle components is the temperature of the powder particles during milling. When impact pressure on the powder particle reaches yield stress, the mechanical energy generated may transform the powder particles resulting in an increase in temperature, thus stabilizes the different structural defects (8). No alloying could be observed when the milling vial was cooled with liquid nitrogen (22). It has been claimed that pure ceramic materials of WC, TiN and TaN can partially be mechanically alloyed at certain conditions (23). In general, the temperature during mechanical alloying is believed to be far below the melting temperatures of WC, TiN and TaN or their diffusion temperatures. The temperature measured after a certain duration of mechanical alloying was only about 393 K. However, because the change in temperature of the powder particles is a dynamic process, it is very difficult to directly monitor the temperature during the alloying process. Davis and co-worker (22) used an indirect method to determine the change in temperature. In their test, a Fe-1.2 wt.% C alloy was prepared by melting electrolytic pure iron and graphite in an induction furnace. The sample was placed in an evacuated quartz tube containing Zr and austenitised at 1,150 K followed by ice-water quenching. Finally about 140 mesh powder with martensitic structure was produced. The resultant powder particles were then milled and their microstructure was analyzed. Since the martensitic structure will decompose into ε-carbide and a lower carbon martensite at 418 K and into cementite and ferrite at 575 K, the temperature of the mechanical alloying process can be evaluated by examination of the remaining convertible cementite in the sample. The results showed that a change in temperature did occur during mechanical alloying, but the maximum temperature was only about 559 K which is far from the normal diffusion temperature. Lu *et al.*[15] measured the thermodynamic properties of nickel and aluminium mixture, and of copper-zinc and aluminium powder mixture using differential scanning colorimeter after different times of mechanical alloying. It was found that the lowest exothermic reaction in copper-zinc-aluminium system was about 473 K corresponding to the recovery and partial release of the stored energy in the particles. This implies that the temperature of the powder particles during mechanical alloying was lower than 473 K otherwise they would have been annealed during the process. differential scanning calorimetry measurement of nickel-aluminium also revealed that the lowest exothermic reaction occurred at about 633 K. Although the maximum temperature during mechanical alloying is dependent on several factors, such as type of ball mill used, speed of rotation employed and properties of the powder particles to be mechanically alloyed, according to the above measurements, it is believed that the global temperature is relatively low. Temperature at certain locations under high energy collision, however, could be higher due to the large amount of plastic deformation and sliding but this heat is very rapidly dissipated by the large mass of the milling vial, balls and powder particles. Nevertheless, the increase in milling temperature certainly plays a very important role in mechanical alloying, not only in the diffusion process but also in controlling the size of the particles. In addition, the change in temperature during reaction milling is very significant due to the exothermic reaction causing local combustion. An abrupt increase in temperature observed may cause dangerous explosive reaction to form new compounds.

As discussed earlier, two major phenomena can contribute to the increase in milling temperature: friction during collision and localized plastic deformation. Schwarz and Koch [24] proposed a model by which the rise in temperature due to localized deformation can be predicted:

$$\Delta T = \frac{\phi_f}{2}\sqrt{\frac{\Delta t}{\pi k_0 \rho_p c_p}} \qquad [8.28]$$

where $\Delta T$ is the increase in milling temperature, $\phi_f$, dissipated energy flux given as $\sigma v$ where $\sigma$ is the normal stress due to a head-on collision and, $v$, the relative velocity of the ball before impact, $\Delta t$, stress state life, $\rho_p$, density of the powder particles, $k_0$, thermal conductivity of the particles and, $c_p$, the specific heat of the powders. If the following values are taken: $k_0$=2.37 $Wcm^{-1}K^{-1}$, $\rho_p$=2.70g $cm^{-3}$, $c_p$=900J $kg^{-1}$ $K^{-1}$, and $\sigma$=170MN $m^{-2}$ (31), the increase in temperature during mechanical alloying can be calculated as a function of collision speed and stress state life using this model. Figure 8.5 shows the change in temperature of aluminium powder at different mechanical alloying conditions. Although strongly dependent on stress state life and collision velocity, it has been pointed out by Davis (22) that the temperature rise in most collision events is about 100-350K.

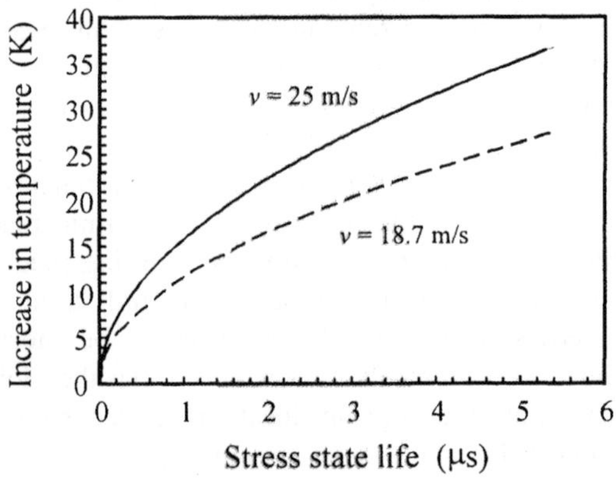


Figure 8.5. Change in temperature at different impact velocities.

Streletskii (25) considered the energy transfer during a single collision. The average energy intensity, $I_m$, is directly proportional to the power, $W$, of the electric motor. This intensity may be measured with the help of the calorimetric method (25-27). The main part of the impact energy $I_m$ is transformed to heat resulting in an increase in milling temperature (25, 28-30). Only a small part of the energy contributed to the structural defects of the substance.

The relationship between impact intensity, $I_m$, and the rate of temperature increment, $\partial T/\partial t$, can be written as (25):

$$I = c_e \sum m_i \frac{\partial T}{\partial t} \qquad [8.29]$$

where $c_e$ is the effective specific heat capacity of the milling tools and powder and $m_i$ is the mass of substance $i$. Eq.[8.29] gives the change in global temperature when heat transfer is isolated. Since impact during milling occurs within a few milliseconds, it may be assumed that the particle being deformed is thermally isolated at the instance of impact with no dissipation of heat. In this case, at the event of impact, the impact intensity, $I_m$, may only lead to an increase in temperature of the powder particles under deformation. Consequently, $m_i$ in Eq.[8.29] will be the mass of the particles under deformation. The effective specific heat capacity, $c_e$, will be the heat capacity, $c_p$, of the powder particles. The initial temperature before impact should therefore be the global temperature. During the impact the temperature of the powder increases and soon after impact, it will be dissipated to the surrounding.

Butyagin [1] considered the influence of crystalline size $d_c$ on the change in temperature in the mechanical milling process. During surface contact, the heat of adhesion, $Q_{ad}$, is transformed into kinetic energy of the atoms causing the temperature to increase. The increase in temperature can be written as:

$$\Delta T = \frac{Q_{ad}}{c_p(d_c)} \qquad [8.30]$$

Besides heating due to plastic deformation, increase in temperature on a microscopic level as a result of sliding friction can also be predicted [32]. The method of prediction considers a system where a body comes into contact with another over a limited area and moves over the surface of the other body at a constant velocity. The contact area is considered to be square. Under these conditions, the microscopic temperature rise is given by:

$$\Delta T = \frac{\mu P V_l}{4.24 l Q_h^*(K_1 + K_2)} \qquad [8.31]$$

where $\mu = 0.6$ is the frictional coefficient, $P$, the load, $l$, half the length of the side of the contact area, $Q_h^*$, the mechanical equivalent of heat, and $K_1$ and $K_2$, the thermal conductivity of the respective bodies.

In addition to temperature change due to localized plastic deformation and friction, combustion [33], reaction of enthalpy [34] and precipitation can also contribute to the increase in temperature, especially reaction of enthalpy.

### *8.2.2 Effect of crystalline size*

High density of defects is generated with prolonged mechanical milling duration, leading to an increase in microhardness of the powder particles. Because of large plastic deformation and repeated fracturing, the crystalline size of the particles can be reduced to nanometer scale as may be evidenced by the change in width at half of maximum X-ray peak. The presence of large surface area of the mechanically milled powder particles implies a large amount of energy being stored in the milled particles. Fecht [17] indicated that the final energies stored during mechanical milling exceeded those resulting from conventional cold working of metals and alloys. During conventional plastic deformation, the excess energy is seldom found to be above 1-2 kJ $mol^{-1}$ whereas in the case of mechanical milling, the energy can reach values typically those of crystallization enthalpies of metallic glasses corresponding to about 40% heat of fusion. The study of Fecht shows that the change in enthalpy can be divided into two regimes depending upon the size of the grains. As shown in Figure 8.6, the stored enthalpy slowly increases with decreasing grain size. The change in enthalpy weakly depends on grain size as a result of the dislocation controlled deformation processes of large grained polycrystals [17]. When grain size is smaller than 25 nm, the stored enthalpy becomes strongly dependent on the grain size. The critical grain size corresponds to the size of nanograins which are formed within the shear bands. The atomic level strain shows an increase with decreasing grain size. Two regimes can also be clearly observed as shown in Figure 8.7. Contribution to stored enthalpy due to grain boundary energy is considerably higher in comparison to that of fully equilibrated higher angle grain boundaries [35-37]. The high stored energy in the mechanically milled powder particles drives the structural change in the materials to minimize their internal energy level.

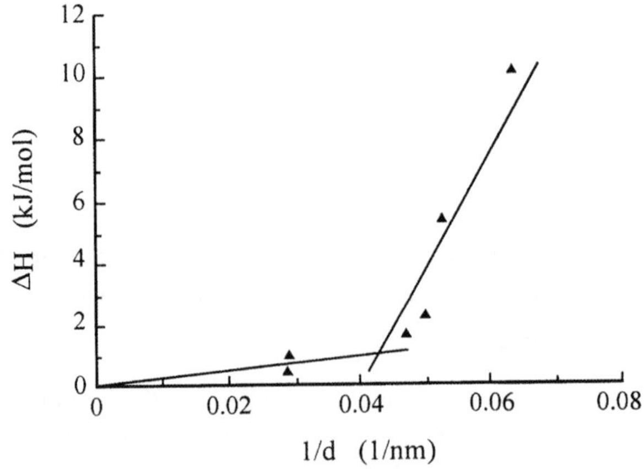


Figure 8.6. Stored enthalpy as a function of the reciprocal of grain size [17].

Microscopically, there are two opposing processes to promote and to limit the growth of the surface area as a result of the dynamic equilibrium between forward and reverse processes. The forward process is mechanical dispergation of particles by mechanical fracture and plastic deformation of the powder particles while the reverse process is the aggregation of powder particles. The structure of aggregates depends on the nature of interatomic bonds [(1)]. In brittle materials, the reversible process is called molecular-dense aggregation. Mechanical milling of the powder mixtures results in aggregation and structural relaxation by the action of chemical forces. If two elemental particles A and B are milled, aggregations may consist of particles A-A, B-B and A-B, depending on the relative strength of the forces of cohesion and/or adhesion. Consequently, the relaxation of the disordered mixture of substances may result in the formation of crystals A, B or AB [(1)].

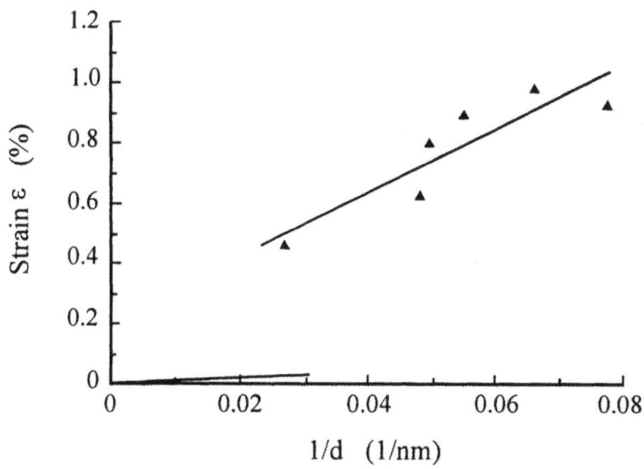


Figure 8.7. Relationship between atomic level strain and the reciprocal of the grain size [(17)].

As reported by Horvath *et al.* [(39)], measurement of self-diffusivity in nanocrystalline copper revealed an enhancement of its self-diffusivity by a factor of about $10^{19}$. This remarkable enhancement may be understood in terms of the large amount of boundaries that provide a connected network of short-circuit diffusion paths. An enhancement of self-diffusivity in comparison to boundary diffusion of a factor of about 100 has been reported [(39)]. Because a large volume fraction of the atoms resides in the grain boundaries, nanocrystalline materials have high reactivity and high diffusivity, thus enable the alloying of normally immiscible metals [(40)].

Furthermore, it is known [13] that at any temperature, the magnitudes of the diffusivity along grain boundary $D_b$ and along free surface $D_s$ relative to the diffusivity through defect-free lattice $D_l$, are such that

$$D_s > D_b > D_l \qquad [8.32]$$

Eq.[8.32] mainly reflects the relative ease with which atoms can migrate along free surfaces because such surfaces tend to be regions of relatively higher disorder and lower activation energy for diffusion than the interior. Figure 8.8 shows the relationship between diffusion via the surface, the grain boundary and the lattice [41]. It can be seen that lattice diffusion has the largest slope while surface diffusion, the smallest. As a result, at low temperatures, surface diffusion dominates over grain boundary and lattice diffusions. As the temperature is increased, however, grain boundary diffusion predominates and, at higher temperature, lattice diffusion becomes the principal mode of diffusion [41]. The change from one predominant mechanism to another depends upon the nature of the grain boundaries and surfaces.

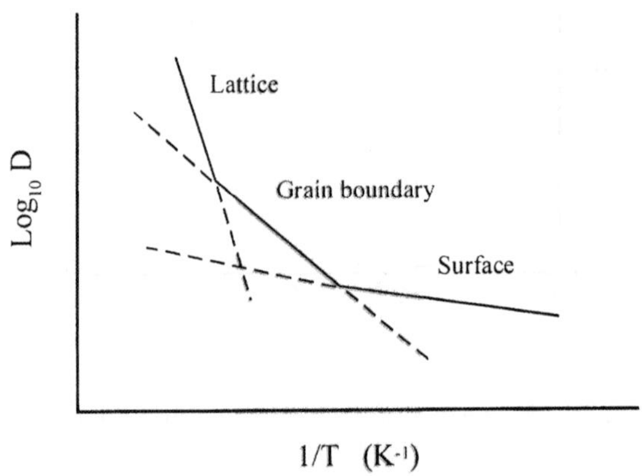


Figure 8.8. Diffusivity of surface, grain boundary and lattice diffusion [41].

The process of mechanical alloying consists of fracturing and welding. Hence, at a microscopic level, the process gives rise to three types of free surfaces for diffusion processes to proceed. The first type of free surface occurs in microcracks. Microcracks can be introduced either by fracture or by the motion of dislocations. Owing to the formation of microcracks, the internal energy is increased by the increase in surface energy. To stabilize the powder particles (minimize internal energy), the atoms would attempt to diffuse along the surface of the microcracks to fill up the crack. As the process is driven by the reduction in surface area, driving force is lowered when the microcracks are filled up by diffused atoms. This process proceeds throughout the entire mechanical alloying process even after a homogeneous material is eventually formed.

Similar to the first type of free surface, the second type of macrocrack-like free surface is due to cold-welding between different particles. The mechanism of diffusion here is similar to the sintering of powder particles. The difference in radii $L$ (radius of particle) and $r$ (radius of a crack tip) at the valley means that the vapor pressure is lower in the valley than over the rest of the particles and thus atoms can be transferred through the gaseous phase from the main surface of the particle to the valley. There is, however, an outward-acting force on the surface of the valley due to surface tension $\gamma$ which gives rise to a stress. According to sintering theory, the magnitude of the stress is approximately equal to $\gamma/r$. If the ratio at the valley is very small, the magnitude of the stress will be very large.

The third type of free surface is pre-cold welding surface which is actually a mechanically bonded surface without real welding. Atoms can diffuse along the rough free surfaces to finally form welded boundaries with the disappearance of mechanical bonding. The three types of free surfaces are generated by fracturing and plastic deformation but will disappear as diffusion proceeds.

In addition, fracturing and cold-welding during mechanical alloying enable different particles to be always in contact with each other with fresh surfaces. The diffused layer between different particles is in turn repeatedly broken leading to a minimized diffusion distance under high hydrostatic pressure which is a major difference from normal diffusion process. The formation of layered structure creates more free surfaces. It is noted that increased collision pressure results in higher compact density of the powder particles. The latter, which occurs during collision event, leads to better interparticle contact which improves inter-diffusion between particles. On the other hand, lower density results in higher surface area in the collision and thus a larger contribution of surface diffusion to homogenization. Therefore, it is believed that diffusivity along free surface $D_s$, dominates the mechanical alloying process.

Schumacher [(42)] found that silver can be diffused into copper at temperature between 303 and 373 K if the crystalline size of copper is in nanometer. This is normally not possible when ordinary copper is used. It implies that the diffusivity of nanocrystalline materials at low temperature is very high in comparison to materials with normal crystalline size. This can be represented using a simple mathematical expression [(41)]:

$$D_{eff} = (1 - F)\, D_l + F D_b \qquad [8.33]$$

where $D_{eff}$ is the effective diffusivity, and $F$, the area fraction of short-circuit paths (i.e. grain boundaries) in a plane perpendicular to the direction of diffusion. It may be understood that the effective diffusivity can be increased by decreasing grain size (i.e. increasing the area of grain boundaries). This situation is well promoted in the mechanical alloying process. If the grain boundary width is $\delta$ and grain size, $d$, then, $F = 2\delta/d$ approximately. Still taking Al-Cu system as an example, if thickness of the grain boundary is $7*10^{-6}$ mm [10], effective diffusivity of this system will be increased from $1.03*10^{-18}$ to $5.70*10^{-15}$ when the grain size is reduced from 50 μm to 100 nm. Figure 8.9 shows the change in diffusivity with crystalline size and temperature. It can be seen that diffusivity is dramatically changed when the crystalline size is reduced to nanosize and that diffusion temperature strongly influences diffusivity at small grain size.

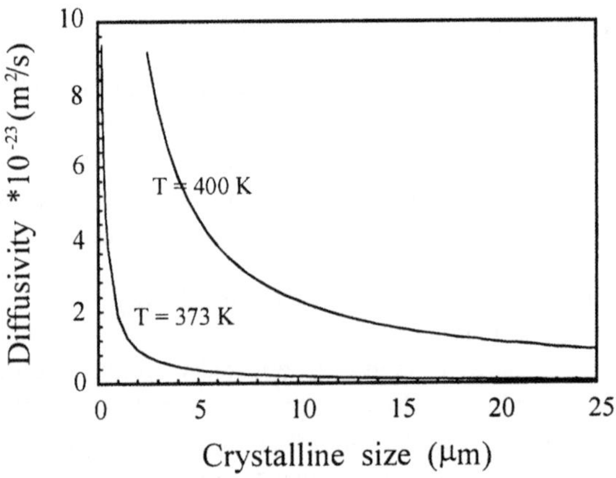


Figure 8.9. Change in diffusivity as a function of crystalline size.

Using Eq.[8.33], it can be observed that grain boundary diffusion takes place if the following expression is satisfied [41]:

$$FD_b > (1 - F)\, D_l \qquad [8.34]$$

For such a situation to occur, the grain size must be small.

For the transition from grain boundary to lattice diffusion, the following criterion should be satisfied:

$$FD_b = (1 - F)D_l. \qquad [8.35]$$

Thus,

$$\frac{D_l}{D_b} = \frac{F}{(1-F)} \qquad [8.36]$$

Since

$$D_l = D_0^l \exp\left(\frac{-\Delta Q}{RT}\right) \qquad [8.37]$$

and

$$D_b = D_0^b \exp\left(\frac{-\Delta Q}{RT}\right), \qquad [8.38]$$

it follows that

$$\frac{F}{1-F} = \frac{D_l}{D_b} = \frac{D_0^l}{D_0^b}\left[\exp\frac{\Delta Q_b - \Delta Q_l}{RT_{trans}}\right]. \qquad [8.39]$$

Taking natural log of Eq.[8.37] (41),

$$\ln\left(\frac{F}{1-F}\frac{D_0^b}{D_0^l}\right) = \frac{\Delta Q_b - \Delta Q_l}{RT_{trans}} \qquad [8.40]$$

or

$$T_{trans} = \frac{Q_b - Q_l}{R\ln\left(\frac{F}{1-F}\frac{D_0^b}{D_0^l}\right)} \qquad [8.41]$$

To evaluate the transition temperature, the following data have been taken in the calculations (10):

$Q_l$ = 126 kJ/mole, $Q_b$ = 84 kJ/mole, $D_0^l = 1.5*10^{-5}$ m²/s, $D_0^b = 15\,D_0^l$, and $\delta \cong 7$ Å.

If $d$ = 50 μm,

$$F = 2\delta/d = 2*7*10^{-10}/50*10^{-6} = 2.8*10^{-5}.$$

If $d = 5$ μm,

$$F = 2\delta/d = 2*7*10^{-10}/5*10^{-6} = 2.8*10^{-4} .$$

Substituting the above values into Eq.[8.41], it can be seen that grain boundary diffusion begins to dominate over lattice diffusion at temperature below 649 K if the grain size is 50 μm, but below 923 K if the grain size is reduced to 5 μm. This explains why for mechanical alloying, only boundary and free surface diffusions are prevalent. Lattice diffusion can be avoided because of the small grain size involved.

It is believed that the diffusivities of stress-free and deforming samples are different. Under high collision stress during mechanical alloying, the lattice parameter at some locations of the deformed particles can be increased, leading to large diffusion path. Consequently, activation energy required by moving atoms is lowered. The solute atoms can easily diffuse into diffusion couple under this high collision stress.

### 8.2.3 *Effect of crystal structure*

Diffusivities, beside being affected by temperature, are also dependent upon crystal structure. The type of crystal structure of the solvent lattice is important. In particular, elements with small atomic diameters that form interstitial solid solutions diffuse very rapidly. Carbon diffuses faster in bcc iron than in fcc iron where diffusivities of bcc and fcc structures are respectively $D_{bcc}=10^{-12}$ $m^2/s$ and $D_{fcc}=5*10^{-15}$ $m^2/s$. The reason for this difference is that the bcc structure has a lower atomic packing factor of 0.68 as compared to that of fcc structure. The passageways of carbon atoms through the unit cell of an fcc structure are tighter for the movement of carbon.

During mechanical alloying, crystal structure of the powder particles can be changed resulting in a change in diffusivity. An example of such occurrence is the mechanically alloyed CuZnAl alloy from fcc crystal structure of α phase CuZn pre-alloyed powders and element Al powder [(49)]. During mechanical alloying, the Al atoms diffuse into the CuZn alloy resulting in the formation of CuZnAl alloy with low Al content. With increase in milling duration, more Al atom diffuses into CuZnAl to form β phase which has a bcc crystal structure. As diffusivity of α phase structure at 700°C is only about $4.5*10^{-11}$ $cm^2$/sec., while in β phase structure at 550°C, it is about $3*10^{-8}$ $cm^2$/sec, diffusion in β phase can be clearly seen to be much faster than that in α phase.

## 8.3 Conclusions

The key factors controlling the formation of new alloys are found to be: activation energy, which is related to the formation of defects during collision of milling balls with the powder particles and the vial; temperature, which is associated with plastic deformation of the powder particles and sliding between powder particles, and between high energetic balls and powder particles, and crystalline size, which is related to the formation of nanometer crystalline structure during mechanical alloying.

Diffusion is a mechanism of formation of new alloys during mechanical alloying. Diffusion process in mechanical alloying, however, differs from thermally induced diffusion process. The latter process is controlled by thermal energy while in the former process formation of new alloys is controlled by both thermal and mechanical energies. During the alloying process, large amount of defects is created by large plastic deformation. According to thermodynamic theory, at a constant temperature, a decease in activation energy can result in an increase in diffusivity. Therefore, decrease in activation energy is equivalent to an increase in temperature. High diffusivity can be achieved by creating large amount of defects through mechanical alloying. It is suggested that the decrease in activation energy due to increase in vacancies gives rise to an increase in diffusivity thereby the diffusion process is eased.

Due to the nature of the mechanical alloying process consisting of fracturing and welding, at least three types of free-surfaces can contribute to surface diffusion mechanisms. Because of the formation of more free surfaces and grain boundaries, surface diffusion which is driven by the reduction in surface area and radius of the crack tip, dominates during mechanical alloying. For thermally induced diffusion process, the driving forces due to surface tension and surface stress always tend to be lower with diffusion duration, while for mechanical alloying induced diffusion process, the opposite trend can be observed due to the formation of microcracks and other defects. By creating nanometer crystalline through repeated fracturing and cold-welding of powder particles, diffusion can easily take place through the grain boundaries. Consequently, elements which are difficult to diffuse may be alloyed using this technique.

Grain boundary and free surface diffusions are always the prevalent modes of diffusion in mechanical alloying process. Although the diffusion temperature is relatively low, diffusivity is very high. Because of small grain size, lattice diffusion becomes significant only at relatively high temperature.

With increase in mechanical alloying duration, temperature of the powder particles is raised by plastic deformation, sliding, chemical, transformation, precipitation and other reactions. All these further enhance the diffusion process.

## 8.4 References


1. P. Yu Butyagin, *InCoMe '93, Fundamentals and Models of Mechanically Stimulated Processes*, Proc. of the 1st Intern. Confer. on Mechanochemistry, Košic, March 23-26, 1993, 27.
2. Gutman, *ibid*, 34.
3. E.L. Goldberg and S.V. Pavlov, *ibid*, 66.

4. N. Lyakhov, *ibid*, 59.
5. V. Jesenak and K. Jesenak, *ibid*, 76.
6. K. Tkacova, G. Paholic, N. Stevulova and V. Sepelak, *ibid*, 79.
7. A.K. Jena and M.C. Chaturvedi, *Phase Transformations in Materials*, Prentic-Hall Inc., (1992), 74.
8. P.Y. Butyagin, *Mechanical Alloying for Structural Applications*, Proc. of the 2nd Intern. Confer. on Structual Appl. of Mechanical Alloying, 20-22 Sep. 1993, Vancouver, British Columbia, Canada, Ed. J.J. deBarbadillo, F.H. Froes and R. Schwarz, Publ. ASM Intern., Mater. Park, Ohio 44073-0002, 385
9. W.F. Smith, *Foundations of Materials Science and Engineering*, 2nd Edition, McGraw-Hill Inter. Ed., (1993), 172.
10. H.J. Frost and M.F. Ashby, *Deformation-Mechanism Maps* -The plasticity and creep of metals and ceramics, Pergamon Press, (1982), 21.
11. A.K. Bhattacharya and E. Arzt, *Scripta Metall.*, Vol. 27 (1992), 635.
12. W.D. Nix and J.C. Gibeling, *Flow and Fracture at Elevated Temperature*, ed. R. Raj, Publ. ASM, Metals Park (1985).
13. D.A. Porter and K.E. Easterling, *Phase Transformations in Metals and Alloys*, Chapmen & Hall, 1993.
14. G.T.Hida and I.L. Lin, *Combusion and Plasma Synthesis of High Temperature Materials*, ed: J.B. Holt and Z.A. Munir, American Ceramic Society, (1988), 246.
15. L. Lu, M.O. Lai and S. Zhang, *J. Mater. Res. Bull.*, Vol.29, 8 (1994), 889.
16. G.B. Schaffer and P.G. McCormick, *Metall. Trans. A*, Vol.22A (1991), 3019.
17. H.J. Fecht, *Nanophase Mater.*, Ed. G.C. Hadjipanayis and R.W. Siegel, Publ. Kluwer Academic Publishers, 1994, 125.
18. Richard. J. Borg and G.J. Dienes, *An Introduction to Solid State Diffusion*, Publ. Academic Press, INC., Harcourt Brace Jovanovich, Publishers, 1988.
19. A.Y. Yermakov, Y.Y. Yurchikov and V.A. Barinow, *Phys. Met. Metall.*, Vol.52 (1981), 50.
20. P.S. Gilman and W.D. Nix, *Metall. Trans. A*, Vol. 12A (1981), 813.
21. R.M. Davis and C.C. Koch, *Script Metall.*, Vol.21 (1987), 305.
22. R.M. Davis, B. McDermott and C.C. Koch, *Metall. Tran. A*, Vol.19A (1988), 2867.
23. S. Zhang, K.A. K.A. Khor , L. Lu, *J. Mater. Proc. Tech.*, Vol.48 (1995), 779.
24. R.B. Schwarz and C.C. Koch, *Appl. Phys. Lett.* Vol.49 (1986), 146.
25. A.N. Streletskii, *Mechanical Alloying for Structural Applications*, Proc. of the 2nd Intern. Confer. on Structual Appl. of Mechanical Alloying, 20-22 Sep. 1993, Vancouver, British Columbia, Canada, Ed. J.J. deBarbadillo, F.H. Froes and R. Schwarz, Publ. ASM Intern., Mater. Park, Ohio 44073-0002, 51.
26. A.R. Kuznetsov, P.Yu. Butyagin and I.K. Pavlichev, *Apparatus & Tech. Experiment*, Vol.6 (1986), 201.
27. I.K. Pavlichev, A.B. Pakovich, A.N. Streletskii and P.Yu. Butyagin, *Proc. Confer. on UDA-Technology*, Tambov, Rassia, (1984), 34.
28. M. Magini, *Mater. Sci. Forum*, Vols.88-90 (1992), 121.
29. M. Magini and A. Iasonna, *Mater. Trans. JIM*, Vol.36 (1995), 123.
30. M. Magini, A. Iasonna and F. Padella, *Scripta Mater.*, Vol.34 (1996), 13.
31. *Metals, Alloys, Compounds, Ceramics, Polymers and Composites -Catalogue* 1994/1995, Goodfellow Cambridge Limited, Cambridge Science Park, Cambridge, CB4 4DJ, England, (1994), 33.
32. H.S. Carslaw and J.C. Jaeger, Oxford University Press, New York (1959), 255.
33. G.B. Schaffer and P.G. McCormick, *Metall. Trans. A*, Vol.21 A (1990), 2789.
34. G.B. Schaffer and P.G. McCormick, *Mat. Forum*, Vol.16 (1992), 91.
35. H.J. Fecht, E. Hellstern, Z. Fu and W.L. Johnson, *Adv. Powder Metall.*, Vol.1 (1989), 111.
36. H.J. Fecht, E. Hellstern, Z. Fu and W.L. Johnson, *Metall. Trans. A*, Vol.21A (1990), 2333.
37. J. Echert, J.C. Holzer, C.E. Krill III and W.L. Johnson, *Mater. Sci. Forum*, Vols.88-90 (1992), 505.

38. P.G. McCormick, V.N. Wharton, M.M. Royhani and G.B. Schaffer, *Microcomposites and Nanophase Materials*, ed. D.C. Van Aken, G.S. Was and A.K. Gosh, TMS, Warrendale, Pennsylvamia (1991), 65.
39. H. Gleiter, *Sci. of Adv. Mater.*, ASM Mater. Sci. Semi., 26-29 Sep. 1988, ASM Intern. Materials Park, Ohio, ed: H. Weidershich and M. Mechii, (1988), 203.
40. C. Suryanarayana and F.H. Froes, *J. Mater. Res.*, Vol. 5 (1990), 1880.
41. H.V. Atkinson and B.A. Rickinson, *Hot Isostatic Processing*, The Adam Hiler Series on New Manufacturing Processes and Materials, ed. J. Wood, Adam Hilger, Bristol, Philadelphia and New York, (1991), 34.
42. S. Schumacher, R. Birringer, R. Straub and H. Gleiter, *Acta Metall.*, Vol.37 (1989), 2485.
43. G.B. Schaffer and P.G. McCormick, *Metall. Trans. A*, Vol. 23A (1992), 1285.
44. G.B. Schaffer and P.G. McCormick, *Metall. Trans. A*, Vol. 22A (1991), 3019
45. M.A. Morris, *J. Mater. Sci.*, Vol. A150 (1991), 1157.
46. C Suryanarayana and L. Schultz, *Mater. Sci. Eng.*, Vol. 93 (1987), 213.
47. B.S. Murty, S. Ranganathan and M. Mohan Rao, *Mater. Sci. Eng.*, Vol. A149 (1991), 231.
48. A. Johson, Topical Discussion, *New Mater. by Mech. Alloying Techs.*, 3-5 October 1988, Calw-Hirsau (FRG), Ed: E. Arzt and L. Schltz, Informationsgesellschaft Verlag, 354.
49. S. Zhang, L. Lu and M.O. Lai, *Mater. Sci. Eng.*, Vol.171 A (1993), 257.
50. W.L. Johson, *Progr. Mater. Sci.*, Vol.30 (1986), 81.
51. H. Bakker and L.M. Di, *Mater. Sci. Forum*, Vols.88-90 (1992), 27.
52. H. Bakker, G.F. Zhou and H. Ynag, *Mater. Sci. Forum*, Vols.179-181 (1995), 47.
53. L.M. Di and H. Bakker, *J. Appl. Phys.*, Vol.71 (1992), 5650.

# 9

# MODELING OF MECHANICAL ALLOYING

## 9.1 Modeling

Mechanical alloying is a process governed by the phenomena of repeated cold welding and fracturing. The degree of cold welding and fracturing and hence the effect of mechanical alloying are directly associated to the energy released from the impact of the balls onto the powder particles. To better understand the mechanics of the mechanical alloying process, different approaches have been employed to develop physical models of the process [(1-5)]. In general, collisions between the balls and the vial are considered to be either perfectly elastic in which there is no loss in kinetic energy or imperfectly elastic in which energy is dissipated [(6, 7)]. Because most collisions are in practice almost perfectly elastic, Hertz's theory of impact can be approximately applied in the study of mechanical alloying process [(7)]. To simulate the alloying process, several key factors have to be considered: (a) kinetic energy of the ball, (b) fraction of the kinetic energy transferred to the powders, and (c) quantity of material entrapped in the impact event [(7, 8)]. This process has been considered to be similar in some respects to hot isostatic pressing and superplastic forming [(9)]. Courtney and Mauric [(10)] have recently divided the modeling of the process into three categories: local modeling, global modeling and synthesis modeling.

## 9.2 Kinetics

### *9.2.1 Planetary ball mill*

A planetary ball mill is a specially designed apparatus for ball milling on a laboratory scale. In the milling process, energy transfer from a flying ball to the powder particles being milled is essentially governed by impact speed and angle of the ball with respect to other objects such as other balls or the wall of the vial. Consequently, the motion and the impact event of the balls are of importance and have mathematically been well studied [(11-13)]. As early as in 1956, Joisel [(14)] had evaluated the trajectories of motion of the ball during ball milling by using a graphic construction. Analytical solutions were later proposed by Rumpf [15] and adopted by John *et al.* [16]. In 1992, Raasch [12] suggested another theory of the motion of the balls in a planetary ball mill by which the cataracting motion of the milling bodies could be predicted. An iterative calculation of time and location of impact is given by evaluation of the trajectories. Based on such formulations, the effects of impact and optimal design of a planetary ball mill can be deduced [11, 12].

Experimentally, it has been observed by using high speed photography that the balls fly along certain trajectory [11]. By using new vials in each experiment, the impact traces after 120 minutes of milling have been clearly identified and are as shown in Figures 9.1 (a) to (d). It is evident that the positions of impact of the balls on the wall of the vials are ball size dependent. Most collisions occur at some certain height. The distance between two traces is more or less equal to the diameter of the ball used if fewer balls are used (Figures 9.1 (a) and (b)). When the number of balls with size of 20 mm diameter is increased, the balls are pulled to the upper side of the vial and the distance between the neighbouring traces decreases (Figure 9.1 (c)). Impact between the balls and the lid of the vial can also be observed. If 40 balls of size 15 mm diameter are used for a milling duration of over 400 hours, four impact bands of about 1 to 2 mm deep have been observed, implying that even though more balls are used, the fly trajectory of the balls can be determined by the size of the ball. The distance between the two neighbouring tracks is about 12 to 13 mm. Based on this measurement, the most probable stacking sequence of the balls is schematically illustrated in Figure 9.2 when 15 mm ball size is used. This kind of stacking sequence gives the distance between two neighbouring tracks to be 13 mm. For 20 mm ball size, a distance of 17.3 mm is obtained, which also corresponds to the track distance as shown in Figures 9.1 (c).

Impact bands are generally very narrow indicating a fixed flying height of repeated collisions. In the case where different sizes of balls are used in one vial, more complicated impact traces would be obtained. It is difficult to find the relationship between the traces and the size of the ball. The impact bands in this case become wider due mainly to the interference of balls of different sizes indicating most likely the occurrence of size effect. If fewer balls are used, no impact trace can be found on the surface of the cover; only one impact trace on the cover could be visible when more balls are used. Logically, therefore, it can be assumed that the flying kinetics of the balls can be simulated by considering the flying trajectory of a single ball with the interference between balls being omitted. Another evidence is the arrangement of the ball after impact. If five balls are used in a planetary ball mill, the original arrangement of the five balls after impact is seen to be kept indicating that there is little interference between the balls [17].

Further hypotheses are needed to simplify the simulation of the mechanical alloying process [13]:

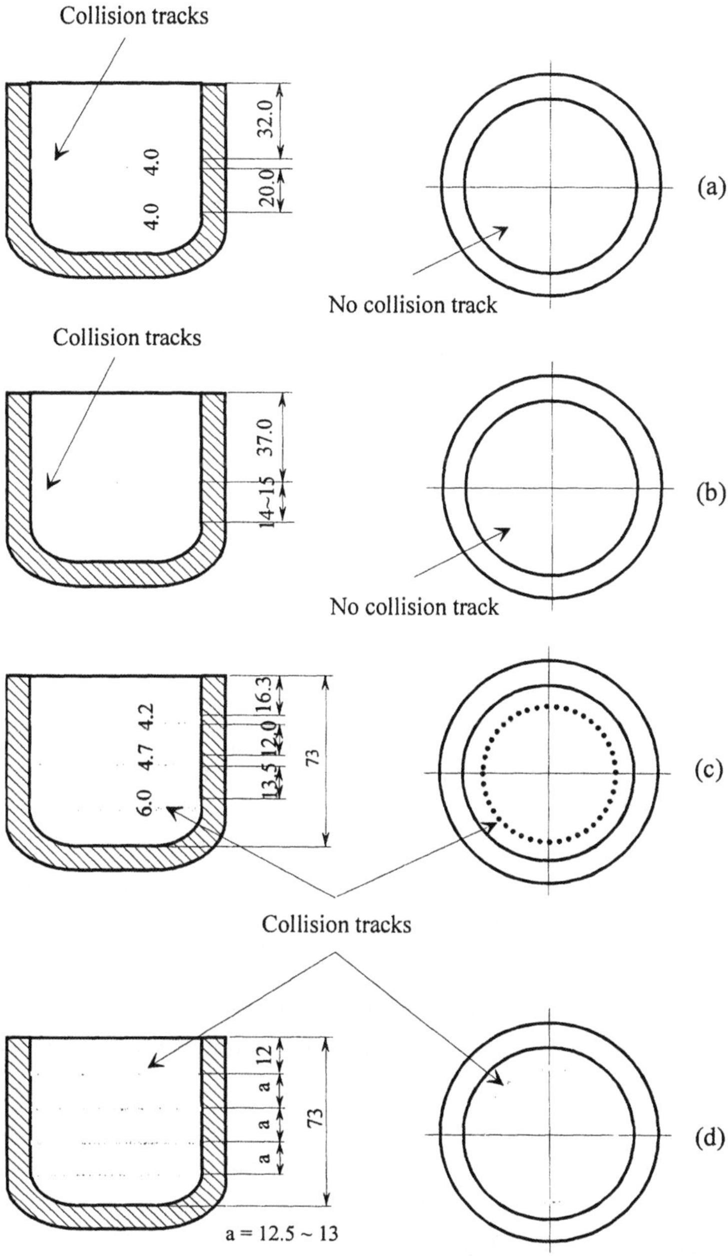


Figure 9.1. Schematic view of impact traces on the surfaces of the vial at 250 rpm by using different number and size of balls: (a) 2 balls of 20 mm, (b) 3 balls of 15 mm, (c) 5 balls of 20 mm, and (d) 40 balls of 15 mm diameter.

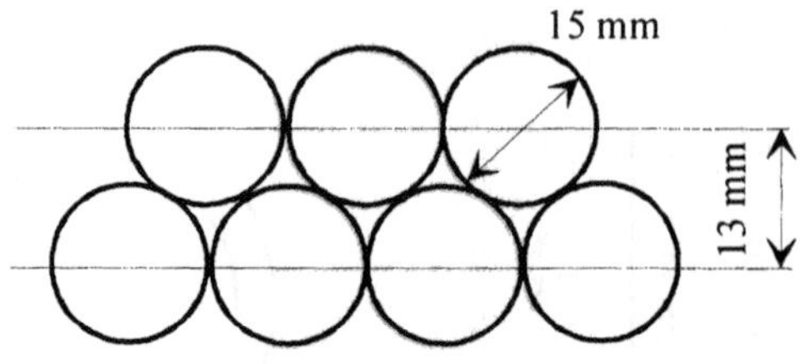


Figure 9.2. Stacking sequence of balls during flying.

(a) There is no relative motion between the ball and the vial wall prior to the point of detachment;

(b) The new attachment point of a ball to the vial is taken to be the impact point without considering the elastic rebounding of the ball.

It is obvious that the two assumptions above will not fully satisfy the actual motions of the balls since there is relative movement between the balls and the vial especially at the beginning of impact and rebounding will occur.

*9.2.1.1 Theoretical considerations*

The motion of a vial in a ball mill involves rotation about two parallel axes, analogous to the revolution of the earth about the sun. The four milling vials with radius $r$ located in the sun disk (sometimes called turn table) rotate about point O with an angular velocity of $\omega_e$ as shown in Figure 9.3. They are centered at point $o_1$, a fixed radius of $r_m$ from point O. They rotate with an angular velocity $\omega_r$ about their own axes at point $o_1$ in the direction opposite to $\omega_e$. The references adopted, XOY, the inertia co-ordinate system and, $xo_1y$, the relative co-ordinate system with respect to the sun disk, are the Cartesian co-ordinate system. In the following mathematics formulation, the terms 'absolute' and 'relative' refer to the parameters calculated with respect to the XOY and $xo_1y$ co-ordinate systems respectively. Consider therefore, a ball of mass $m$ and radius $r_b$ moving along the wall of the milling vial.

The forces exerted on the ball in the milling vial are the centrifugal forces acting outwards from the centre of the sun disk and from the centre of the vial denoted as $F_e$ and $F_r$ respectively, the frictional force $F_f$ resulted from the interaction between the balls and the vial, the force $F_c$ arising from the Corilis effect, and the gravitational force. In this approach, the ball is considered as a point mass represented by the movement of its centre of gravity. Hence, friction between the ball and the wall of the milling vial is not treated in this approach. The formulation therefore simplifies to an analysis of the movement of a ball in a planar section in which the effect of gravity may be neglected. According to Newton's second law of motion, the resultant of the external forces acting on the ball is equal to the product of its mass and its acceleration. Hence, transport force towards O, centrifugal force towards $o_1$, and centrifugal acceleration towards $o_1$ can respectively be written as:

$$\vec{F}_e = -m\vec{a}_e \quad [9.1]$$

$$\vec{F}_r = -m\vec{a}_r \quad [9.2]$$

$$\vec{F}_c = -m\vec{a}_c \quad [9.3]$$

where $m$ is the mass, and $\vec{a}_e, \vec{a}_r, \vec{a}_c$ are respectively the transport acceleration towards O, centrifugal acceleration towards $o_1$, and Corollas acceleration. The latter can be written as:

$$\vec{a}_e = \vec{\omega}_e \mathrm{x} \overrightarrow{\mathrm{OA}} \quad [9.4]$$

$$a_e = \overline{\mathrm{OA}}\,\omega_e^2 = \frac{r_m + r\cos\varphi}{\cos\alpha}\omega_e^2 \quad [9.5]$$

$$\vec{a}_r = \vec{\omega}_r \mathrm{x} \vec{r} \quad [9.6]$$

$$a_r = r\omega_r^2 \quad [9.7]$$

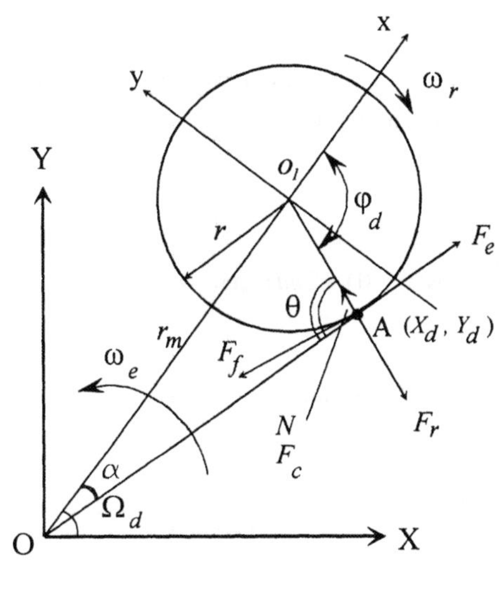


Figure 9.3. Configuration of a planetary ball mill.

and

$$\vec{a}_c = 2\vec{\omega}_e \mathrm{x} \vec{V}_r \qquad [9.8]$$

$$a_c = 2\omega_e\omega_r r \qquad [9.9]$$

From Figure 9.3, the following equations can be deduced:

$$\sum \vec{F} = m(\vec{a}_r + \vec{a}_e + \vec{a}_c) \qquad [9.10]$$

$$N = F_r - F_c - F_e \cos(\pi - \theta) \qquad [9.11]$$

where $N$ is the force normal to the surface of the vial.

It follows from Figure 9.3 that

$$F_e \cos(\pi - \theta) = ma_e \cos(\pi - \theta) = m\omega_e^2 \frac{r_m + r\cos\varphi}{\cos\alpha} \cos(\pi - \theta) \qquad [9.12]$$

$$F_r = mr\omega_r^2 \qquad [9.13]$$

$$F_c = 2m\omega_e\omega_r r \qquad [9.14]$$

It can be seen that when $N$ is zero, the ball will detach itself from the surface of the vial. This critical condition can be written as:

$$N = 0;\ \varphi = \varphi d \qquad [9.15]$$

$$m\omega_e^2 \frac{r_m + r\cos\varphi_d}{\cos\alpha} \cos(\pi - \theta) + 2mr\omega_e\omega_r = mr\omega_r^2 \qquad [9.16]$$

By dividing $m$ and $\omega_e^2$, it follows that

$$-\frac{r_m + r\cos\varphi_d}{\cos\alpha}\cos\theta + 2r\frac{\omega_r}{\omega_e} = r\left(\frac{\omega_r}{\omega_e}\right)^2 \qquad [9.17]$$

From Figure 9.3, it can be seen that $\theta = \varphi_d - \alpha$. Therefore, Eq.[9.17] can be rewritten as:

$$\frac{r_m + r\cos\varphi_d}{\cos\alpha}\cos(\varphi_d - \alpha) - 2r\frac{\omega_r}{\omega_e} = -r\left(\frac{\omega_r}{\omega_e}\right)^2 \tag{9.18}$$

Tangent angle $\alpha$ can be written as:

$$\tan\alpha = \frac{r\sin(\pi - \varphi_d)}{r_m - r\cos(\pi - \varphi_d)} = \frac{r\sin\varphi_d}{r_m + r\cos\varphi_d} = k \tag{9.19}$$

Since

$$\cos\alpha = \frac{1}{\sqrt{1+\tan^2\alpha}} = \frac{1}{\sqrt{1+k^2}} \tag{9.20}$$

$$\sin\alpha = \frac{\tan\alpha}{\sqrt{1+\tan^2\alpha}} = \frac{k}{\sqrt{1+k^2}} \tag{9.21}$$

Therefore,

$$\begin{aligned}\frac{r_m + r\cos\varphi_d}{\cos\alpha}\cos\theta &= -\frac{r_m + r\cos\varphi_d}{\cos\alpha}\cos(\varphi_d - \alpha)\\ &= (r_m + r\cos\varphi)\sqrt{1+k^2}\left[\left(\cos\varphi\frac{1}{\sqrt{1+k^2}} + \sin\varphi\frac{k}{\sqrt{1+k^2}}\right)\right]\\ &= (r_m + r\cos\varphi)(\cos\varphi + k\sin\varphi)\\ &= (r_m\cos\varphi + r)\end{aligned} \tag{9.22}$$

Substituting all the components of the acceleration and Eq.[9.22] into Eq.[9.17], it can be shown that

$$r_m\cos\varphi_d + r - 2r\frac{\omega_r}{\omega_e} = -r\left(\frac{\omega_r}{\omega_e}\right)^2 \tag{9.23}$$

Let

$$R = \frac{\omega_r}{\omega_e} \quad [9.24]$$

After the appropriate manipulations, Eq.[9.23] can be reduced to

$$r_m \cos\varphi_d = -r\left(R^2 - 2R + 1\right) \quad [9.25]$$

If the sun disk rotates in the anti-clockwise direction while the vial clockwise, the detach angle can be found to be:

$$\varphi_d = \arccos\left(-\frac{r(1-R)^2}{r_m}\right) \quad [9.26]$$

If the sun disk and the vial rotate in the same anticlockwise direction, the detach angle becomes:

$$\varphi_d = \arccos\left(-\frac{r(1+R)^2}{r_m}\right) \quad [9.27]$$

The values of $\varphi_d$ and $2\pi - \varphi_d$ define a domain on the surface of the milling vial where the resultant of the forces on the ball is oriented towards the interior of the mill. Hence, the ball will not be in a stable condition between the angles $\varphi_d$ and $2\pi - \varphi_d$. If the radius $r$ of the vial is varied under the condition of fixed $r_m$ and rotational ratio $R$, the locus of the point of detachment can be represented by the Davis circle [12, 18] whose diameter, $D_d$, depicted in Figure 9.4, is given by:

$$D_d = \frac{r_m}{(1+R)^2} \quad [9.28]$$

With the detach angle calculated, the absolute position of detachment ($X_d$, $Y_d$) and the two components of the detached velocity $v_{dx}$, $v_{dy}$ in the X and Y directions respectively can be written as:

$$X_d = r_m \cos\Omega_d + r\cos(\varphi_d - \Omega_d) \quad [9.29]$$

$$Y_d = r_m \sin\Omega_d - r\sin(\varphi_d - \Omega_d) \quad [9.30]$$

$$v_{dx} = -r_m\omega_e \sin\Omega_d - r(\omega_r - \omega_e)\sin(\varphi_d - \Omega_d) \quad [9.31]$$

$$v_{dy} = r_m\omega_e \cos\Omega_d - r(\omega_r - \omega_e)\cos(\varphi_d - \Omega_d) \quad [9.\ 32]$$

Once a ball detaches from the surface of the vial, it is expected to travel with its detach velocity in a linear manner regardless of the influence of gravity. The kinetics of rigid bodies is used to describe the free motion of the ball until it comes in contact again with the vial. Figure 9.5 shows the motion of a ball from a detachment event up to an impact event, where the subscripts *d* and *c* represent the various parameters at detachment and impact events respectively. The trajectory of the ball inside the milling vial at time $\tau$ between the first detachment event and the first impact event can be fully described by:

$$X_\tau = v_{dx}\tau + X_d \quad [9.33]$$

$$Y_\tau = v_{dy}\tau + Y_d \quad [9.34]$$

Here, $\Omega_\tau = \omega_e\tau$, the absolute angular position of the milling vial at time $\tau$, and ($X_\tau$, $Y_\tau$) is the absolute position of the ball at time $\tau$ between the first detachment event and the first impact event. The ball then remains in contact with the surface of the wall and rotates until the detach angle $\varphi_d$ is reached again.

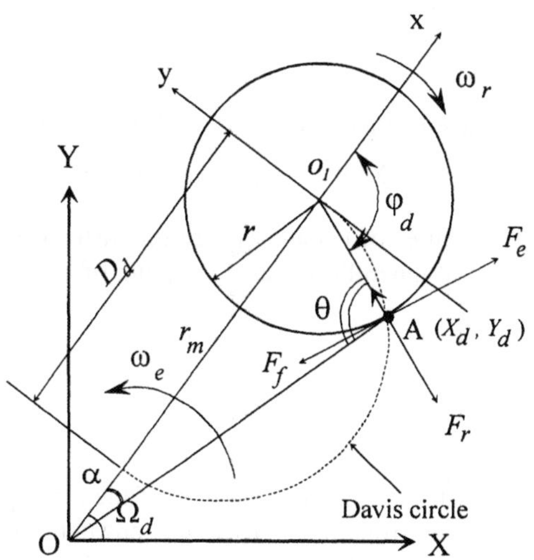


Figure 9.4. Locus of points of detachment.

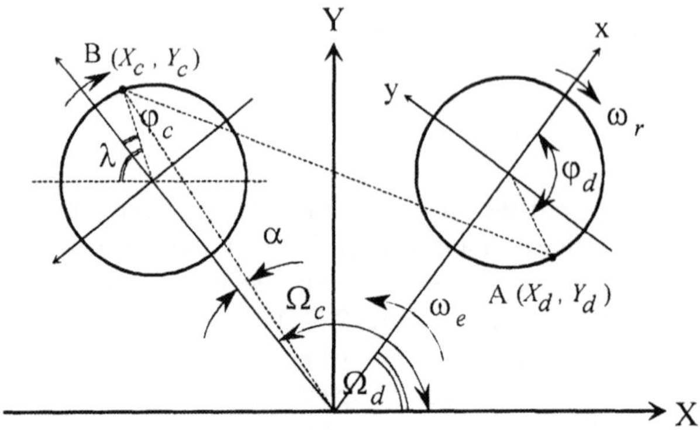


Figure 9.5. Motion of ball from detachment to impact.

With the time from the detachment event to the impact event, $\tau$, calculated, the absolute velocities of the milling vial at the point of impact in the X and Y directions can be evaluated.

$$v_{vx} = -\frac{r_m + r\cos\varphi_c}{\cos\alpha}\omega_e \sin(\lambda - \varphi_c + \alpha) + r\omega_r \sin\lambda \qquad [9.35]$$

$$v_{vy} = -\frac{r_m + r\cos\varphi_c}{\cos\alpha}\omega_e \cos(\lambda - \varphi_c + \alpha) + r\omega_r \cos\lambda \qquad [9.36]$$

where $\Omega_c = \omega_e T_1$, the absolute angular position of the vial at the instant of impact, and $v_{vx}$ and $v_{vy}$ are respectively the velocity of the vial in X and Y directions.

The absolute impact velocities of $v_{ix}$ and $v_{iy}$ are the difference between the detach velocity and the velocity of the milling vial at the point of impact.

$$v_{ix} = v_{dx} - v_{vx} \qquad [9.37]$$

$$v_{iy} = v_{dy} - v_{vy} \qquad [9.38]$$

The angle at which the ball impacts onto the wall of the milling vial determines the amount of energy released from the ball to the powders. The effective impact velocity that generates the impact energy is the normal component of the impact velocity resolved in the radial direction of the milling vial. Therefore, the effective impact energy released per hit during the mechanical alloying process can be calculated as follows:

$$v_i = v_{ix} \cos\lambda + v_{iy} \sin\lambda \quad [9.39]$$

$$E_i = \frac{1}{2} m v_i^2 \quad [9.40]$$

where $v_i$ is the effective impact velocity, $E_i$ represents the impact energy released per hit and $\lambda$ is the geometrical parameter shown in Figure 9.5. The tangential component of the impact velocity, $v_f$, gives rise to $E_f$, the frictional energy generated during the milling process.

$$v_f = \sqrt{v_{ix}^2 + v_{iy}^2 - v_i^2} \quad [9.41]$$

and

$$E_f = \frac{1}{2} m v_f^2 \quad [9.42]$$

Abdellaou *et al.* [10] found that in a planetary ball mill, the collision energy is almost a parabolic function of the angular velocity of the sun disk but is less sensitive to the angular velocity.

The impact frequency is the number of collisions per second of a ball onto the vial. Thus, a knowledge of the total time taken for the ball to travel from the first detachment event to the next detachment event is required to calculate the frequency of impact. Each cycle may be decomposed into two periods, $T_1$, time taken by a ball to go from the first detachment point to the first impact point, and $T_2$, time required to meet the second detachment event after the first impact event. $T_1$ is calculated based on the detach angle. To obtain $T_2$, the impact angle $\varphi_c$ relative to the moving reference frame has to be determined by assuming that the ball detaches at the same detach angle as that when there is no slip between the moving ball and the surface of the vial. The absolute difference between the detach angle and the impact angle in the relative frame gives the angular position that the ball has to travel before it reaches the second detachment point. $T_2$ can then be computed as follows:

$$T_2 = \frac{\varphi_d - \varphi_c}{\omega_r} \quad [9.43]$$

As period of the cycle is the sum of the two component periods $T_1$ and $T_2$, the impact frequency $f$ can be found. The impact frequency, which is the number of impact per second may be obtained by the inverse of the cycle period [13].

$$f = \frac{1}{T_1 + T_2} \qquad [9.44]$$

where $f$ is the impact frequency.

This impact frequency corresponds to the impact of a single ball inside the milling vial. To take into account the effect of the number of balls in the milling process, the effective impact frequency, $f_{eff}$, taken to be equal to the product of the impact frequency of a ball and the number of balls [13]. The power released, $P$ is the product of the effective impact frequency and the impact energy released during the mechanical alloying process.

$$P = f_{eff} E_i \qquad [9.45]$$

This parameter gives a comparison of the rate of reaction. A higher power released by the ball to the powders means that a shorter milling time is required for the milling process. Experimentally, $P$ is associated to the weight ratio of ball to powder.

*9.2.1.2 Effect of speed ratio, R*

Detach and impact positions are functions of the size of the vial, $r$, the position of the vial on the sun disk $r_m$ as well as the ratio of the rotational speed $R$. When $r$ and $r_m$ are fixed, detachment is dependent only on $R$. Based on the trajectories of the balls, the motion of the balls can be categorized into three modes by altering $R$ [11], namely, chaotic mode, impact and friction mode and pure friction mode. The state of the balls in these three modes is schematically shown in Figures 9.6 (a), (b) and (c) respectively.

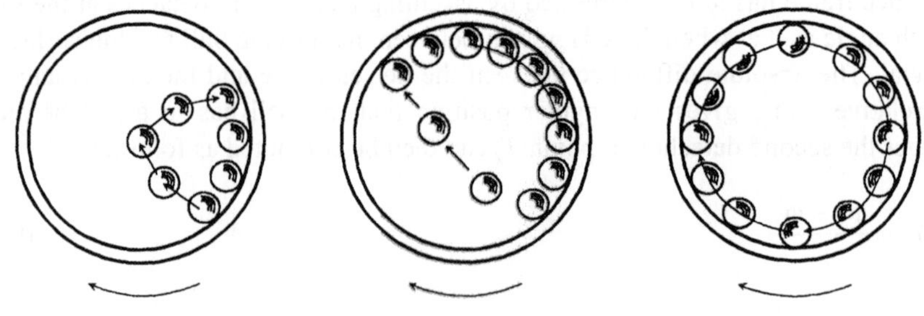

(a) (b) (c)

Figure 9.6. Modes of ball motion at different $R$ ratios. Direction of rotation of sun disk, counter-clockwise.

Chaotic mode (for $R < R_{limit}$) [11]

At this speed ratio, the sun disk rotates faster than the milling vial. As a result, the detach angle $\varphi_d$ and impact angle $\varphi_c$ are close to each other. The milling vial comes in contact with the balls immediately when the balls detach themselves from the surface of the vial at $\varphi_d$. A chaotic situation arises due to the interactions between the milling balls such that the effect of the impact onto the wall becomes insignificant. The vial chases after the balls rather than the balls impacting onto its surface as illustrated in Figure 9.6 (a). The limiting rotational speed ratio at this condition is definded as $R_{limit}$. Since impact energy of the balls is very low, any $R$ value smaller than $R_{limit}$ may be considered to have no effect on milling.

Pure friction mode (for $R > R_{critical}$) [11]

For speed ratio greater than a value of $R_{critical}$, the resultant force experienced by the ball is not oriented towards the interior of the milling vial. The model foresees that the ball will always be stuck to the wall of the vial. Theoretical calculations show no detachment event and so there is no energy loss either in the form of impact energy or frictional energy. In practice, however, interactions between the ball and the wall of the vial will occur when the ball slides on the wall surface of the vial even though this gives rise only to frictional energy (Figure 9.6 (c)). For this reason, it is known as pure friction mode and the corresponding rotational ratio $R$ is defined as $R_{critical}$ [11].

Impact and friction mode (for $R_{limit} \leq R \leq R_{critical}$) [11]

In this mode as shown in Figure 9.6 (b), the trajectory of the ball is well described by the fundamental principles of dynamics as shown in Figure 9.7. The energy at the time of impact can be resolved into two components, namely, the normal component which gives rise to the effective impact energy released to the powder particles and the tangential component which results in frictional energy. They have been calculated for the planetary ball mill with a speed ratio modification capability and the evolution of the two energy components is presented as a function of the milling rotational ratio $R$.

"Fritsch Pulverisette 5" planetary ball mill (P5) has a velocity ratio of $R = 2.17$, revolution radius $r_m = 125$ mm and radius of the milling vial, $r = 50$ mm. Results from simulation work show that altering the impact energy via the change in the angular velocity of the sun disk enables the impact velocity and impact frequency to be changed without varying the trajectory of the ball and the impact angle. Figure 9.8 shows the time taken for the ball to travel a one cycle period from the first detachment event to the second detachment event. Both the flying time of the ball from the point of first detachment to the point of first impact, $T_1$ and the 'sticking' time from the point of first impact to the point of second detachment, $T_2$ decrease exponentially as the angular velocity of the sun disk increases. This implies that as the angular velocity is increased the balls take shorter time to complete one cycle period. With a combination of increased detach velocity and a shorter flying time, the impact velocity thus increases with the angular velocity of the sun disk. This impact velocity is an important parameter that determines how much energy is released to the powder particles during the mechanical alloying process. Figure 9.9 shows the plot for both impact energy and frictional energy as a function of the angular velocity of the sun disk. The impact energy released per hit and the resultant frictional energy can be seen to increase as the revolution speed of the milling vial increases. The results show that purely raising the angular velocity of the sun disk may not necessarily be a better method to increase the impact energy released to the powders. The increase in frictional energy may generate a higher frictional force leading to frictional wear of the milling tools.

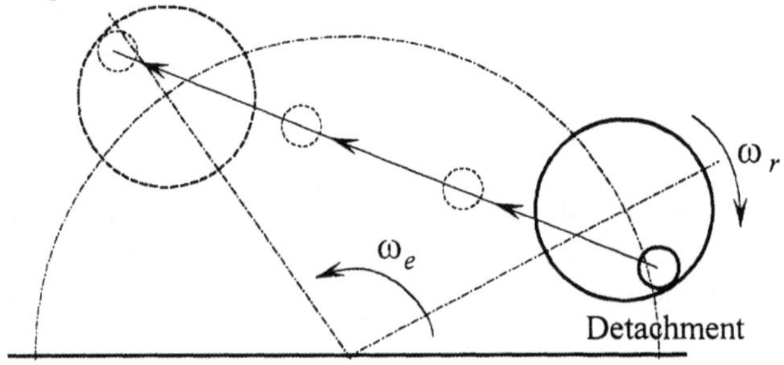


Figure 9.7. Movement of ball after detachment when $R_{\text{limit}} \leq R \leq R_{\text{critical}}$.

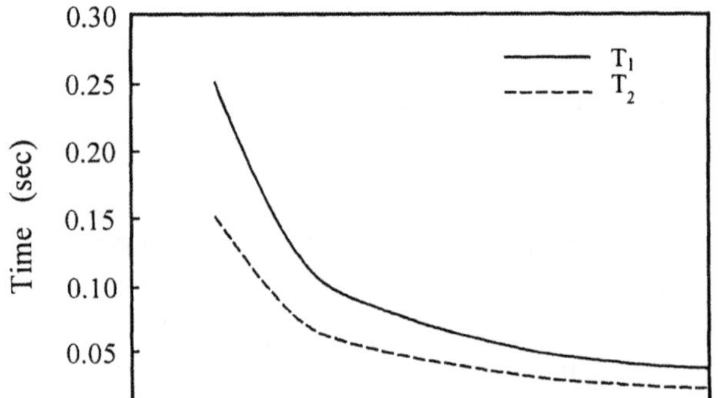


Figure 9.8. Time taken for one cycle as a function of angular velocity.

*9.2.1.3 Effect of ball size*

The effect of the ball size on impact energy released is shown in Figure 9.10. It can be seen that higher impact energy can be transferred to the powder particles when a bigger milling ball is used in a single impact due to the increase in mass of the ball. It can also be inferred that in order to release the same amount of impact energy per hit to the powder particles using different ball sizes, the mechanical alloying process has to be carried out with the sun disk rotation at different angular velocities.

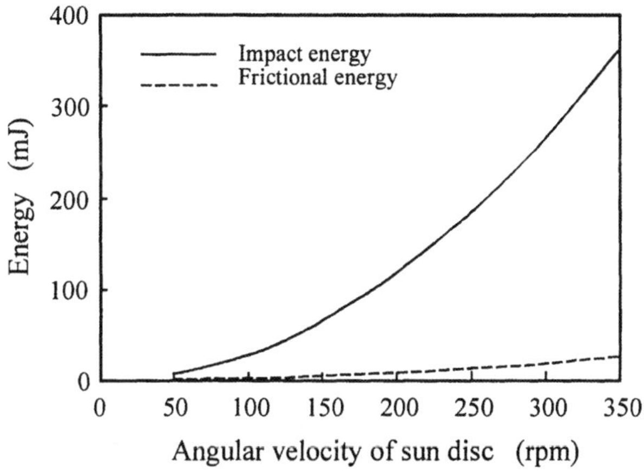


Figure 9.9. Impact and frictional energies as a function of angular velocity.

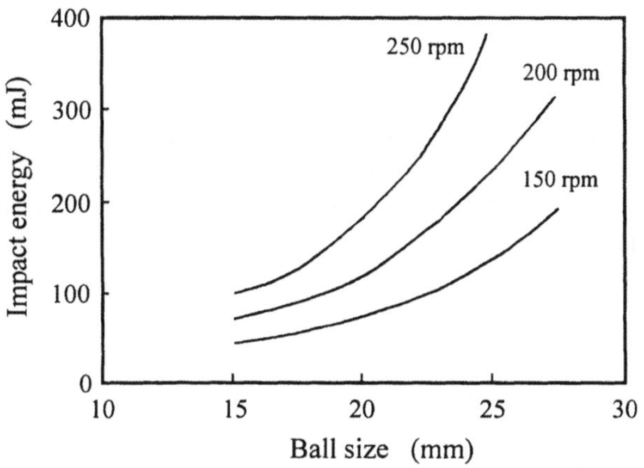


Figure 9.10. Influence of ball size on impact energy.

Since the ball to powder weight ratio is one of major factors influencing the milling process, to keep this ratio constant when different sets of ball sizes are used, the number of balls employed in each operation has to be adjusted according to the total mass. When the number of balls used is taken into account, the effective frequency which is the product of the impact frequency and the number of balls used, is found to increase when smaller balls are used but the impact energy is lower compared to that using bigger balls. Figure 9.11 shows an example of the relationship between effective frequency and velocity of the sun disk with different size balls for ball to powder weight ratio of 20:1.

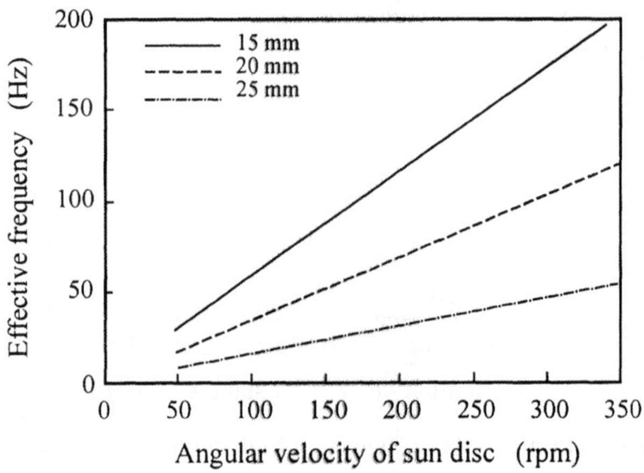


Figure 9.11. Effective frequency as a function of angular velocity of the sun disk.

Another parameter to describe the mechanical milling process is the power released. This is defined as the product of impact energy and impact frequency. Taking the same example as that considered in Figure 9.11 it is found that the power released by different sets of ball exactly fit into one single line as shown in Figure 9.12 indicating that the number of balls does not influence the power generated as long as the ball to powder weight ratio is constant. However, at the same power, the impact energy is different for the three sets of balls.

*9.2.1.4 Influence of ball size under the same impact energy and power*

Experimental results have shown that the formation of new materials via mechanical alloying depends upon both the impact energy per hit and the total impact power absorbed [(21-23)]. According to the above simulation, if different sets of ball size are used at the same angular velocity, the impact energies will be different. In order to study the effect of ball size at identical impact energy as well as total energy released, the angular velocity has to be adjusted accordingly. Based on simulation works, when ball sizes of 15, 20 or 25 mm diameter are used, impact energy of 118 mJ will be generated if the angular velocities of the ball milling are set at 285, 200 and 125 rpm respectively. By considering equal amount of total energy input to the powders from the above three cases, the time taken for the three cases can be calculated from the frequency of impact. Figure 9.13 shows the x-ray diffraction spectra of Ti-58at.%Al powder mixture that has been mechanically alloyed at the same impact energy per hit and the same total impact energy. The results show that super-saturated solid solution has taken place only if 15 mm ball size is used, while Ti and Al diffraction peaks are clearly visible if 20 or 25 mm ball size is used. Table 9.1 gives in detail the calculated data, the temperature measured and the end products. The table suggests that other factors may need to be considered, for example, the rise in temperature during impact. For the same impact energy per impact, the smaller the ball size, the higher is the impact frequency since higher angular velocity is required to generate the same impact energy as that using bigger ball size. Consequently, heat is accumulated faster than that with low impact frequency, leading to an increase in milling temperature.

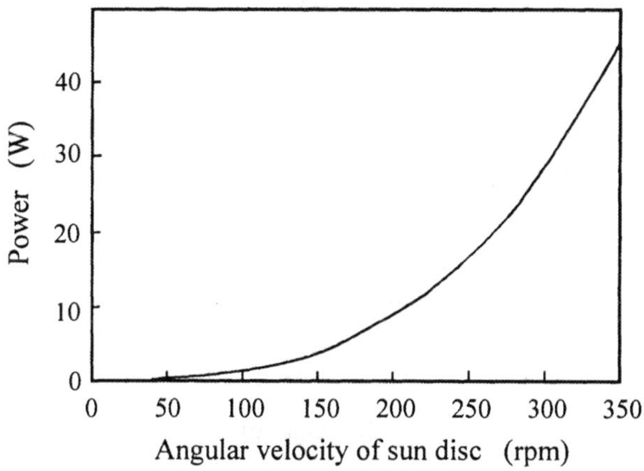


Figure 9.12. Power released as a function of angular velocity of the sun disk.

Intensive theoretical and experimental studies have been carried out to find the relationship between ball milling parameters and the end products [13, 24, 21]. McCormick *et al.* [17] proposed a so-called slip factor that takes into account of the sliding phenomenon between the balls and the wall of the vial since significant slip occurs between them. The slip factor has been defined as:

$$f_s = 1 - \frac{\omega_B}{\omega_r - \omega_e} \qquad [9.46]$$

where $\omega_B$ is the relative angular velocity of the balls in contact with the wall of the vial. When $f_s = 0$, there is no friction between the balls and the vial while if $f_s = 1$, the balls slip completely. Therefore, in general, the value of $f_s$ is between 0 and 1.

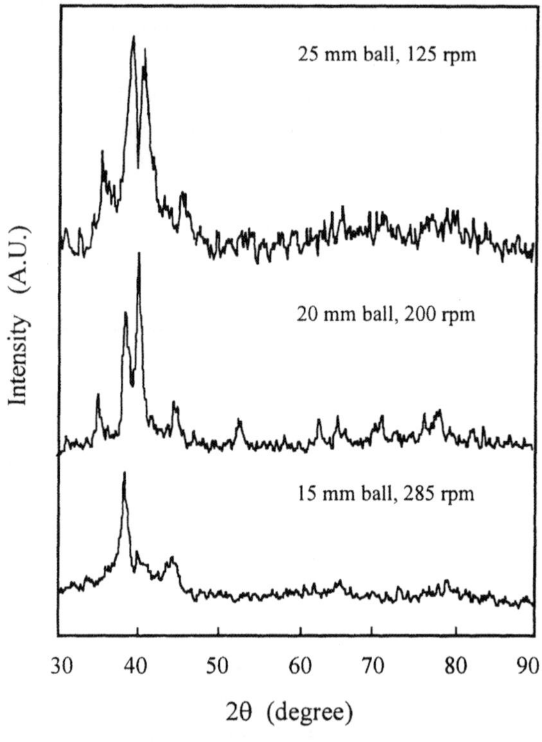


Figure 9.13. X-ray diffraction spectra of mechanically alloyed Ti-Al mixtures subjected to identical impact energy per hit and total impact energy absorbed.

McCormick [17] has investigated the motion of the ball by using a standard planetary ball mill (Fritsch Pulversette 7). It was found that the slip factor was 0.8 which remained approximately constant over a rotation speed of the sun disk from 100 to 300 rpm. This range of angular velocity generally covers almost all ball milling situations. Since the speed of the balls along the rotational direction of the vial is almost zero when they impact onto the wall, they have to be accelerated from a minimum to a maximum speed value within a very short time period. The acceleration is mass dependent where higher mass may lead to slower acceleration.

Therefore, the speed of the ball within the attached area is slower than the motion of the inner surface of the vial. According to Burgio *et al.* [19], if ball to ball interaction is considered, the energy per impact is reduced by a factor which is essentially the fraction of vial space occupied by the balls.

Table 9.1 Effects of impact energy, power released and temperature measured on the structure of mechanically alloyed Ti-Al mixture.

| Ball size (mm) | angular velocity (rpm) | impact energy (mJ/hit) | Power released (W) | Milling duration (minute) | Tempera-ture (K) | End product |
|---|---|---|---|---|---|---|
| 15 | 285 | 118.0 | 1.992 | 264 | 318 | SS* |
| 20 | 200 | 117.9 | 8.163 | 750 | 310 | Ti and Al |
| 25 | 125 | 117.5 | 23.405 | 3105 | 304 | Ti and Al |

* solid solution

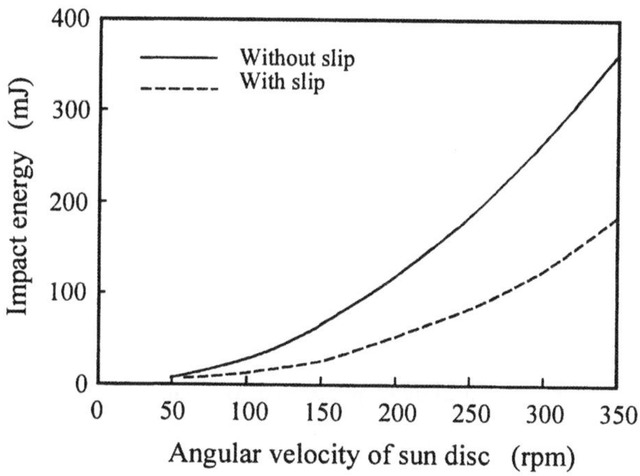


Figure 9.14. Impact energies at different values of $f_S$.

An example of the influence of slip factor is illustrated in Figure 9.14 in which two cases based on Fritsch Pulversette 5 are considered. The ball mill has an angular velocity ratio of $R$=2.17 and the ball size used is 20 mm. If a slip factor of 0.35 is considered, the resultant effective angular velocity ratio is found to be equal to $R$=1.76. In the figure, the solid line represents $f_S$=0, namely no slip while the dashed line $f_S$=0.35. In the case of $f_S$=0.8, the effective angular velocity ratio is equal to $R$=0.434. At this value of $R$, the ball motion is in the chaotic mode.

*9.2.1.5 Modification of impact energy*

The final structure of a mechanically alloyed or milled powder may be controlled by the energy input to the powder during the mechanical alloying process. If the energy ratio between impact and friction can be controlled, the structure of the powder may be modified. To achieve this, an attraction force may be introduced, for example, by the introduction of magnet fixed on the sun disk near the vial as shown in Figure 9.15.

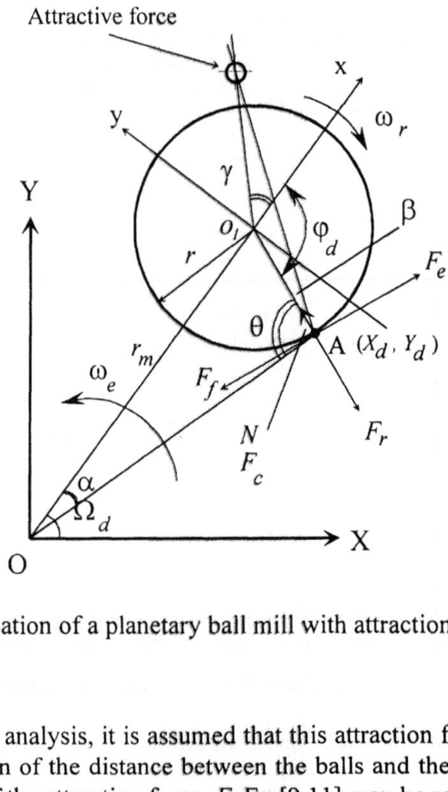


Figure 9.15. Modification of a planetary ball mill with attraction force.

For simplicity in the analysis, it is assumed that this attraction force is constant, that is, it is not a function of the distance between the balls and the source of the force. With the presence of the attraction force, $F$, Eq.[9.11] may be rewritten as:

$$N + F\cos\beta = F_r - F_c - F_e\cos(\pi - \theta) \qquad [9.47]$$

The rest of the force components remains the same as discussed earlier. For trigonometry, cosβ can be solved as follows:

$$\cos\beta = \frac{(r/b) - \cos(\varphi_d + \gamma)}{\sqrt{1 + \frac{r^2}{b^2} - \frac{2r}{b}\cos(\varphi_d + \gamma)}} \qquad [9.48]$$

where $b$ is the distance between the centre of the attraction force and $o_1$.

The critical condition in which the ball detaches from the wall of the vial requires N = 0. Therefore, the Eq.9.11 can be further rewritten as:

$$\Psi r_m \frac{(r/b) - \cos(\varphi_d + \gamma)}{\sqrt{1 + \frac{r^2}{b^2} - \frac{2r\cos(\varphi_d + \gamma)}{b}}} = r(1 - R)^2 + r_m \cos\varphi_d \qquad [9.49]$$

where $\Psi = \frac{F}{m r_m \omega_e^2}$ is a dimensionless force constant.

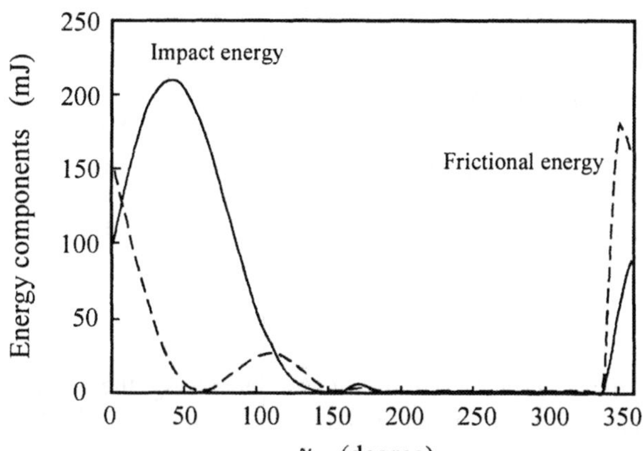


Figure 9.16. Variation in impact energy released per impact as a function of location of attraction force ($\Psi = 1.0$; $\omega_e = 200$ rpm; ball diameter = 20 mm).

Figure 9.16. shows the change in impact energy as a function of location of the attraction force. It can be observed that the impact energy can be increased while the frictional energy decreased by changing angle $\gamma$. The attraction force may be seen to serve two functions, namely, increasing flying velocity of the balls and changing the flying trajectory of the balls. With a change in the trajectory, the angle of impact can be adjusted to yield different impact to friction energy ratios. A maximum impact energy released is seen to exist at the location of $\gamma$=40°. This maximum impact energy can be further increased by increasing the value $\Psi$. If the attraction force is placed at 60°, a minimum frictional energy is experienced. To obtain maximum impact energy with the lowest frictional energy possible, $\gamma$ may be optimized at about 50°. The advantage of incorporating this attraction force is hence to better control the ratio of impact to frictional energies so that energy distribution may be modified according to different needs. A comparison of impact and frictional energies between a normal and a modified planetary ball mill is given in Figures 9.17 and 9.18.

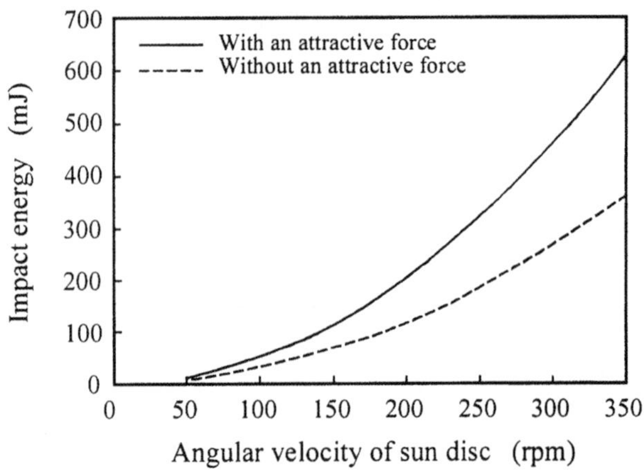


Figure 9.17. Effect of attraction force on impact energy ($\Psi$ = 1.0; ball diameter = 20 mm).

Impact energy may be increased by simply increasing the value of $\Psi$. Under constant $m$, $\rho$ and $\omega_e$, $\Psi$ can be increased by increasing force $F$. To obtain similar impact energy, the angular velocity may be lowered if the attraction force is increased. The lowering of the angular velocity may have the advantage that the impact frequency is decreased and the effect of heat accumulation may be reduced. In addition, it may reduce the reaction rate such that crystallization during the milling process may be avoided. Figure 9.19 shows the influence of $\Psi$ on impact energy. It is clear that under the condition of similar impact energy, the angular velocity can dramatically be lowered.

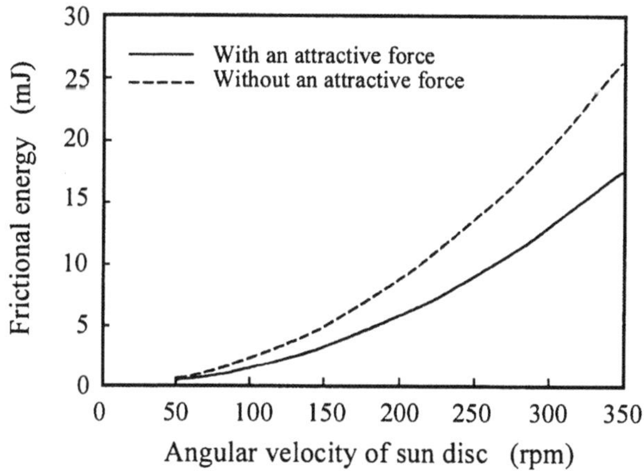


Figure 9.18. Effect of attraction force on frictional energy ($\Psi$ = 1.0; ball diameter = 20 mm).

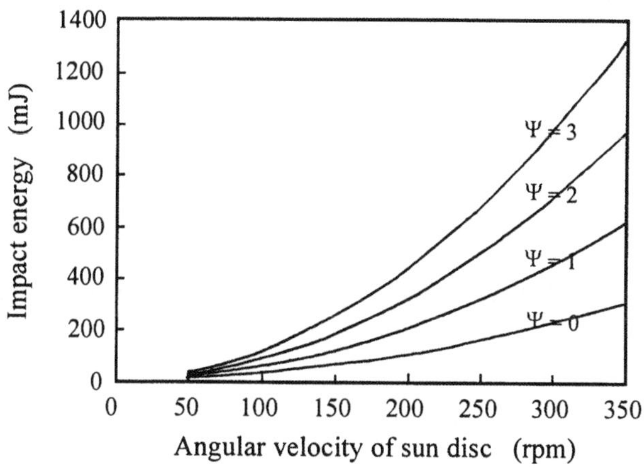


Figure 9.19. Influence of $\Psi$ on impact energy.

### *9.2.2 Horizontal ball mill*

*9.2.2.1 Theoretical considerations*

Consider the mechanical alloying process using a conventional horizontal ball mill illustrated in Figure 9.20. In this system, the balls or cylindrical rods and powders are placed in a vial which rotates at a speed of $\omega$ about its horizontal axis. The balls and the powders will move along the internal surface of the vial if the vial rotates fast enough. Assuming that the balls and the powders move at the same speed as that of the vial, the normal acceleration of the $i_{th}$ ball can be written as:

$$a_i = \frac{D-d}{2}\omega^2 \qquad [9.50]$$

where $D$ is the inner diameter of the vial and, $d$, the diameter of the ball.

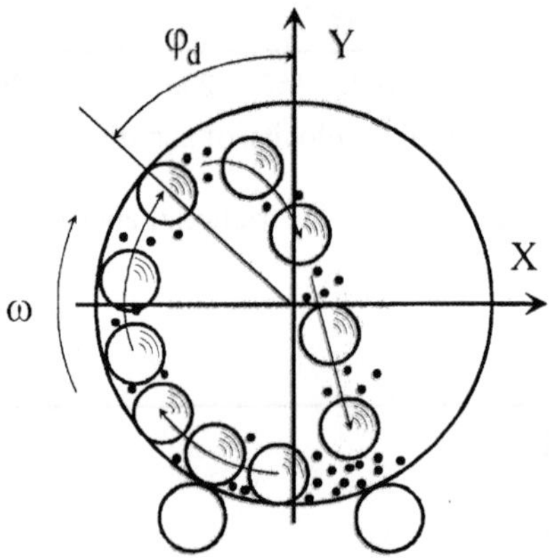


Figure 9.20. Horizontal ball mill.

According to Newton's law, the ball should satisfy the following conditions during motion [26]:

$$\sum_{j=1}^{N} f_{ij} = m_i a_i \qquad [9.51]$$

where $f_{ij}$ is the force acting on the $i_{th}$ ball, $m_i$, and $a_i$ are respectively the mass and the acceleration in the normal direction of the $i_{th}$ ball. Three forces, namely, gravity $W_b$, friction force between the inner wall and the ball, $F$ and the normal force $N$, exert on a ball during milling. From Figure 9.20, the resultant force in the radial direction can be written as:

$$N + W_b \cos\varphi \qquad [9.52]$$

By substituting Eq.[9.52] into Eq.[9.51], it can be shown that:

$$N + W_b \cos\varphi = \frac{W_b}{g} a_i \qquad [9.53]$$

where $g$ is the gravitational acceleration. Substituting Eq.[9.50] into Eq.[9.53], Eq.[9.53] can be written as:

$$N = \frac{W_b}{g} \frac{D\text{-}d}{2} \omega^2 - W_b \cos\varphi \qquad [9.54]$$

$N$ is a function of $\varphi$, if $\omega$ is constant during milling. The essential condition for milling is that the balls must fall down to impact the powders. This critical condition can be described by $N = 0$ when the balls separate from the inner surface of the vial and fall down. The critical angle $\varphi_d$ at which this critical condition occurs can be obtained from Eq.[9.54]:

$$\cos\varphi_d = \frac{(D-d)\omega^2}{2g} \qquad [9.55]$$

It can bee seen from Eq.[9.55] that the critical angle is a function only of the inner diameter and the angular velocity of the vial. For different values of ($D$ - $d$) at a given $\omega$, different detach angles can be obtained. The locus of the point of detachment may also be presented by a Davis circle [(12)]:

$$D_d = \frac{g}{\omega^2} \qquad [9.56]$$

After the ball has separated from the wall of the vial, it travels along a parabolic trajectory which can be expressed as:

$$x = v_0 t \cos\varphi_d - \frac{D-d}{2} \sin\varphi_d \qquad [9.57]$$

$$y = v_0 t \sin\varphi_d - \frac{1}{2} g t^2 + \frac{D-d}{2} \cos\varphi_d \qquad [9.58]$$

where $v_0$ is the speed when the ball just separates from the wall and, $t$, time after the separation. The ball travels as a free body until it impacts onto the wall (or other balls) below. The height $h$ through which the ball falls can be calculated by solving Eqs.[9.57], [9.58] as well as the following:

$$\frac{(D-d)^2}{4} = x^2 + y^2 \quad [9.59]$$

The speed of the ball before impacting can be shown to be:

$$v_x = v_0 \cos\varphi_d \quad [9.60]$$

$$v_y = v_0 \sin\varphi_d - gt \quad [9.61]$$

$$v = \sqrt{v_x^2 + v_y^2} \quad [9.62]$$

where $v_x$, $v_y$ and $v$ are the speed components of the ball in the x and y and resultant directions respectively. The impact energy, $E_i$, is given by:

$$E_i = \frac{1}{2} m v^2 \quad [9.63]$$

### *9.2.3 Vibration ball mill*

#### *9.2.3.1 Theoretical considerations*

Gavrilov *et al.* [25] proposed a model to simulate a vibration or shaker ball mill. The essential feature of this 3-D model is that it takes into account several parameters including geometry and interference of the balls. Since it is a 3-D simulation, a closer estimation to the reality of the ball milling process may be obtained. In this model, the following hypotheses are assumed:

(a) Collisions between the balls and between the balls and the vial are centered and the tangential velocity of the ball is conserved.

(b) The velocity of the vial is not influenced by the impact events as the vial is assumed to have infinite mass.

(c) The impact event between ball to ball is instantaneous and on a one-to-one basis only.

(d) The vial moves with constant acceleration during each half-period of the oscillations

The trajectory of the ball motion in a gravitational field follows a parabolic line given by:

$$u(t) = u_0 + v_0 t + \frac{gt^2}{2} \quad [9.64]$$

$$v(t) = v_0 + gt \quad [9.65]$$

where $u$, $v$ and $g$ are the position, velocity and gravitational vectors respectively.

The vibration of the vial is a harmonic motion in the vertical Z direction. From hypothesis (d), the position of the vial at time $t$ can be written as:

$$z(t) = v_0 t + \frac{a_v t^2}{2}$$

where $z(t)$ is the position of the vial in the Z direction since vibration takes place in the vertical direction and $a_V$, the acceleration of the vial during each half-period of oscillation.

Consider an impact between two balls, since linear momentum is conserved, the linear momentum of the balls should be the same before and after the impact. The energy balance is given by:

$$m_1 v_{10} + m_2 v_{20} = m_1 v_1 + m_2 v_2 \quad [9.66]$$

Figure 9.21 shows three events of two balls before, during and after impact. The ratio of the relative velocities between the two balls after and before is defined as coefficient of restitution, $\Pi$, which can be written as:

$$\Pi = \frac{v_2 - v_1}{v_{10} - v_{20}} \quad [9.67]$$

If the velocity before impact and the coefficient of restitution are known, the velocity after impact can be calculated:

$$v_1 = \frac{(m_1 - \Pi m_2) v_{10} + m_2 (1 + \Pi) v_{20}}{m_1 + m_2} \quad [9.68]$$

$$v_2 = \frac{m_1 (1 + \Pi) v_{10} + (m_2 - \Pi m_1) v_{20}}{m_1 + m_2} \quad [9.69]$$

In comparison to the mass of a ball, the mass of a vial can be considered infinitely large. Therefore, from Eq.[9.68], the velocity of a ball after impact with the vial can be expressed as:

$$v_1 = v_{20} + \Pi(v_{20} - v_{10}) \quad [9.70]$$

where $v_{20}$ refers to the velocity of the vial, and $v_{10}$ and $v_1$, the velocities of ball before and after impact with the vial.

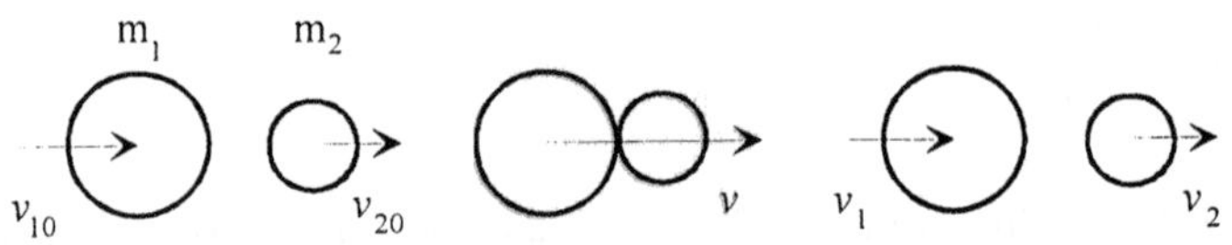


Figure 9.21. Schematic view of ball motion before, during and after impact.

This model essentially considers a one-to-one impact. In the case where there are more than two objects, the impact event may be handled sequentially in pairs. The impact event between two objects is firstly calculated. Then calculation is carried out between the next object with either one of the first two objects. According to statistics from their simulation work, Gavrilov *et al.* (25) found that the cases of simultaneous impacts between three or more objects are very rare.

## 9.3 Local modeling

Local modeling is performed to simulate deformation behaviour of the powder particles during impact events. Mauroce *et al.* (7) were the first to attempt to simulate such event by considering the geometry, mechanics and physics of the mechanical alloying process. Detailed study based on Hertzian contacts between the milled media which entrap a certain volume of material between the impacting surfaces has been carried out (7). The key factor is to consider the geometry which essentially defines the volume of powder particles affected per impact. From the volume, material properties and impact energy, milling duration and other parameters can be predicted.

### *9.3.1 Volume of material per collision*

It is essential to know the volume of powder cluster during an impact. According to Gilman *et al.* (27), powder particles entrapped between two impacting media are in the order of 1000 during a single impact. If the influence of atmospheric force which resists the motion of the balls and the powder particles is taken into consideration, the associated frictional drag force has been shown to be dependent on the Reynolds number [7]:

$$\mathrm{Re} = D_m \upsilon_{air} \rho / \eta \qquad [9.71]$$

where $D_m$ is the diameter of the vial, $\upsilon_{air}$, the average velocity of the air (1.94m/s for Spex), $\rho$, the density and $\eta$, the viscosity of the air (1.8*10-5 kg/m.s).

Because of the resistance of air, the flying body, either balls or powder particles will be decelerated. This deceleration can be expressed as:

$$a = \frac{3\upsilon_{air}^2 f}{8r} \qquad [9.72]$$

where $f$ is the frictional factor, $f = 0.079\mathrm{Re}^{-0.25}$.

Eq.[9.72] indicates the deceleration speed as a function of the size of a flying body. The smaller the size, the higher the deceleration speed. Therefore, Maurice *et al.* [7] proposed a uniform distribution of powder particles over the length of the vial in a Spex mill. When the balls fly from one end to the other end of a Spex mill, some powder particles fly away from the front of the balls while others are pushed ahead by the balls as schematically shown in Figure 9.22. It is assumed that as the ball approaches the surface of the vial, fewer and fewer powder particles "escape" from the path of the ball. Accordingly, the volume (called the swept volume) of powder entrapped in the process can be considered as the volume of a cone with length $L - 2r_b$, where $L$ is the length of the vial and $r_b$, the radius of the flying ball. The radius of the base of the cone can be defined by the Hertz radius [7].

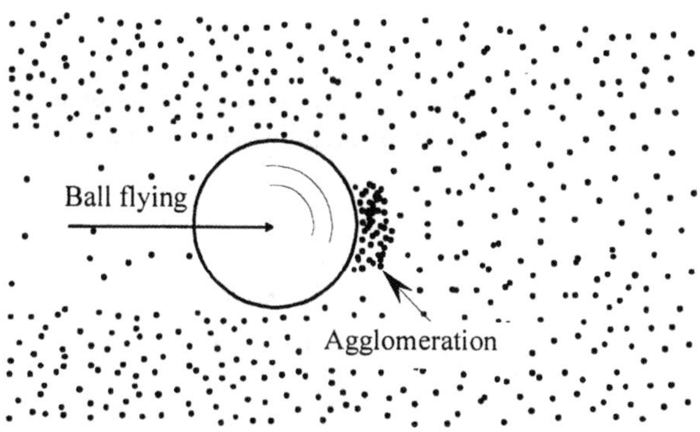


Figure 9.22. Schematic view of agglomeration of powder particles ahead of a flying ball.

Based on the above analysis, two swept volumes have been formulated [7]. The swept volume for a Spex mill $V_{\text{Spex}}$ is

$$V_{\text{Spex}} = \pi(L - 2r_b)r_h^2 / 3 \tag{9.73}$$

where $r_h$ is the bass radius of the cone, and $r_b$, radius of the ball.

The swept volume $V_S$ for an attritor $V_{\text{Attritor}}$ is

$$V_{\text{Attritor}} = \vartheta\pi r_b r_h^2 / 3 \tag{9.74}$$

where $\vartheta$ is the packing parameter (= 0.8. For more details, refer to reference 7).

Only the powder particles within the corn can have the possibility of being entrapped between two impacting bodies. It is, therefore, important to evaluate the volume of the powder particles within the cone or the density of powder distribution in the vial [7]. Since the free volume of the vial $V_{\text{free}}$ is the difference between the volume of the vial and volume of all the balls, the density of the powder distribution in the vial is the volume ratio of powder particles $V_{\text{powder}}$ to $V_{\text{free}}$. The value of $V_{\text{free}}$ can be written as [7]:

$$_{\text{free}} = \pi\left(\frac{LD_m^2}{4} - \frac{4n_b r_b^3}{3}\right) \tag{9.75}$$

The volume of the powder particles can be written as a function of the ball to powder weight ratio [7]:

$$_{\text{powder}} = \frac{4}{3}\frac{\pi n_b \rho_b r_b^3}{\rho_p C_R} \tag{9.76}$$

where $C_R$ is the ball to powder weight ratio, $\rho_b$ and $\rho_p$, the density of the ball and the powder respectively. The density of the powder distribution $f_p$ can therefore be written as [7]:

$$f_p = \frac{V_{\text{powder}}}{V_{\text{free}}} = \frac{\frac{4}{3}\frac{\pi n_b \rho_b r_b^3}{\rho_p C_R}}{\pi\left(\frac{LD_m^2}{4} - \frac{4n_b r_b^3}{3}\right)} = \frac{16\frac{n_b \rho_b r_b^3}{\rho_p C_R}}{3LD_m^2 - 16n_b r_b^3} \tag{9.77}$$

Volume of the cluster of powder to be impacted between two bodies, $V_C$, is a fraction of the swept volume, namely [7],

$$V_C = f_P V_s \qquad [9.78]$$

while the height of the cluster, $h_0$, is given by [7]:

$$h_0 = \frac{V_c}{r_h^2} \qquad [9.79]$$

For an attritor, $f_p$ can be obtained accordingly. For a planetary ball mill however, the value of $V_s$ may not be the same as that of $V_s$ in Eq.[9.73] since the flying trajectory of the ball relative to the vial is not a straight line. The powder particles swept in front of the ball at the early stage of flying of the ball may escape when it gradually changes its relative flying direction. Another consideration is the influence of the impact angle. Since the flying direction of the ball is usually not normal to the vial, the swept powder particles may not be fully impacted between the balls and the wall of the vial. Hence, the actual volume of powder to be impacted is less than that swept. This hypothesis is schematically depicted in Figure 9.23.

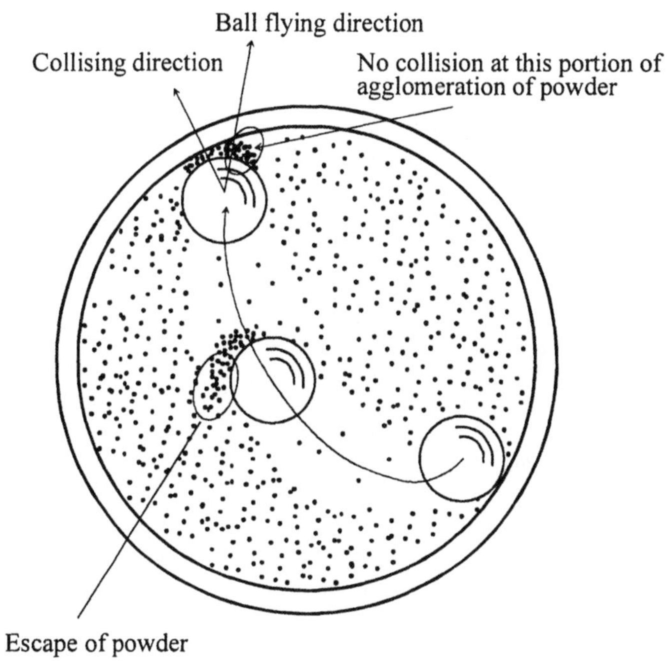


Figure 9.23. Ball flying and colliding in a planetary ball mill.

Based on this hypothesis, Eq.[9.77] can be rearranged by introducing an escaping and collision factor $\xi < 1$ as:

$$f_{\text{planetary}} = \xi f_p \qquad [9.80]$$

The factor $\xi$ is dependent upon the flying trajectory and the collision angle, namely, the angular velocity of the ball mill.

### 9.3.2 Impact Event

During mechanical alloying, four possible impact events may take place [6]: (a) normal impact between balls and inner surface of the vial, (b) impact with sliding displacement between balls and inner surface of the vial, (c) normal impact between balls and (d) impact with sliding displacement between balls. The effectiveness of impact between balls is not great since the balls normally move in the same direction. The more efficient impact is the direct impact between the balls and the inner surface of the vial. Because the powder particle aggregates trapped in between colliding bodies are much smaller than the colliding bodies themselves, the deformation of the powders between two balls or between ball and inner wall of the vial can be likened to that in an upset forging between two parallel plates. If the change in velocity of the objects after impact is taken into account, the energy dissipation during impact can be evaluated by:

$$\Delta E = \left(\frac{1}{2}m_1 v_{10}^2 + \frac{1}{2}m_2 v_{20}^2\right) - \left(\frac{1}{2}m_1 v_1^2 + \frac{1}{2}m_2 v_2^2\right) \qquad [9.81]$$

Since

$$v_{10} - v_1 = (1+\Pi)\frac{m_2}{m_1 + m_2}(v_{10} - v_{20}) \qquad [9.82]$$

$$v_{20} - v_2 = -(1+\Pi)\frac{m_1}{m_1 + m_2}(v_{10} - v_{20}) \qquad [9.83]$$

Energy dissipation during impact, which is due to heat, sound, plastic deformation and fracture of the particles, can be directly written as a function of the velocities of the objects and the coefficient of restitution,

$$\Delta E = (1 - \Pi^2)(v_{10} - v_{20})^2 \frac{m_1 m_2}{2(m_1 + m_2)} \qquad [9.84]$$

It can be seen that when $\Pi = 1$, there is no energy dissipation since $\Delta E = 0$. However, if $\Pi = 0$, Eq.[9.84] can be simplified to be,

$$\Delta E = \left(v_{10} - v_{20}\right)^2 \frac{m_1 m_2}{2\left(m_1 + m_2\right)} \qquad [9.85]$$

In general, the greater the amount of dissipated energy, the larger will be the plastic deformation and fracture. The former is dependent upon both velocity and mass of the objects. The efficiency of impact can be determined from the ratio of dissipated energy to impact energy of the ball,

$$\Theta = \frac{\Delta E}{E_i} = \frac{1 - \Pi^2}{1 + m_1 / m_2} \qquad [9.86]$$

There are two possible impact events during milling, namely, impact between the balls and that between the ball and the vial. Since the mass of the vial $m_2$ is much larger than that of the ball, impact between the balls and the vial is more efficient, such that the plastic deformation and fracture as a result of impact between ball and ball can be omitted.

Magini [(8)] simplified the impact event into two different elemental mechanical actions: collision when milling is conducted with a limited number of balls and attrition when milling is carried out in a container filled with balls. The kinetic energy dissipated during the impact event is mainly transferred into heat which increases the temperature of the powder and the milling tools. A minor fraction of it is stored in materials as structural disorder [(28)]. By introducing a collision energy transfer coefficient $K_a$, the energy transferred in each collision can be obtained [(29)].

$$\Delta E = K_a E_i \qquad [9.87]$$

$E_i$ can be obtained from Eqs.[9.40] and [9.63]. If $K_a = 0$, the collision is perfectly elastic, that is no energy transfer takes place. If $K_a = 1$, the collision is perfectly inelastic [(29)].

The energy transfer coefficient $K_a$ may be evaluated by a free ball falling test [(8)] where two parameters are recorded: the height at which the ball is launched and the height of rebound of the ball.

$$\zeta = \frac{h'}{h} \qquad [9.88]$$

where $\zeta = 1 - K_a$, and $h'$ and $h$, the heights of the ball at launch and rebound respectively.

The energy released during a collision is the kinetic energy of the ball minus the energy restituted $E_r$ [8, 28].

$$\Delta E = E_i - E_r = (1 - \zeta)E_i \quad [9.89]$$

Figure 9.24 shows the relationship between impact energy and energy transfer coefficient in two cases. In the first case, represented by open symbols in the figure, the ball freely falls from a given height onto a flat surface of the same material. The energy transfer coefficient represents a loss of impact energy as a result of an inelastic collision. In the second case, as shown by the filled symbols, the free falling test was conducted with a ball that had been firstly used to mill with Ti-Al powder. After the ball milling process, the ball was covered with a thin layer of Ti-Al powder mixture. The absorption of impact energy by the powder has been taken into account in the measurement of energy transfer coefficient. It can be seen from Figure 9.24 that the yield of the ball varies from about 0.9 for the lowest impact energy and tends towards to a plateau value of about 0.65 [28].

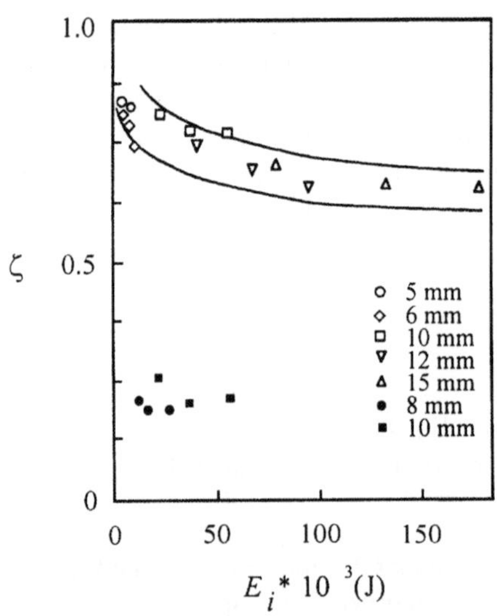


Figure 9.24. Measurement of energy transfer coefficient by free falling ball test. Open symbols refer to balls while the filled symbols, Ti-Al coated balls [8].

Since impact energy contributes only partially to the plastic deformation involved in an impact event [7], it is necessary to evaluate the effect of the elastic energy component via an equivalent Young's modulus, $E_{eq}$, which is a function of the Young's moduli of the ball and the powder compact [34].

$$E_{eq} = \frac{E_c E_b}{E_c + E_b} \qquad [9.90]$$

where $E_c$ and $E_b$ are respectively the moduli of the particle compact and the ball. The elastic energy is proportional to $E_{eq}^{0.2}$ [34].

### 9.3.3 *Deformation of powder particles*

#### 9.3.3.1 *Maurice and Courtney model* [7]

In this model, impacts between two balls having the same radius are considered. Assuming the maximum pressure, $p_{max}$, to be acting on a cluster of powder, the elastic energy, $u_u$, per unit volume of the ball may be given as:

$$u_u = \frac{1}{2}\varepsilon_e p_{max} = \frac{p_{max}^2}{2E_b} \qquad [9.91]$$

where $E_b$ is the Young's modulus. If volume associated with the elastic deformation is approximated as $\pi r_b^2 \delta_{max} / 3$, the elastic energy, $U_e$, of two balls can be expressed as:

$$U_e = \frac{p_{max}^2}{2E_b}\frac{2\pi r_b^2 \delta_{max}}{3} = \frac{2\pi r_b r_h^2 p_{max}}{3E_b} \qquad [9.92]$$

since $\delta_{max} = r_h^2 / 2r_b$. The ratio of total plastic deformation energy, $U_p$, to elastic energy, $U_e$, can be obtained by considering the plastic deformation of a volume of $\pi r_h^2 h_0$ with a unit plastic deformation energy of $u_p$:

$$\frac{U_p}{U_e} = \frac{3h_0 u_p E_b}{2r_b p_{max}^2} \qquad [9.93]$$

Maurice showed that the deformation strain of the powder depended upon the value of the centre-to-centre approach of the balls. This function of $\alpha(r_x)$ can be written as a function of the distance $r_x$ from the centre of contact by:

$$\alpha(r_x) = r_b v \left( \frac{\rho_b}{H_v} \right)^{1/2} - \frac{r_x^2}{r_b} \qquad [9.94]$$

where $\rho_b$ is the density of the balls, $v$, collision velocity and $H_V$, hardness.

The deformation strain of the powders is therefore:

$$\varepsilon(r) = -\ln\left\{1 - \frac{\alpha(r_x)}{h_0}\right\} \qquad [9.95]$$

When $r_x = 0$, maximum strain may be obtained by:

$$\varepsilon(0) = -\ln\left\{1 - \frac{r_b v}{h_0} \left( \frac{\rho_b}{H_v} \right)^{1/2} \right\} \qquad [9.96]$$

Two criteria have been developed based on the observation that successful mechanical alloying is a balance between cold welding and fracturing [(9)]. The criterion for welding is that the weld tensile force must be equal or large than the sum of the square of the normal elastic recovery force and the square of the comparable shear force. The criterion for fracturing is that strain is greatest at the circumference of the particle where edge fracture initiates and that strain is greatest at the centre of contact where crack initiation takes place in the case of mechanical alloying of a brittle material.

More details could be obtained from the following references (7, 30, 31).

*9.3.3.2 Schwarz model* [(32)]

Consider the elastic head-on collision of two balls with relative velocity $v_i$, the compressive stress at the point of contact can be approximated as:

$$\sigma_c = Z_b E_b \qquad [9.97]$$

where $\sigma_c$ is the compressive stress and $Z_b$ is equal to:

$$Z_b = \sqrt{\rho_b E_b} \qquad [9.98]$$

where $\rho_b$ is the density of the material for the ball. To achieve plastic deformation during mechanical alloying, it is essential to satisfy the following requirement,

$$v_i > \frac{\tau}{\sqrt{\rho_b \Gamma}} \qquad [9.99]$$

where $\tau$ is the compressive flow stress of the powder material. If the velocity of the balls satisfy the above condition, plastic deformation takes place in powder particles which will lead to the formation of layered structure and finally, the new materials.

*9.3.3.3 Bhattacharya and Arzt model* [34]

Based on porous constitutive, Bhattacharya and Artz [33] assumed the similarity of powder deformation during a ball to ball collision event to the deformation of hollow powder particles. This model assume that a homogeneous medium containing homogeneously and isotropically distributed pores, namely, the agglomeration of the powder particles, is subjected to head-on impact. Pores are spatially isolated so that interaction between them can be neglected. The initial porosity is simply due to the stacking of powder particles as shown in Figure 9.25. From this stacking sequence, pore size can be seen to be 0.46 times the average particle size. The porosity is 0.476. Pore size changes during milling and porosity accordingly reduces with prolonged milling. If the porous material is deformed such that the porosity is reduced from $\phi_0$ to $\phi$, the volume fraction of the pore, the energy per unit mass, $E_p$, for the deformation of hollow powder with inner of diameter $2a_0$ and outer diameter $2b_0$ can be expressed as

$$E_p = \frac{2\sigma_y(1-\phi)}{3\rho_p} \ln\left[\frac{\alpha_0^{\alpha_0}(\alpha-1)^{\alpha-1}}{\alpha^{\alpha}(\alpha_0-1)^{\alpha_0-1}}\right] \qquad [9.100]$$

where $\alpha_0 = 1/(1 - \phi_0)$, $\alpha = 1/(1 - \phi)$, $\rho_p$, density of the powder and $\sigma_y$, the flow stress of the powder. $\alpha_0$ may also be written as:

$$\alpha_0 = \frac{b_0^3}{b_0^3 - a_0^3} \qquad [9.101]$$

By introducing a linear relationship between yield stress and deformation strain for the powder, the linear relationship of yield stress can be expressed as:

$$\sigma_y = \sigma_0 + \frac{H\bar{\varepsilon}}{2} \quad [9.102]$$

where $\sigma_0$ is the flow stress at the beginning of the impact event and $H$, the hardening constant. Substituting Eqs.[9.102] and [9.103] into Eq.[9.100], impact deformation can be evaluated.

$$\bar{\varepsilon} = \frac{1}{3}\ln\left[\frac{\alpha(\alpha-1)}{(\alpha_0-1)^2}\right] \quad [9.103]$$

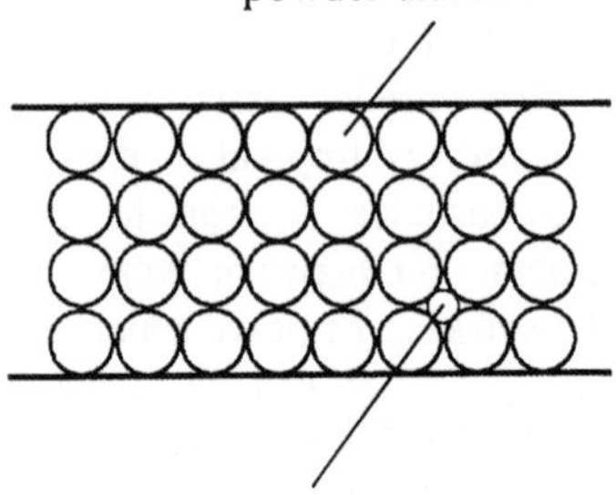


Figure 9.25. Stacking sequence of powder at initial stage of milling.

*9.3.3.4 Upset forging model*

Energy required to obtain deformation $dh$ can be written as:

$$E_i = -\int_{h_0}^{h} P dh = \int_{h}^{h_0} P dh \quad [9.104]$$

where $h$ is the instantaneous height of the powder cluster, and $P$, the force acting on the powder cluster during one impact. $P$ can be further written as a function of volume of the cluster and impact pressure $p$:

$$P = p\frac{V_{cluster}}{h} \quad [9.105]$$

If upset forging model is adapted to evaluate powder forging, the mean flow stress needed for the deformation of a cluster of powders having height $h$ would be

$$\overline{\sigma} \approx 2k\varsigma\left(1+\frac{d_c\mu}{3h}\right) \quad [9.106]$$

where $\overline{\sigma}$ is the mean stress during upset forging, $k$, the shear stress at yield, $d_c$, diameter of a cluster of powders to be equal to $2r_h$, $\mu$, the friction parameter between the powder and the balls and $\varsigma$, density factor which is always $\leq 1$. Because of work hardening of the powders, the yielding shear stress is actually a function of deformation strain. For simplicity, however, it is assumed to be constant during a single impact. From Eq.[9.106], it can be seen that the mean flow stress during impact varies according to $d_c$, $h$ and $\varsigma$. The change of mean flow stress is normally simplified by the following rule:

$$\overline{\sigma}_a = \frac{\overline{\sigma}_0 + 2\overline{\sigma}_1}{3} \quad [9.107]$$

where $\overline{\sigma}_a$ is the average flow stress, $\overline{\sigma}_0$, mean flow stress at the beginning of impact and $\overline{\sigma}_1$, mean flow stress at the end of impact.

Substituting Eq.[9.107] into Eq.[9.104], the deformation energy, $E_i$, for a cluster of powder having volume $\pi d_0^2 h/4$ can be obtained:

$$E_i = \overline{\sigma}_a \frac{\pi d_0^2 h_0}{4}\left[\ln\frac{h_0}{h} - \frac{1}{9}\left(\frac{d}{h} - \frac{d_0}{h_0}\right)\right] \quad [9.108]$$

The above equation is adopted from the analysis of upset forging. It can be used as an approximation of powder forging during impact. Since the major difference is the porosity, the flow stress will increase not only due to work hardening but also due to the change in porosity during each single impact period. Based on Maurice and Courtney model [(7)], sweeping core contains loss powder particle. Therefore, it can be assumed that every single impact starts from the deformation of porous material (or loss particles) but ends with the deformation of dense material (all particles are cold welded together to form one particle or fractured from one particle). The change in density of a cluster of particles and in flow stress are schematically given in Figure 9.26.

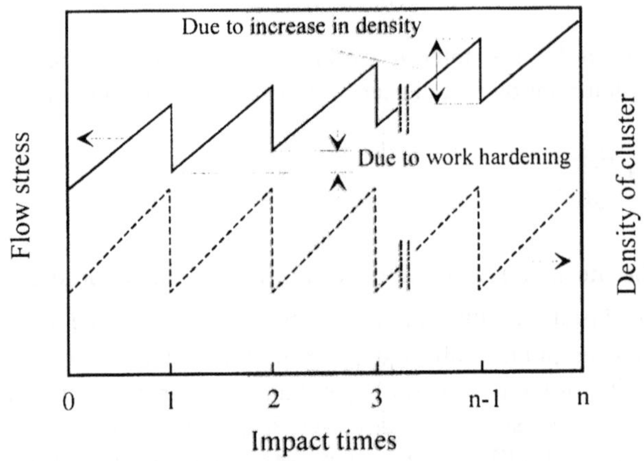


Figure 9.26. Schematic diagram showing the change in density and flow stress of a cluster before and after impact.

Assuming $\overline{\sigma}_a = 100$ MPa, and $h = 1.2$ mm, the relationship between deformation $\Delta h$ and energy $E_i$ can be evaluated. Figure 9.27 shows the relationship between impact energy required for the deformation of a cluster of powder with different initial height and deformation. The energy consumption is very low unless the deformation approaches the height of the powder cluster.

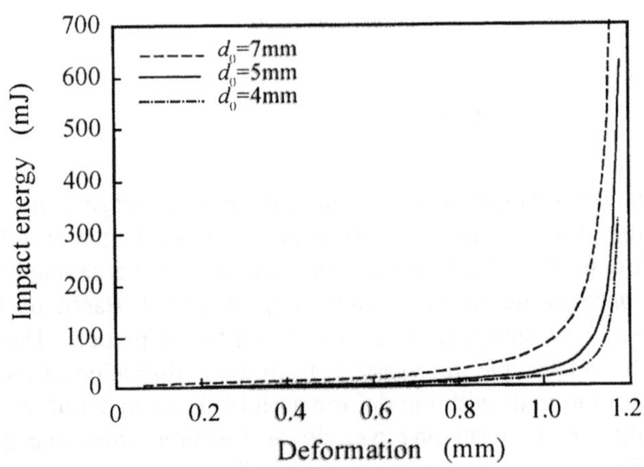


Figure 9.27. Impact energy required for the deformation of powder cluster with different heights.

The above energy calculation considers only the energy for the deformation of powder cluster without taking into account energy storage in the particles and in the milling tools. The latter part of the energy will be released after impact. Depending upon the properties of the materials to be milled, according to Maurice and Courtney [7], the ratio of the plastic energy to the elastic energy is only about 0.1 to 1% for an attritor mill and 0.3 to 9% for a Spex mill.

According to cold welding theory, there exists a critical deformation strain which is described by a minimum bonding. For instance, the minimum deformation strains for Pb, Sn, Al, Cu and Zn are about 9, 13, 40, 42, and 55% respectively [35]. Using the data employed in Figure 9.27, the energy of deformation can be determined. Figure 9.28 shows an example of energy required for the cold welding of Pb, Al and Zn.

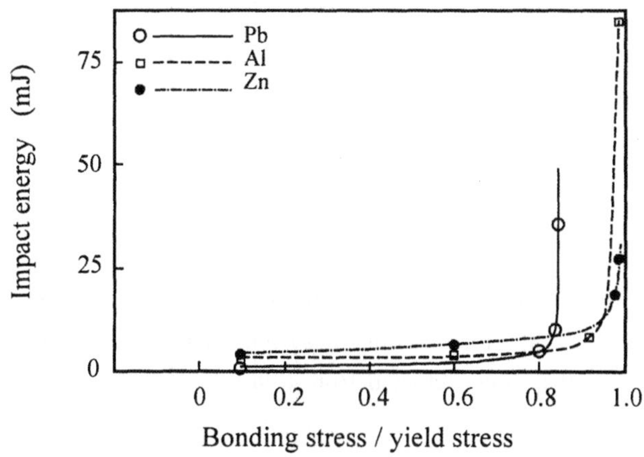


Figure 9.28. Minimum energy required for cold welding.

### *9.3.4 Fracture of particles*

Owing to high density of defects and improperly welded locations in the powder particles, fracture may occur in these weak locations during impact. The strain energy released during high speed impact will exceed the energy associated with the formation of new surface. The surplus energy may be dissipated onto other cracks in the powder particles causing propagation, thus breaking up particle into several smaller ones. Griffith [36] criterion for crack propagation has been employed for the analysis of particle refinement during milling process [37].

According to Griffth, the specific surface energy $\Re_S$, can be written as a function of crack length and stress $\sigma$ at fracture:

$$\Re_s = \frac{\pi\sigma^2 a_c}{2E_p} \quad [9.109]$$

and

$$\sigma = \sqrt{\frac{2E_p \Re_s}{\pi a_c}} \quad [9.110]$$

where $a_c$ is one-half the crack length.

The Griffith relation was derived for a material containing a very sharp crack. Since plastic deformation is generally involved in fracturing, Orowan modified the Griffith relation by introducing an energy term that is associated with plastic deformation:

$$\sigma = \sqrt{\frac{2E_p(\Re_s + \Re_p)}{\pi a_c}} = \sqrt{\frac{2E_p \Re_s}{\pi a_c}\left(1 + \frac{\Re_p}{\Re_s}\right)} \quad [9.111]$$

where $\Re_p$ is the plastic deformation energy.

When the crack driving force $\wp = (\Re_s + \Re_p)$ reaches a critical value $\wp_c$, fracture occurs in the particles. $\wp_c$ may be obtained from fracture test. In reality, however, $\wp_c$ is smaller than the value obtained from a large material sample as a result of imperfect bonding.

## 9.4 Conclusions

Since mechanical alloying or mechanical milling is a process in which energy is transferred from the high energetic milling medium to the powder particles, it is essential to understand this process of energy transfer. A better understanding of the kinetics of motion of the ball in the ball mill provides the necessary knowledge on the energy of the milling medium. Several models have been proposed for the simulation of a planetary ball mill, conventional ball mill and vibration mill. They provide means to predict the trajectories of the ball. From these models, the influence of major parameters such as size and mass of the ball, ball velocity and impact frequency can be better understood. The simplicity of these models enables clearer understanding to the relationships between the different parameters, hence allowing these operating parameters to be controlled. However, the simplicity also has limited the reality of the models. On the basis of slip factor between the balls and the vial, results of the simulation become more realistic. By considering the interference between the balls and between the balls and the vial, the combination of using different ball size in one milling process can be optimized. Experimental observations have shown certain regular pattern for the trajectory of the ball flying in a planetary ball mill, the interference between balls may consequently be very little. The global modeling of the milling process can further provide useful information to the new generations of ball mills.

By introducing Hertz theory to the impact medium, the difficulty in identifying the impact area and the volume of deformed powder during impact event has been solved although it is only an approximation. However, as a result of the difficulty in obtaining information on the change in density and average yield stress of the powder cluster during milling, the understanding on local modeling is still currently limited. Since local modeling may provide more information on modeling the synthesis, it is important to fully understand the physical mechanisms of energy transfer during impact event and the influence of various operating parameters.

## 9.5 References

1. T.H. Courtney, J.C.M. Kampe, J.K. Lee and D.R. Maurice, *Diffusion Analysis & Applications*, Ed. A.D. Romog, Jr. and M.A. Dayananda, Publ. TMS, Warrendale, PA, (1989), 225.
2. T.H. Courtney and D.R. Maurice, *Solid State Powder Processing*, Proc. of The Confer., Indianapolis, IN, 1-5 Oct. 1989, Ed. A.H. Clauer and J.J. deBarbadillo, Publ. TMS, Warrendale, PA, (1990), 3.
3. H. Hashimoto and R. Watanabe, *Mater. Trans., JIM*, Vol.31 (1990), 219.
4. H. Hashimoto and R. Watanabe, *Mater. Sci. Forum*, Vol.88-90 (1992), 89.
5. D. Basset, P. Matteazzi and F. Miani, *Nanophase Materials: Synthesis-Properties-Applications*, Ed. G.C. Hadjipanayis and R.W. Siegel, Nato-Asi Series E: Applied Sci., Vol.260 (1994), Kluwer Acad. Publ., Dordrecht, The Netherlands, 149.
6. T.H. Courtney and D.R. Maurice, *Solid State Powder Processing*, Proc. of a Symp., Indianapolis, Indiana, 1-5 October, 1989, Ed: A.H. Clauer and J.J. deBarbadillo, Publ: The Minerals, Metals & Mater. Soc., (1990), 3-9.
7. D.R. Maurice and T.H. Courtney, *Metall. Trans.*, Vol.21A (1990), 289.
8. M. Magini and A. Iasonna, *Mater. Trans. JIM*, Vol.36 (1995), 123.
9. T.H. Courtney, *Mater. Trans. JIM*, Vol.36 (1995), 110.
10. T.H. Courtney and D. Maurice, *Scripta Mater.*, Vol.34 (1996), 5.
11. P.Le Brun, L. Froeyen and L. Delaey, *Mater. Sci. Eng.*, A161 (1993), 75.
12. J. Raasch, *Chem. Eng. Techn.*, Vol.15 (1992), 245.
13. M. Abdellaoui and E. Gaffet, *Acta Metall. Mater.*, Vol.43 (1995), 1087.
14. A. Joisel, *Rev. Matér. Constr. Trav. Publics* (1956), No.493, 1.
15. H. Rumpf, *Vorlesungen über Mechanische Verfahrenstechnik*, Univ. Karlsruhe (1965/64).
16. G. John and F. Vock, *Chem.-Ing.-Tech.*, Vol.37 (1965), 411.
17. P.G. McCormick, H. Huang, M.P. Dallimore, J. Ding and J. Pan, *Mechanical Alloying for Structural Applications*, Ed. J.J. de Barbadillo, F.H. Froes and R.B. Schwarz, Publ. ASM., (1993), 45.
18. E. W. Davis, *Trans. AIMME*, Vol61 (1919), 250.
19. N. Burgio, A. Lassona, M. Magini, S. Martelli and F. Padella, *Il Nuovo Cimento*, Vol.13D (1991), 459.
20. M. Abdellaoui and E. Gaffet, *J. Phys. IV*, Vol.4 (1994), C3-291.
21. J. Eckert, L. Schultz and E. Hellstern, *J. Appl. Phys.*, Vol.64 (1988), 3224.
22. A.W. Webber, W.J. Haag, A.J.H. Wester and H. Bakker, *J. Less Common Met.*, Vol.140 (1988), 119.
23. E. Gaffet and M. Harmelin, *Mater. Sci. Eng.*, A119 (1989), 185.
24. E. Gaffet and L. Yousfi, *Mater. Sci. Forum*, Vol.88-90 (1992), 51.

25. D. Gavrilov, O. Vinogradov and W.J.D. Shaw, *Proc. of ICCM-10*, Whistler, BC, Canada, 14-18 August 1995, Ed. A. Poursartip and K. Street, Publ. Woodhead Publ. Ltd., Vol.3 (1995), 11.
26. D.J. McGill and W.W. King, *An Introduction to Dynamics, 2nd* edition, PWS-KENT Publishing Company, Boston, 1989.
27. P.S. Gilman and J.S. Benjamin, *Annu. Rev. Mater. Sci.*, Vol.13 (1983), 279.
28. M. Magini, *Mater. Sci. Forum*, Vols.88-90 (1992), 121.
29. M. Magini, A. Iasonna and F. Padella, *Scripta Mater.*, Vol.34 (1996), 13.
30. D.R. Maurice and T.H. Courtney, *JOM*, Vol44 (1992), 437.
31. D.R. Maurice and T.H. Courtney, *Metall. Mater. Trans.*, Vol.25A (1994), 147.
32. R.B. Schwarz, *Scripta Mater.*, Vol.34 (1996), 1.
33. M.M. Carroll and A.C. Holt, *J. Appl. Phys.* Vol.34 (1972), 1626.
34. A.K. Bhattachary and E. Artz, *Scripta Metall. Mater.*, Vol.28 (1993), 395.
35. J.M. Alexander and R.C. Brewer, *Manufacturing properties of Materials*, Van Nostrand Co. Ltd, London, S.W.1, (1993), 369.
36. A.A. Griffith, *Phil. Trans.* Royal Soc. A, 221 (1920), 163.
37. C.L. Prasher, *Crushing and Grinding Process Handbook*, John Wiley & Sons Ltd, 1987.

# INDEX

*E*



*F*



*G*



*H*



*I*



*K*



*L*



*M*

MIX
Papier aus verantwortungsvollen Quellen
Paper from responsible sources
FSC® C105338

If you have any concerns about our products,
you can contact us on
**ProductSafety@springernature.com**

In case Publisher is established outside the EU,
the EU authorized representative is:
**Springer Nature Customer Service Center GmbH**
**Europaplatz 3, 69115 Heidelberg, Germany**

Printed by Libri Plureos GmbH
in Hamburg, Germany